213847

89 U2
56.85

INTERSCIENCE TRACTS
IN PURE AND APPLIED MATHEMATICS

Editors: L. BERS · R. COURANT · J. J. STOKER

1. *D. Montgomery and L. Zippin*—**Topological Transformation Groups**
2. *Fritz John*—**Plane Waves and Spherical Means Applied to Partial Differential Equations**
3. *E. Artin*—**Geometric Algebra**
4. *R. D. Richtmyer and K. W. Morton*—**Difference Methods for Initial-Value Problems.** Second Edition
5. *Serge Lang*—**Introduction to Algebraic Geometry**
6. *Herbert Busemann*—**Convex Surfaces**
7. *Serge Lang*—**Abelian Varieties**
8. *S. M. Ulam*—**A Collection of Mathematical Problems**
9. *I. M. Gel'fand*—**Lectures on Linear Algebra**
10. *Nathan Jacobson*—**Lie Algebras**
11. *Serge Lang*—**Diophantine Geometry**
12. *Walter Rudin*—**Fourier Analysis on Groups**
13. *Masayoshi Nagata*—**Local Rings**
14. *Ivan Niven*—**Diophantine Approximations**
15. *S. Kobayashi and K. Nomizu*—**Foundations of Differential Geometry.** In two volumes
16. *J. Plemelj*—**Problems in the Sense of Riemann and Klein**
17. *Richard Cohn*—**Difference Algebra**
18. *Henry B. Mann*—**Addition Theorems: The Addition Theorems of Group Theory and Number Theory**
19. *Robert Ash*—**Information Theory**
20. *W. Magnus and S. Winkler*—**Hill's Equation**

Additional volumes in preparation

INTERSCIENCE TRACTS
IN PURE AND APPLIED MATHEMATICS

Editors: L. BERS · R. COURANT · J. J. STOKER

Number 4

DIFFERENCE METHODS FOR INITIAL-VALUE PROBLEMS
Second Edition

Robert D. Richtmyer and K. W. Morton

INTERSCIENCE PUBLISHERS
a division of John Wiley & Sons
New York • Chichester • Brisbane • Toronto

DIFFERENCE METHODS
FOR
INITIAL-VALUE PROBLEMS

Second Edition

ROBERT D. RICHTMYER
University of Colorado,
Boulder, Colorado

K. W. MORTON
United Kingdom Atomic Energy Authority,
Culham Laboratory,
Abingdon, England

INTERSCIENCE PUBLISHERS
a division of John Wiley & Sons
New York • Chichester • Brisbane • Toronto

20 19 18 17 16 15

First edition copyright © 1957 by Interscience Publishers, Inc.
Second edition copyright © 1967 by John Wiley & Sons, Inc.

All Rights Reserved

Reproduction or translation of any part of this work beyond that permitted by Sections 107 or 108 of the 1976 United States Copyright Act without the permission of the copyright owner is unlawful. Requests for permission or further information should be addressed to the Permissions Department, John Wiley & Sons, Inc.

Library of Congress Catalog Card Number 57-13249
Printed in the United States of America

ISBN 0 470 72040 9

Preface to the Second Edition

In the ten years since the publication of the first edition of this book, the literature on the theory of difference methods for partial differential equations has increased by several orders of magnitude, and the reader might suppose that the main purpose of the revision of the book is to present this body of theory (or that part of it that is relevant to initial-value problems), but this is not our main purpose. Although certain important theoretical advances are described in some detail, especially in Chapters 4, 5, and 6, the book is intended primarily for users, not for numerical analysts. As in other parts of mathematics, the abstract theory and the applications have tended to grow apart, and a new branch of pure mathematics has come into being in this area. Although this new branch has already produced numerous results and concepts of value to the users, most practical problems in physical science are far too complex, in one way or another, to be yet covered by the theorems that have been published. In practical computation, it is just as necessary to utilize physical intuition, heuristic reasoning, and trial-and-error procedures as it was when von Neumann started the subject, nearly twenty-five years ago.

We have presented certain parts of the theory that we feel are useful or significant for practical applications, but we have also tried to retain the practical approach that was emphasized in the first edition. The principal theoretical advances are (*1*) the rounding-out or completion of the theory for pure initial-value problems with constant coefficients by the general sufficient conditions for stability obtained by Buchanan and Kreiss, and (*2*) the rigorous stability theory for certain classes of problems with variable coefficients, of mixed initial-boundary-value problems, and of quasi-linear problems. Among the ideas that we believe should be of value to people engaged in the solution of practical problems are (*1*) the notion of dissipative difference schemes, (*2*) the Lax-Wendroff method for systems of conservation laws, (*3*) the alternating-direction methods for multidimensional parabolic problems, (*4*) practical stability criteria for cases in which stability as normally defined is inadequate, and the use of energy methods in the analysis of stability.

The recent theoretical advances have generally used mathematical methods and concepts which are likely to be difficult or completely inaccessible to persons who are not specialists in numerical analysis, and this has posed a difficult problem for us in presenting the new material, where the proofs may be rough going for many readers, in spite of attempts to simplify them. As far as we can see, this is unavoidable, and we feel that the importance of the results warrants inclusion of the proofs.

We wish to pay particular tribute to the stimulus provided by Heinz Kreiss and to the valuable discussions with Peter Lax, Gilbert Strang, Beresford Parlett, and Sam Schechter. We wish to thank Valerie Breeze and Jenny Apeland for valuable assistance in the difficult mathematical typing.

Preface to the First Edition

It is perfectly obvious that a revolution of some sort became due in the field of numerical methods with the advent of modern computing machines. Numerical calculation is being used today in fields where it was unheard of fifteen years ago. Although electronic computers perform basically just the same operations as can be performed by hand, their speed and capacity make procedures feasible which are entirely out of the question for hand calculation. Finite-difference methods for solving partial differential equations were discussed in 1928 in the celebrated paper of Courant, Friedrichs, and Lewy but were put to use in practical problems only about fifteen years later under the stimulus of wartime technology and with the aid of the first automatic computers. At Los Alamos, for example, the calculation of certain time-dependent fluid flows played an important part in the wartime work of the laboratory and much time and effort was devoted to these calculations, which were performed on the punch-card machines available at that time. Immediately after the war still more complicated problems involving several simultaneous partial differential equations were solved numerically on the ENIAC in Aberdeen by members of the Los Alamos staff, and very soon problems involving fluid dynamics, neutron diffusion and transport, radiation flow, thermonuclear reactions, and the like were being solved on various machines all over the United States.

However, the expected revolution in numerical methods has been rather slow in coming. A few new ideas have emerged, but many of the basic methods are essentially the same as have been used for many years. The Gauss elimination method (in one form or another) is still among the best methods for solving simultaneous linear equations and the Runge-Kutta method is still among the best methods for solving ordinary differential equations. It is likely that such basic methods of classical numerical analysis will continue to be stock-in-trade for some time, and perhaps they will somehow become incorporated in, rather than displaced by, the methods of the future. However, it seems clear on combinatorial grounds alone that the high-speed, large-capacity computing machines are inherently capable of doing things in countless ways that have not yet even been dreamed of. Probably many powerful methods await discovery, both for problems that can be handled with existing techniques and for those that cannot. Methods now in use are, for the most part, generated by simply pyramiding methods designed for hand computation and for much smaller problems; surely we have not yet come even close to utilizing the inherent power of the machines.

The part of numerical analysis which has been most changed so far is probably the solution of partial differential equations, especially those describing non-steady physical phenomena, by difference methods—and that is the main subject of this book. The equations encountered in practical problems are often very complicated. They generally have variable coefficients or are even nonlinear; there are often coupled systems of differing types (say hyperbolic and parabolic); often there are several space variables; and integro-differential equations are also encountered. There has been a rapid growth of new calculational algorithms for handling these problems.

This development of numerical methods for automatic machines has differed in one important respect from the previous development of methods for hand computation. Because of the complexity of the problems, because of the rapid accumulation of new algorithms, and because the people who have worked with electronic computers have generally not been expert in the classical foundations of numerical analysis, the new development has been based more on empiricism and intuition and less on a mathematical basis than the classical development. One should not condemn the new development for this because if we were to wait for convergence proofs and error estimates for the new methods, most of the computers now in use in technology and industry would come grinding to a halt. Error estimates and convergence proofs become rapidly more difficult to obtain as the problems become more complicated. One consequence of this new development is a gap between the mathematician and the practical man (e.g., physicist) over these questions. It is my conviction that the revolution we are looking for can be achieved only by a collaboration between the two. To narrow the gap between them, we must persuade the mathematician to be a little more flexible about accepting the empirical intuitive approach, and we must persuade the practical man to attach more importance to a real basic understanding of the methods he uses.

In the fall of 1953, Dr. Leslie Peck and the author, who were visiting the Institute of Mathematical Sciences at New York University, organized a seminar at the request of Professor Courant to discuss those parts of numerical analysis that are the joint property of the mathematician and the computing machine man. Many members and visitors at the Institute collaborated, including John Curtiss, Joel Franklin, Wallace Givens, Eugene Isaacson, Fritz John, Peter Lax, and Milton Rose. Other similar seminars and courses on related subjects have been held subsequently at NYU, with many additional participants, including a large fraction of the Institute faculty and of the members of the AEC Computing Facility, which is a part of the Institute, and also including a few long-term visitors, particularly John and Olga Todd and George Forsythe.

The present volume presents part of the lore on certain subjects in numerical analysis that has accumulated at Los Alamos and other places during the last dozen years, somewhat modified and interpreted by the discussions at the seminars and courses referred to above. The ideas in the theoretical part of the book are primarily those of John von Neumann and Peter Lax and, to some extent, of the author; but it is clear that if a different approach had been followed the contributions of certain other people would also have appeared prominently. An attempt has been made to tie together the mathematical and the practical; although the revolution seems to be as far away as ever, it is hoped that we are indeed inching along in the direction of knowing what to do with the modern computing machines and of understanding the results that come out of them.

R. D. RICHTMYER

New York University, New York

Contents

Part I
GENERAL CONSIDERATIONS

1 Introduction
- 1.1 Initial-Value Problems 3
- 1.2 The Heat Flow Problem 4
- 1.3 Finite-Difference Equations 7
- 1.4 Stability 9
- 1.5 Implicit Difference Equations 16
- 1.6 The Truncation Error 19
- 1.7 Rate of Convergence 22
- 1.8 Comments on High-Order Formulas and Rounding Errors . . . 24
- 1.9 Outline of the Remainder of the Book 25

2 Linear Operators
- 2.1 The Function Space of an Initial-Value Problem . . . 28
- 2.2 Banach Spaces 30
- 2.3 Linear Operators in a Banach Space 33
- 2.4 The Extension Theorem 34
- 2.5 The Principle of Uniform Boundedness 34
- 2.6 A Fundamental Convergence Theorem 37
- 2.7 Closed Operators 37

3 Linear Difference Equations
- 3.1 Properly Posed Initial-Value Problems 39
- 3.2 Finite-Difference Approximations 42
- 3.3 Convergence 44
- 3.4 Stability 45
- 3.5 Lax's Equivalence Theorem 45
- 3.6 The Closed Operator A' 49
- 3.7 Inhomogeneous Problems 52
- 3.8 Change of Norm 57
- 3.9 Stability and Perturbations 58

4 Pure Initial-Value Problems with Constant Coefficients
- 4.1 The Class of Problems 60
- 4.2 Fourier Series and Integrals 61

4.3	Properly Posed Initial-Value Problems	63
4.4	The Finite-Difference Equations	65
4.5	Order of Accuracy and the Consistency Condition	67
4.6	Stability	68
4.7	The von Neumann Condition	70
4.8	A Simple Sufficient Condition	72
4.9	The Kreiss Matrix Theorem	73
4.10	The Buchanan Stability Criterion	80
4.11	Further Sufficient Conditions for Stability	83

5 Linear Problems with Variable Coefficients; Non-Linear Problems

5.1	Introduction	91
5.2	Alternative Definitions of Stability	95
5.3	Parabolic Equations	100
5.4	Dissipative Difference Schemes for Symmetric Hyperbolic Equations	108
5.5	Further Results for Symmetric Hyperbolic Equations	119
5.6	Non-Linear Equations with Smooth Solutions	124

6 Mixed Initial-Boundary-Value Problems

6.1	Introduction	131
6.2	Basic Ideas of the Energy Method	132
6.3	Simple Examples of the Energy Method: Stable Choice of Approximations to Boundary Conditions and to Non-Linear Terms	137
6.4	Coupled Sound and Heat Flow	143
6.5	Mixed Problems for Symmetric Hyperbolic Systems	146
6.6	Normal Mode Analysis and the Godunov-Ryabenkii Stability Criterion	151
6.7	Application of the G-R Criterion to Mixed Problems	156
6.8	Conclusions	164

7 Multi-Level Difference Equations

7.1	Notation	168
7.2	Auxiliary Banach Space	169
7.3	The Equivalence Theorem	171
7.4	Consistency and Order of Accuracy	174
7.5	Example of Du Fort and Frankel	176
7.6	Summary	179

Part II

APPLICATIONS

Preface to Part II 183

8 Diffusion and Heat Flow

8.1 Examples of Diffusion 185
8.2 The Simplest Heat-Flow Problem 186
8.3 Variable Coefficients 193
8.4 Effect of Lower Order Terms on Stability 195
8.5 Solution of the Implicit Equations 198
8.6 A Non-Linear Problem 201
8.7 Problems in Several Space Variables 206
8.8 Alternating-Direction Methods 211
8.9 Splitting and Fractional-Step Methods 216

9 The Transport Equation

9.1 Physical Basis 218
9.2 The General Neutron Transport Equation 219
9.3 Homogeneous Slab: One Group 222
9.4 Homogeneous Sphere: One Group 223
9.5 The "Spherical Harmonic" Method 224
9.6 Slab: Difference System I for Hyperbolic Equations . . 228
9.7 A Paradox 230
9.8 Slab: Difference System II (Friedrichs) 232
9.9 Implicit Schemes 233
9.10 The Wick-Chandrasekhar Method for the Slab 233
9.11 Equivalence of the Two Methods 235
9.12 Boundary Conditions 237
9.13 Difference Systems I and II 237
9.14 System III: Forward and Backward Space Differences . 238
9.15 System IV (Implicit) 239
9.16 System V (Carlson's Scheme) 240
9.17 Generalization of the Wick-Chandrasekhar Method . . 243
9.18 The S_n Method of Carlson (1953) 244
9.19 A Direct Integration Method 246

10 Sound Waves

10.1 Physical Basis 259
10.2 The Usual Finite-Difference Equation 260
10.3 An Implicit System 263
10.4 Coupled Sound and Heat Flow 264
10.5 A Practical Stability Criterion 269

11 Elastic Vibrations

11.1 Vibrations of a Thin Beam 271
11.2 Explicit Difference Equations 273

	11.3	An Implicit System	274
	11.4	Virtue of the Implicit System	275
	11.5	Solution of Implicit Equations of Arbitrary Order	275
	11.6	Vibration of a Bar Under Tension	282

12 Fluid Dynamics in One Space Variable

12.1	Introduction	288
12.2	The Eulerian Equations	289
12.3	Difference Equations, Eulerian	290
12.4	The Lagrangean Equations	293
12.5	Difference Equations, Lagrangean	295
12.6	Treatment of Interfaces in the Lagrangean Formulation	298
12.7	Conservation-Law Form and the Lax-Wendroff Equations	300
12.8	The Jump Conditions at a Shock	306
12.9	Shock Fitting	308
12.10	Effect of Dissipation	311
12.11	Finite-Difference Equations	317
12.12	Stability of the Finite-Difference Equations	320
12.13	Numerical Tests of the Pseudo-Viscosity Method	324
12.14	The Lax-Wendroff Treatment of Shocks	330
12.15	The Method of S. K. Godunov	338
12.16	Magneto-Fluid Dynamics	345

13 Multi-Dimensional Fluid Dynamics

13.1	Introduction	351
13.2	The Multi-Dimensional Fluid-Dynamic Equations	354
13.3	Properly and Improperly Posed Problems	356
13.4	The Two-Step Lax-Wendroff or L-W Method	360
13.5	The Viscosity Term for the L-W Method	365
13.6	Piecewise Analytic Initial-Value Problems	368
13.7	A Program for the Development of Methods	374
13.8	Characteristics in Two-Dimensional Flow	375
13.9	Shock Fitting in Two Dimensions	378
13.10	The Problem of the Atmospheric Front	383

References 389
Subject Index 401

PART I
GENERAL CONSIDERATIONS

CHAPTER 1

Introduction

1.1. Initial-Value Problems

The problems to be considered in this book are initial-value problems that arise in various branches of continuum physics, such as heat flow, diffusion, fluid dynamics, magneto-fluid dynamics, acoustics, electromagnetism, wave mechanics, radiation transfer, neutron transfer, and elastic vibrations. A steady state is not assumed to exist, and we are led to partial differential or integro-differential equations in which one of the independent variables, called t, has the role of time. The equations are of such a nature that if a state of the physical system is arbitrarily specified at some initial time $t = t_0$, a solution exists for $t \geqq t_0$ and is uniquely determined by the equations together with boundary conditions or other auxiliary conditions.

The specific subject of the book is finite-difference methods for the approximate numerical solution of problems of this kind. In Part I a general discussion of these methods is given for linear problems in terms of the theory of linear operators, the purpose of this discussion being to crystallize the main concepts relevant to stability and convergence. General results are derived for pure initial-value problems with constant coefficients; some results for special classes of equations with variable coefficients and for some mixed initial-boundary-value problems are given in subsequent chapters. Part II contains descriptions of the principal finite-difference methods currently in use, with automatic digital computers, for the solution of initial-value problems in various fields of applied physics. As far as possible, the discussions are based on the theoretical material of Part I and of certain published works, but the existing mathematical theory is still often inadequate (mainly because of non-linearities), and we then have recourse to a combination of intuition and experimental evidence.

1.2. The Heat Flow Problem

In this chapter, the main ideas are introduced by considering the familiar example of linear heat flow or diffusion in one dimension. If x denotes a coordinate along the length of a thin insulated rod in which heat can flow, or a coordinate perpendicular to the sides of a large slab with uniform temperature on each face, and if $u = u(x, t)$ denotes the temperature at position x, time t, the temperature satisfies the differential equation

$$(1.1) \qquad a\frac{\partial u}{\partial t} = \frac{\partial}{\partial x}K\frac{\partial u}{\partial x},$$

where $a = a(x)$ or $a(x, t)$ is the heat capacity of the material per unit volume and $K = K(x)$ or $K(x, t)$ is the thermal conductivity.

This equation is linear, although it would be non-linear if we allowed the material properties denoted by a and K to vary with the temperature as well as with x and possibly t. Furthermore, we shall limit the discussion in this chapter to the case of a homogeneous material, so that a and K are constants, and we suppose that $K/a = \sigma > 0$.

The differential equation is to be supplemented by boundary conditions at say $x = 0$ and $x = L$, which denote the ends of the rod or the sides of the slab; for example, the temperature is fixed, say $u = u_0$ at $x = 0$, if the end of the rod is in contact with a large heat reservoir, while $\partial u/\partial x = 0$ at $x = 0$ if the end is insulated.

Generally, the problems considered will consist of one or more partial differential or integro-differential equations together with boundary and initial conditions. There may also be internal boundary conditions: for example, if $K = K(x)$ has a simple discontinuity across an interface in the above problem, then the heat flow laws require that u and $K\partial u/\partial x$ be continuous across it. Similarly, at a shock discontinuity in the flow of a compressible fluid, the differential equations are replaced, at the shock front, by the Rankine-Hugoniot jump conditions (see Chapter 12).

The important but difficult questions of the existence and uniqueness of solutions will not be discussed in detail; instead, the problems dealt with are mostly assumed to be properly posed from the outset, in such a way that existence and uniqueness are ensured under physically reasonable assumptions (such as piecewise differentiability of coefficients and initial data).

In the stepwise numerical solution, one encounters the questions of construction of finite-difference systems, methods of solving them, their stability, and their accuracy. In the next few sections of this chapter these questions will be discussed for the simple heat flow problem mentioned above with even further simplifications as follows: we suppose the unit of length so chosen that $L = \pi$, and we assume that the boundary condition is $u = 0$ at each end, and that the initial function $\varphi(x)$ has whatever kind of smoothness we find convenient. The initial-value problem is:

$$(1.2) \quad \frac{\partial}{\partial t} u = \sigma \frac{\partial^2}{\partial x^2} u \quad \text{for} \quad \begin{cases} \sigma = \text{constant} > 0 \\ 0 \leq x \leq \pi \\ t \geq 0 \end{cases}$$

where $u = u(x, t)$,

$$(1.3) \quad u(x, 0) = \varphi(x) \text{ (given)} \quad \text{for } 0 \leq x \leq \pi,$$

$$(1.4) \quad u(0, t) = 0, \quad u(\pi, t) = 0 \quad \text{for } t > 0.$$

For illustration, one may think of an electric blanket stretched out flat and sandwiched between thick layers of ordinary blankets, and one may suppose that the bottom surface of the bottom blanket and the top surface of the top one are always at room temperature, which we take as $u = 0$ by suitable choice of thermometric scale. If the electric blanket is suddenly turned off after having been on for a long time (long enough to allow a steady state of heat flow to be established), we have the above problem with an initial temperature given by

$$(1.5) \quad \varphi(x) = \begin{cases} Cx & \text{for } 0 \leq x \leq \frac{\pi}{2} \\ C(\pi - x) & \text{for } \frac{\pi}{2} \leq x \leq \pi \end{cases}$$

x being now a vertical coordinate and C a constant. This function and typical temperature distributions $u(x, t)$ at later times are shown graphically by the solid curves of Figure 1.1.

For comparison with the results of the calculations with difference equations, we note that the exact solution of this initial-value problem can be obtained by the familiar Fourier series method. This may be written as a sine-series; or alternatively, if we define $\varphi(x)$ as $-\varphi(-x)$ for $-\pi \leq x \leq 0$, the solution may be written as a series of complex exponentials:

$$(1.6) \qquad u(x, t) = \sum_{-\infty}^{\infty} {}_{(m)} A_m \exp(imx - m^2 \sigma t),$$

where

$$(1.7) \qquad A_m = \frac{1}{2\pi} \int_{-\pi}^{\pi} \varphi(x) \exp(-imx) dx.$$

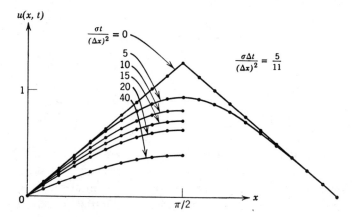

Fig. 1.1. Solution of the one-dimensional heat flow problem. The curves show the exact solution and the dots show the solution of the finite-difference equations (1.9), (1.10), and (1.11). Since each curve is symmetric, it is only necessary to show half of it.

The series (1.6) is sometimes regarded as a generalized solution of the initial-value problem even if it merely converges for $t > 0$; but in this chapter we assume that the Fourier series for $\varphi(x)$ itself is convergent, and in fact absolutely convergent.

For the initial data given by (1.5) for the electric blanket problem, we have

$$(1.8) \qquad A_m = \begin{cases} 0 & \text{if } m \text{ is even} \\ \dfrac{2iC}{\pi m^2}(-1)^{(m+1)/2} & \text{if } m \text{ is odd.} \end{cases}$$

1.3. Finite-Difference Equations

Obviously the Fourier series method is applicable only to a restricted class of problems. Indeed, its success, at least in the form presented above, depends on the linearity of the differential equation, the constancy of its coefficients, and the fact that the boundary conditions are equivalent to reflection and periodicity conditions. This is true also for the other partial differential equations we shall consider and for problems in more dependent and independent variables. The finite-difference methods, on the other hand, are by no means so restricted, although the analysis we shall give of their stability and convergence is often strictly valid only under these restrictions.

Let Δx and Δt be increments of the variables x and t, where $\Delta x = \pi/J$, J being an integer. The set of points in the x, t-plane given by $x = j\Delta x$, $t = n\Delta t$, where $j = 0, 1, 2, \ldots, J$, and $n = 0, 1, 2, \ldots$, is called a *net* (or *grid* or *lattice*) whose mesh size is determined by Δx and Δt; Δx and Δt are thought of as small increments, and in fact we shall later consider limiting processes in which they approach zero. The approximation to $u(j\Delta x, n\Delta t)$ is denoted by u_j^n. One of the simplest difference equations approximating the differential equation (1.2) is

$$(1.9) \qquad \frac{u_j^{n+1} - u_j^n}{\Delta t} = \sigma \frac{u_{j+1}^n - 2u_j^n + u_{j-1}^n}{(\Delta x)^2}, \quad \begin{array}{l} j = 1, 2, \ldots, J-1, \\ n = 0, 1, \ldots. \end{array}$$

The boundary condition is interpreted as

$$(1.10) \qquad u_0^n = 0, \; u_J^n = 0, \quad n = 0, 1, \ldots,$$

and the initial condition as

$$(1.11) \qquad u_j^0 = \varphi(j\Delta x), \quad j = 0, 1, \ldots, J.$$

Clearly, these equations can be used recursively to determine all the u_j^n for $0 \leq j \leq J$, $n \geq 0$.

Calculation with these equations is illustrated in Figures 1.1, 1.2a, b, c for the problem with initial data as given by (1.5). Two calculations are presented, both with $J = 20$, and therefore with $\Delta x = \pi/20$, but differing slightly as to the value at Δt used, viz.,

$$\frac{\sigma \Delta t}{(\Delta x)^2} = \begin{cases} 5/11 \text{ for Figure 1.1} \\ 5/9 \text{ for Figure 1.2a, b, c.} \end{cases}$$

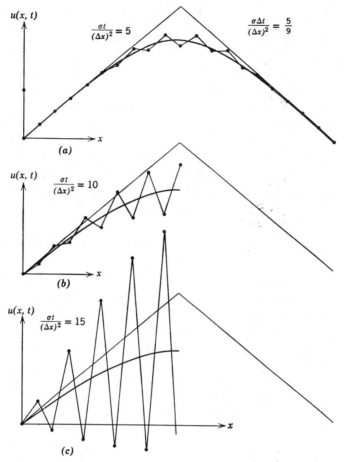

Fig. 1.2a, b, c. Solution of the same problem as for Figure 1.1 but calculated with a slightly larger value of Δt. Figures 1.2a, b, c correspond to the second, third, and fourth curves from the top in Figure 1.1.

The curves were obtained from the exact solution[1] (1.6) with the coefficients (1.8) and the dots from the finite-difference equations (1.9),

---

[1] Like most of the illustrative examples in this book, the numerical values were obtained in the course of studies on the Univac and IBM 7090 at the AEC Computing and Applied Mathematics Center at New York University. In this example, which is of course quite simple, the problem was coded in such a way that the machine printed out tables of the approximate and exact solutions in parallel columns at selected values of n. The exact solution was obtained by summing the Fourier series to an accuracy of around eight places of decimals.

(1.10), (1.11). It is seen that the calculation with the smaller value of Δt is quite satisfactory, whereas in the other one errors of some sort have accumulated and been amplified to an unacceptable degree after a relatively small number of cycles. The phenomenon appearing in the second calculation has nothing to do with round-off errors, which are completely negligible in the present example, but is a property of the difference equation system (1.9), (1.10), (1.11) and is called *instability*. Furthermore, the errors would only be made worse by taking a smaller value of Δx, unless Δt is also suitably reduced—these facts were pointed out by Courant, Friedrichs, and Lewy (1928).[2] One of the main aims of the studies presented in this book is the analysis and prevention of instability.

1.4. Stability

If $u(x, t)$ is the exact solution of the initial-value problem, and u_j^n is the solution of the finite-difference equations, the error of the approximation is $u_j^n - u(j\Delta x, n\Delta t)$ and one may ask two questions:

1. What is the behavior of $|u_j^n - u(j\Delta x, n\Delta t)|$ as $n \to \infty$ for fixed $\Delta x, \Delta t$?
2. What is the behavior of $|u_j^n - u(j\Delta x, n\Delta t)|$ as the mesh is refined (i.e., as $\Delta t, \Delta x \to 0$) for a fixed value of $n\Delta t$?

We shall see that the answer to the second question depends on the relative rates at which Δx and Δt go to zero. We regard this question as the more significant one because the basic notion of an approximation is that of a scheme whereby the error can be made as small as one pleases, and one hopes that the error will really tend to zero in the limit.

For both questions the number of cycles of calculation becomes infinite in the limit, and there is a possibility of unlimited amplification of errors. In most pure initial-value problems, the two questions lead to quite similar conclusions as to what restrictions must be imposed on the difference equations to keep the errors small, but for problems in a bounded domain, the results can be very different.

To answer these questions for the present simple problem, we note that

---

[2] Original works are referred to by author and year. A list of references will be found at the end of the book.

an explicit solution of the difference equations can also be written as a Fourier series. To see this, let A, ξ and m be constants, where m is an integer. If the expression $A\xi^n e^{imj\Delta x}$ is substituted for u_j^n into the difference equations (1.9), it satisfies these equations provided ξ is given by

(1.12) $$\xi = \xi(m) = 1 - \frac{2\sigma \Delta t}{(\Delta x)^2}(1 - \cos m\Delta x).$$

Therefore, the equation

(1.13) $$u_j^n = \sum_{-\infty}^{\infty}{}_{(m)} A_m e^{imj\Delta x} [\xi(m)]^n,$$

where the A_m are given by (1.7), gives the exact solution of the difference equation (1.9) and the boundary condition (1.10) and the initial condition (1.11), for (a) the series is absolutely convergent, because the series for $\varphi(x)$ is absolutely convergent and $\xi(m)$ is a bounded function of m, (b) each of its terms satisfies the difference equations and therefore the sum does also, (c) for $n = 0$ it reduces to the series for $\varphi(j\Delta x)$ and hence satisfies the initial condition, (d) the oddness of $\varphi(x)$ implies that A_{-m} is the complex conjugate of A_m which implies the oddness, in j, of u_j^n which in turn implies $u_0^n = 0$, and (e) a similar argument applies at the boundary $j = J$; therefore, all the equations are satisfied.

It may be noted in passing that since the function given by the Fourier series in (1.13) is relevant only at the net points, that is, at $x = j\Delta x$, there is a great deal of arbitrariness in the choice of the coefficients. The choice made above is convenient, but we could equally well have written

$$u_j^n = \sum_{-J}^{J}{}_{(m)} B_m e^{imj\Delta x} [\xi(m)]^n,$$

where the B_m are to be obtained from the initial conditions regarded as $2J + 1$ linear equations[3] in the $2J + 1$ unknowns B_{-J} to B_J.

The interpretation of (1.13) is this: because of the separability of variables in the present problem, each harmonic grows (or decays) independently of the others, as time goes on; $\xi(m)$ is the growth factor (or decay factor) for the amplitude of the mth harmonic for a time interval Δt. In the exact solution (1.6), all harmonics decay, and the decay factor

---

[3] The initial function $\varphi(x)$ has been extended into the interval $(-\pi\ 0)$; hence, j now runs from $-J$ to $+J$ in (1.11).

corresponding to (1.12) is exp $(-m^2\sigma \Delta t)$. If this exponential and the cosine in (1.12) are expanded in power series, it is found that the two growth factors agree through first-order terms:

$$\begin{aligned}(1.14) \qquad \xi(m) &= 1 - m^2\sigma\Delta t + \tfrac{1}{12}m^4\sigma\Delta t(\Delta x)^2 - \cdots, \\ e^{-m^2\sigma\Delta t} &= 1 - m^2\sigma\Delta t + \tfrac{1}{2}m^4\sigma^2(\Delta t)^2 - \cdots.\end{aligned}$$

For any harmonic, the two growth factors can be made to agree, to any desired accuracy, by taking Δx and Δt sufficiently small; thus there is hope that the solution of the difference equations may be a good approximation to the solution of the differential equation. On the other hand, it is clear from equation (1.12) that no matter how small Δx and Δt are, there is always some harmonic (some sufficiently high value of m) for which the two growth factors disagree seriously. This may or may not invalidate the approximate solution. The Fourier series for the initial function $\varphi(x)$ was assumed to be absolutely convergent, so that the true solution (1.6) is absolutely convergent for all non-negative t. Therefore, when one has decided on a certain degree of accuracy for the calculation, all harmonics above one of a certain suitably high order, say $m = m_0$, are negligible at all times in the true solution. It does no harm for these higher harmonics to be falsified in the approximate solution provided only that they do not become amplified to such an extent as to be no longer negligible. It is intuitively evident, therefore, that the stability condition is

$$(1.15) \qquad \text{Max}_{(m)}\,|\xi(m)| \leq 1.$$

If this condition is satisfied, no harmonic is amplified at all, whereas if it is violated, there is some harmonic that is amplified without limit as n increases.

Of course it may happen that for very special choices of the initial function the initial amplitudes A_m of the higher harmonics referred to above $(m > m_0)$ are exactly zero, in which case they will remain zero after any amount of amplification, provided that all the arithmetic steps in the solution of the difference equation are performed with infinite precision (e.g., with no round-off allowed), but the condition (1.15) may be regarded as necessary in practice.

We are now in a position to answer the first question raised at the beginning of this section: *The error $u_j^n - u(j\Delta x, n\Delta t)$ remains bounded as $n \to \infty$ for fixed $\Delta x, \Delta t$, for a general initial function with absolutely convergent Fourier series, if and only if* (1.15) *is satisfied.* By way of proof, note first that since the true solution $u(x, t)$ is bounded as $t \to \infty$, the error will be bounded if and only if u_j^n is bounded as $n \to \infty$.[4] But

$$|u_j^n| = |\sum_{(m)=-\infty}^{\infty} A_m e^{imj\Delta x}[\xi(m)]^n| \leq \sum_{(m)=-\infty}^{\infty} |A_m| \cdot |\xi(m)|^n,$$

so that if $|\xi(m)| \leq 1$ for all m as required by (1.15), we have

$$|u_j^n| \leq \sum_{(m)=-\infty}^{\infty} |A_m|.$$

Furthermore, we assumed at the beginning that $\varphi(x)$ has an absolutely convergent Fourier series; therefore, the last summation converges and u_j^n is bounded. On the other hand, if $|\xi(m)| > 1$ for some m, say $m = m_1$, it suffices to take $\varphi(x) = \sin m_1 x$ to exhibit an unbounded solution, for in this case u_j^n is simply $[\xi(m_1)]^n \sin m_1 j\Delta x$.

To apply the stability condition (1.15) to the present example, note from (1.12) that the growth factor $\xi(m)$ is real for all real m and never exceeds $+1$. The condition is therefore that the greatest negative value of $\xi(m)$ be not less than -1. The greatest negative value is attained when $\cos m\Delta x = -1$. Since $\Delta x = \pi/J$, this happens when m is an odd multiple of J. *The condition that $\xi(m) \geq -1$ for such values of m is that*

(1.16) $$2\sigma\Delta t/(\Delta x)^2 \leq 1;$$

this is the stability condition for the difference equation system (1.9), (1.10), (1.11).

For the present simple example, the sufficiency of (1.16) can be obtained by much more elementary methods, such as by observing that the difference equations can be written as $u_j^{n+1} = Au_{j+1}^n + Bu_j^n + Cu_{j-1}^n$, where $A = C = \sigma\Delta t/(\Delta x)^2$, $B = 1 - 2\sigma\Delta t/(\Delta x)^2$, so that if condition (1.16) is satisfied, A, B, C, are positive and have sum $= 1$; hence,

$$\text{Max}_{(j)}|u_j^{n+1}| \leq (A + B + C) \text{Max}_{(j)}|u_j^n| = \text{Max}_{(j)}|u_j^n|$$
$$\leq \text{Max}_{(j)}|u_j^{n-1}| \leq \cdots \leq \text{Max}_{(j)}|u_j^0|.$$

The solution therefore is bounded.

---

[4] We make no attempt here to prove the bound of the error is in any sense small.

SEC. 1.4] STABILITY 13

This argument has the virtue that it applies in a slightly modified form even if the coefficients of the differential equation are variable; but the Fourier series method of analyzing stability, introduced by J. von Neumann, can be applied to a wider variety of types of difference equations, where the more elementary methods fail. Furthermore, it gives insight into what happens in practice, as illustrated in Figures 1.1 and 1.2, where one sees that when instability occurs, those harmonics whose wavelengths are comparable to about twice the mesh spacing become unacceptably amplified.

A further method, which we shall study in some detail in Chapter 6, lies somewhere between these two in its range of applicability. This is the so-called energy method, the name being derived from the fact that the argument used is often closely related to an energy conservation principle for the differential equation. A note of warning should be made, however, for in many cases the conserved quantity that one uses is not the physical energy, and then the name can give rise to considerable confusion. The present problem is a good example: the physical energy in the system is proportional to

$$\int_{-\pi}^{\pi} u \, dx,$$

but instead we consider

$$\int_{-\pi}^{\pi} u^2 \, dx.$$

We multiply each side of equation (1.2) by u and integrate with respect to x, performing an integration by parts on the right-hand side. The boundary terms are also zero because of the boundary condition (1.4) and we obtain

$$\frac{\partial}{\partial t} \int_{-\pi}^{\pi} u^2 \, dx = -\sigma \int_{-\pi}^{\pi} \left(\frac{\partial u}{\partial x}\right)^2 dx \le 0.$$

Similarly, for the difference scheme (1.9) we introduce

$$\|u^n\|^2 = \Delta x \sum_{j=1}^{J-1} (u_j^n)^2$$

and try to show that this quantity remains bounded as $n \to \infty$ if (1.16) is satisfied. To do this we multiply (1.9) by $(u_j^{n+1} + u_j^n)$ and sum over j, using the "summation by parts" formula

$$\sum_{j=1}^{J-1} u_j(v_{j+1} - 2v_j + v_{j-1}) = u_J(v_J - v_{J-1}) - u_1(v_1 - v_0)$$
$$- \sum_{j=1}^{J-1} (u_{j+1} - u_j)(v_{j+1} - v_j).$$

Taking the boundary condition (1.10) into consideration, we get

(1.17) $\quad \|u^{n+1}\|^2 - \|u^n\|^2 =$

$$-\tfrac{1}{2} L[(u_1^n + u_1^{n+1})u_1^n \Delta x + \|\Delta_+ u^n\|^2 + \Delta x \sum_{j=1}^{J-1} \Delta_+ u_j^n \cdot \Delta_+ u_j^{n+1}],$$

where $L = 2\sigma \Delta t/(\Delta x)^2$, and we have used the abbreviation $\Delta_+ u(x)$ for the *forward difference* $u(x + \Delta x) - u(x)$. Now by Schwarz's inequality,

$$|\Delta x \sum_{j=1}^{J-1} \Delta_+ u_j^n \cdot \Delta_+ u_j^{n+1}| \leq \|\Delta_+ u^n\| \cdot \|\Delta_+ u^{n+1}\|$$
$$\leq \tfrac{1}{2}(\|\Delta_+ u^n\|^2 + \|\Delta_+ u^{n+1}\|^2).$$

Hence, if we define S_n by

(1.18) $\quad S_n = \|u^n\|^2 - \tfrac{1}{4} L[(u_1^n)^2 \Delta x + \|\Delta_+ u^n\|^2)],$

we find that

(1.19) $\quad S_{n+1} - S_n = -\tfrac{1}{4} L[(u_1^n + u_1^{n+1})^2 \Delta x + \|\Delta_+ u^n\|^2 + \|\Delta_+ u^{n+1}\|^2$

$$+ 2\Delta x \sum_{j=1}^{J-1} \Delta_+ u_j^n \cdot \Delta_+ u_j^{n+1}] \leq 0,$$

so that S_n is non-increasing. It is clear that $S_0 \leq \|u^0\|^2$ and so from (1.19) that $S_n \leq \|u^0\|^2$; hence, the final step is to show that $\|u^n\|^2 \leq$ const. S_n. This is where (1.16) has to be used. For since $u_J^n = 0$,

$$(u_1^n)^2 + \sum_{j=1}^{J-1} (u_{j+1}^n - u_j^n)^2 \leq 4 \sum_{j=1}^{J-1} (u_j^n)^2$$

and, therefore, if $L \leq 1 - \varepsilon, \varepsilon > 0, S_n \geq \|u^n\|^2 - (1 - \varepsilon)\|u^n\|^2 = \varepsilon \|u^n\|^2$. Thus, the solution u^n is bounded as $n \to \infty$.

More important than the question of boundedness is the second question raised at the beginning of the section—the question of the behavior of the error for a given point (x, t) as the mesh is made finer and finer—for we want the error to $\to 0$ in the limit. Suppose that the point (x, t) is a net-point in a net with spacing $\Delta_1 t$, $\Delta_1 x$ with which a finite-difference calculation has been made, and that one wishes to improve the calculation by repeatedly subdividing the net and thereby repeating the calculation of u with successively finer meshes. The value of n corresponding to the fixed time t will of course go to infinity, and it is intuitively clear that one must avoid growth factors greater in absolute value than 1, just as in the previous argument. Therefore, consider another net with increments given by $\Delta x = \Delta_1 x / K$, $\Delta t = \Delta_1 t / K^2$ (K = integer) and assume that $\Delta_1 x$ and $\Delta_1 t$ are such that the quantity L given by

$$L = \frac{2\sigma \Delta t}{(\Delta x)^2} = \frac{2\sigma \Delta_1 t}{(\Delta_1 x)^2}$$

is less than or equal to 1 so that $|\xi(m)| \leq 1$ for all m. Let $u^n(x, t)$ denote the approximate value of $u(x, t)$ obtained at the given point (x, t), where $t = n \Delta t$, by solving the finite-difference equations. We assert: *Under the foregoing assumptions, $u^n(x, t) - u(x, t) \to 0$ as $K \to \infty$.*

Proof: Let m_0 be a positive integer (arbitrary for the moment). For given K,

$$(1.20) \quad |u^n - u| = \left| \left(\sum_{\substack{(m) \\ |m| \leq m_0}} + \sum_{\substack{(m) \\ |m| > m_0}} \right) A_m e^{imx} \cdot [\xi^{t/\Delta t} - (e^{-m^2 \sigma \Delta t})^{t/\Delta t}] \right| = |\Sigma^1 + \Sigma^2|,$$

where Σ^1 and Σ^2 stand for the sums occurring in the second member of the equation. The second sum satisfies

$$|\Sigma^2| \leq 2 \sum_{\substack{(m) \\ |m| > m_0}} |A_m|,$$

because $|\xi| \leq 1$ and $|e^{-m^2 \sigma \Delta t}| \leq 1$; therefore, Σ^2 can be made as small as one wishes by choosing m_0 sufficiently large, because the Fourier series for $\varphi(x)$ is absolutely convergent. For estimating the first sum, note that the quantity in brackets in (1.20) has the form $\xi^n - \eta^n$, where ξ and η are the growth factors. It can be written as the product $(\xi - \eta)(\xi^{n-1} +$

$\xi^{n-2}\eta + \cdots + \eta^{n-1}$), whose second factor is not greater than n since each of its n terms is not greater than 1. Furthermore,

$$\frac{\xi - e^{-m^2\sigma\Delta t}}{m^4(\Delta t)^2}$$

is an analytic function of $m^2\Delta t$ in some neighborhood of the origin, according to the power series expansions in equations (1.14), and is obviously bounded for all real $m^2\Delta t$ greater than any fixed $\varepsilon > 0$, from which it follows that it is bounded for all non-negative $m^2\Delta t$. Let the bound be B. Then,

$$|\Sigma^1| \leq \sum_{\substack{(m) \\ |m| \leq m_0}} m^4(\Delta t)^2 B \frac{t}{\Delta t} |A_m| \leq m_0^4 B t \Delta t \sum_{-\infty}^{\infty} {}_{(m)}|A_m|.$$

Having chosen m_0 large to make Σ^2 small, one now chooses Δt small to make Σ^1 small. Thus the error can be made arbitrarily small by choosing K sufficiently large, and the stated convergence is proved. A similar proof has been given by Hildebrand (1952).

In summary, two points of view on stability have been presented. Whether one thinks of letting t go to infinity with fixed Δt, or of letting Δt go to 0 with fixed t (or, for that matter, both), it is necessary to observe the restriction (1.16) to prevent errors (truncation errors, round-off errors, or errors of any kind) from becoming so amplified as to make gibberish of the whole calculation. In practice, of course, one does not go to either limit, for one has only a finite number of cycles of calculation with a finite mesh, but the symptoms of instability show up in a relatively small number of cycles if condition (1.16) is not adhered to. Whenever finite-difference equations are used for the approximate solution of an initial-value problem, one must know the conditions under which they are stable. The conditions usually amount to a restriction on the permissible size of Δt in terms of the sizes of the other increments, but there are also examples of unconditionally stable and of unconditionally unstable difference equations.

1.5. Implicit Difference Equations

The stability condition (1.16) arrived at for the simple difference system considered in the foregoing section has the unfortunate consequence that if Δx is chosen rather small in the interest of accuracy, the allowed Δt may turn out to be so very small that an enormous number of cycles of calculation is required to complete a problem. The limit imposed on Δt by (1.16) is proportional to the square of Δx. The situation is different with the system

$$(1.21) \qquad \frac{u_j^{n+1} - u_j^n}{\Delta t} = \sigma \frac{u_{j+1}^{n+1} - 2u_j^{n+1} + u_{j-1}^{n+1}}{(\Delta x)^2},$$

which is the same as (1.9) except for the superscripts in the right member. One says that (1.9) and (1.21) utilize *forward* and *backward* time differences, respectively (meaning relative to the time t at which space differences are expressed).

We shall see that (1.21) is stable under all circumstances, but it is more instructive to consider a general family of difference systems obtained by making the right member a weighted average of the right members of (1.9) and (1.21). If $f(x)$ is any function of x, we denote by δf_j or $(\delta f)_j$ the *central difference* $f((j + \tfrac{1}{2})\Delta x) - f((j - \tfrac{1}{2})\Delta x)$, where j may be an integer or an integer plus $\tfrac{1}{2}$, so that $\delta^2 f_j$ or $(\delta^2 f)_j$ denotes $f((j + 1)\Delta x) - 2f(j\Delta x) + f((j - 1)\Delta x)$, etc.; if f depends also on t, there may also be superscripts such as n, $n + 1$, and the like. With this notation, we wish to consider the difference system.

$$(1.22) \qquad \frac{u_j^{n+1} - u_j^n}{\Delta t} = \sigma \frac{\theta(\delta^2 u)_j^{n+1} + (1 - \theta)(\delta^2 u)_j^n}{(\Delta x)^2}$$

where θ is a real constant, generally thought of as lying in the interval $0 \le \theta \le 1$.

When $\theta = 0$, as in the preceding section, the system is called *explicit*. Each equation of an explicit system gives one of the unknowns u_j^{n+1} directly in terms of the known quantities u_j^n. If θ is not 0, one must solve a set of simultaneous linear equations to obtain the u_j^{n+1}, and the system is called *implicit*.

On substituting the expression $Ae^{imj\Delta x}\xi^n$ in place of u_j^n into (1.22), for given A and m, one finds the growth factor is given by

$$(1.23) \qquad \xi = \xi(m) = \frac{1 - (1 - \theta)L(1 - \cos m\Delta x)}{1 + \theta L(1 - \cos m\Delta x)}$$

where L is again an abbreviation for $2\sigma\Delta t/(\Delta x)^2$. Like (1.12), (1.23) has the consequence that $\xi(m)$ is real for all real m, and never exceeds $+1$. In Figure 1.3 the value of $\xi(m)$ is plotted as a function of the argument $y = L(1 - \cos m\Delta x)$ for various values of θ. As y increases through positive values, the value of $\xi(m)$ falls monotonically from 1 to $-(1 - \theta)/\theta$. If $\frac{1}{2} \leq \theta \leq 1$, the asymptote is not less than -1, hence the difference equations are always stable. But if $0 \leq \theta < \frac{1}{2}$, y must be restricted, for

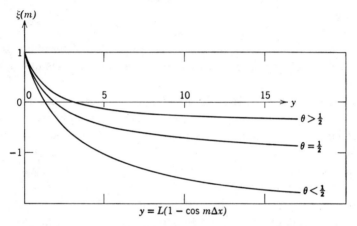

Fig. 1.3. Growth factor for the implicit difference equation (1.22), see also equation (1.23).

stability, by the value at which the curve intersects the line $\xi = -1$. Consequently, the stability condition is

$$(1.24) \qquad \frac{2\sigma\Delta t}{(\Delta x)^2} \leq \frac{1}{1 - 2\theta} \quad \text{if } 0 \leq \theta < \frac{1}{2},$$

$$\text{no restriction} \quad \text{if } \frac{1}{2} \leq \theta \leq 1.$$

It is also a simple matter to show that this is sufficient for stability by the energy method. After operating on equation (1.22) as on (1.9), we obtain an expression for $\|u^{n+1}\|^2 - \|u^n\|^2$ similar to (1.17). Then, defining S_n as

$$(1.25) \qquad S_n = \|u^n\|^2 - \tfrac{1}{4}(1 - 2\theta)L[(u_1^n)^2\Delta x + \|\Delta_+ u^n\|^2]$$

gives the same relation (1.19) as in the special case $\theta = 0$. Hence, $\|u^n\|^2 \leq \text{const.} \, S_n \leq \text{const.} \, S_0 < \text{const.} \, \|u^0\|^2$, if $(1 - 2\theta)L < 1$ which, apart from the limiting case of equality, is the same as (1.24).

The favorite choices for θ are 0, $\tfrac{1}{2}$, 1. The first gives the explicit equation (1.9); the other two give implicit equations which are unconditionally stable. The choice θ = 1 gives the equation (1.21). Using them, at the expense of solving systems of simultaneous equations (which does not turn out to be very costly in practice—see Chapter 8) one can avoid all stability worries and choose a value of Δt based solely on such considerations as desired accuracy.

In Section 8.6 of Chapter 8, a numerical example is given in which the difference system (1.22) is adapted to the non-linear equation

$$\frac{\partial}{\partial t} u = \frac{\partial}{\partial x^2} u^5.$$

Here, the variable quantity $5u^4$ plays the role of the present σ. The results indicate that (1.24) is still the correct stability condition if σ is replaced by $5u^4$, as would be expected.

1.6. The Truncation Error

According to the assertion of Section 1.4, the solution of the simple explicit difference equation actually converges to the true solution of the initial-value problem if Δt and Δx go to zero in such a way that the stability condition (1.16) is satisfied. Corresponding statements hold for the implicit equations. For practical purposes, the *rate* of convergence is also of great interest; in this section it will be discussed in terms of the truncation error. Let $\tilde{u}(x, t)$ be an exact solution of (1.2) with continuous partial derivatives of the orders appearing below. Denote $\tilde{u}(j\Delta x, n\Delta t)$ by $\tilde{u}_j^n$ etc., even for non-integer j and n. By Taylor's series with remainder, we have the expansions

$$\tilde{u}_j^{n+1} = \tilde{u}_j^n + \Delta t \left(\frac{\partial \tilde{u}}{\partial t}\right)_j^n + \tfrac{1}{2}(\Delta t)^2 \left(\frac{\partial^2 \tilde{u}}{\partial t^2}\right)_j^{n+\theta_1},$$

$$\tilde{u}_{j+1}^n = \tilde{u}_j^n + \Delta x \left(\frac{\partial \tilde{u}}{\partial x}\right)_j^n + \tfrac{1}{2}(\Delta x)^2 \left(\frac{\partial^2 \tilde{u}}{\partial x^2}\right)_j^n + \tfrac{1}{6}(\Delta x)^3 \left(\frac{\partial^3 \tilde{u}}{\partial x^3}\right)_j^n$$

$$+ \tfrac{1}{24}(\Delta x)^4 \left(\frac{\partial^4 \tilde{u}}{\partial x^4}\right)_{j+\theta_2}^n$$

and a similar expansion for $\tilde{u}_{j-1}^n$; θ_1 and θ_2 are numbers between 0 and 1. Since $\tilde{u}$ satisfies the differential equation, $\partial \tilde{u}/\partial t$ can be replaced by $\sigma \partial^2 \tilde{u}/\partial x^2$; then, by substituting, we find

$$\frac{\tilde{u}_j^{n+1} - \tilde{u}_j^n}{\Delta t} - \sigma \frac{\tilde{u}_{j+1}^n - 2\tilde{u}_j^n + \tilde{u}_{j-1}^n}{(\Delta x)^2} = $$
$$\tfrac{1}{2}\Delta t \left(\frac{\partial^2 \tilde{u}}{\partial t^2}\right)_j^{n+\theta_1} - \frac{\sigma}{24}(\Delta x)^2 \left[\left(\frac{\partial^4 \tilde{u}}{\partial x^4}\right)_{j+\theta_2}^n + \left(\frac{\partial^4 \tilde{u}}{\partial x^4}\right)_{j+\theta_3}^n\right].$$

The coefficients of Δt and of $(\Delta x)^2$ appearing in the right member of the equation are bounded, because of the continuity of the partial derivatives. This fact is expressed in symbolic notation as follows:

$$(1.26) \quad \frac{\tilde{u}_j^{n+1} - \tilde{u}_j^n}{\Delta t} - \sigma \frac{(\delta^2 \tilde{u})_j^n}{(\Delta x)^2} = O(\Delta t) + O[(\Delta x)^2] \quad \text{as } \Delta t, \Delta x \to 0.$$

This equation is to be interpreted as meaning that there exist two positive constants K_1 and K_2, depending on $\tilde{u}$, such that the absolute value of the left member is less than or equal to $K_1 \Delta t + K_2 (\Delta x)^2$ for all sufficiently small Δt and Δx. When there can be no confusion, the qualifying phrase "as $\Delta t, \Delta x \to 0$" is omitted from the equation.

We shall call the left member of (1.26) the *truncation error* of the difference system (1.12). Some authors multiply it by Δt and call the result the truncation error, but we prefer to take the above expression as it stands.

Equation (1.26) tells how closely a solution of the differential equation satisfies the difference equations as the mesh is made finer and finer. From it, one can derive information on how rapidly the solution of the difference system approaches that of the differential equation. In the present example this is particularly simple: if we denote by ε_j^n the difference $u_j^n - \tilde{u}_j^n$, by subtracting (1.26) from (1.9), we get

$$\varepsilon_j^{n+1} = \tfrac{1}{2}L\varepsilon_{j-1}^n + (1 - L)\varepsilon_j^n + \tfrac{1}{2}L\varepsilon_{j+1}^n + O[(\Delta t)^2] + O[(\Delta x)^2 \Delta t].$$

Hence, if the stability condition $L \le 1$ is satisfied,

$$\text{Max}_{(j)} |\varepsilon_j^{n+1}| \le \text{Max}_{(j)} |\varepsilon_j^n| + K_1(\Delta t)^2 + K_2(\Delta x)^2 \Delta t.$$

Therefore, for $n\Delta t \le t$,

$$\text{Max}_{(j)} |\varepsilon_j^n| \le t[K_1 \Delta t + K_2(\Delta x)^2],$$

i.e., $|\varepsilon_j^n|$ is bounded by t times the bound for the local truncation error. The fact that this tends to zero as $\Delta x \to 0$, with L fixed, proves convergence.

When further terms are included in (1.26), the terms of order Δt and of order $(\Delta x)^2$ are

$$\frac{\Delta t}{2} \frac{\partial^2 \tilde{u}}{\partial t^2} - \sigma \frac{(\Delta x)^2}{12} \frac{\partial^4 \tilde{u}}{\partial x^4}.$$

But since $\tilde{u}$ satisfies the heat flow equation, it also satisfies the equation

$$\frac{\partial^2 \tilde{u}}{\partial t^2} = \sigma^2 \frac{\partial^4 \tilde{u}}{\partial x^4},$$

and the dominant terms become

$$\left(\frac{\sigma^2 \Delta t}{2} - \sigma \frac{(\Delta x)^2}{12} \right) \frac{\partial^4 \tilde{u}}{\partial x^4}.$$

Therefore, if Δt and Δx go to zero in such a way that

$$\sigma \Delta t / (\Delta x)^2 = 1/6,$$

the above expression vanishes, and the truncation error is now $O[(\Delta t)^2]$, or, what is the same thing, $O[(\Delta x)^4]$.[5] In this special case, the error goes to zero even faster than is indicated by (1.26), and one would expect that the solution of the difference equation would approach the solution of the differential equation with special rapidity. This is indeed true and has been made use of by some authors to obtain accurate numerical solutions of the simple heat flow problem. Unfortunately, there does not appear to be any simple way of generalizing this trick to the general heat flow or diffusion problem with variable coefficients. However, certain other schemes for obtaining higher order accuracy can be generalized to some extent. Two such schemes are given for the heat flow problem in Section 8.3.

Generally, the smaller the truncation error, the faster the convergence of the numerical solution to the true one. However, it should be noted that (1.26) continues to hold even if the stability condition is violated, and in fact the essence of instability can be described in the pseudo-paradoxical form: if the mesh is refined, but in such a way as to violate

---

[5] Note also that in this case the expansions (1.14) agree to second-order terms.

the stability condition, the exact solution of the differential equation comes closer and closer to satisfying the difference equations, but the exact solution of the difference equations departs, in general, more and more from the true solution of the initial-value problem.

1.7. Rate of Convergence

An estimate of the rate of convergence for the simple difference equation (1.9) can also be obtained from an examination of the two partial sums Σ^1 and Σ^2 into which the error was broken in the proof of convergence in Section 1.4, see equation (1.20). There the error was taken as the difference between the solution u^n of the difference equation and the solution u of the differential equation at some fixed point (x, t) common to an infinite sequence of nets with progressively finer subdivisions; we desire more information on how rapidly this error tends to zero along the corresponding sequence of calculations than was obtainable by the methods of the previous section. We shall find that the result depends on the asymptotic behavior of the Fourier coefficients of the initial function $\varphi(x)$, which in turn depends on the smoothness of this function, in accordance with the following lemma: *If $\varphi(x)$ and its first $p - 1$ derivatives are continuous, and the pth derivative has bounded variation, then the asymptotic behavior of the Fourier coefficient is given*[6] *by*

$$A_m = O\left(\frac{1}{m^{p+1}}\right), \quad \text{as } m \to \infty.$$

A bound for Σ^2 follows immediately from this lemma: for $p > 0$,

(1.28) $$|\Sigma^2| \leq 2 \sum_{|m| > m_0}^{(m)} |A_m| = O\left(\frac{1}{m_0^p}\right).$$

A bound for Σ^1 also follows, for if we write $A_m \leq C_0/m^{p+1}$, then the estimate for Σ^1 found in Section 1.4 becomes

$$|\Sigma^1| \leq \sum_{|m| \leq m_0} m^4 (\Delta t)^2 B \frac{t}{\Delta t} |A_m| \leq C_0 B t \Delta t \sum_{1 \leq |m| \leq m_0} m^{4-p-1}.$$

---

[6] See Zygmund (1952), page 18, for proof of the case $p = 0$; for $p > 0$ the result is obtained by p-fold term-by-term integration.

If $p \leq 3$, the last term in the sum is the largest and may be used as a bound for the others; if $p = 4$, then the sum diverges only logarithmically, as $m_0 \to \infty$, while, for $p > 4$, it converges. Therefore,

$$|\Sigma^1| \leq \begin{cases} C_1 m_0{}^{4-p}\Delta t & \text{if } p \leq 3 \\ C_2 (\ln m_0)\Delta t & \text{if } p = 4 \\ C_3 \Delta t & \text{if } p > 4 \end{cases}$$

where C_1, C_2, C_3 are constants, by which is meant, in this context, numbers that may depend on x, t, p, σ, etc., but not on Δt or m_0. The parameter m_0, which is arbitrary, is now so chosen as to minimize the resulting estimate of $|\Sigma^1| + |\Sigma^2|$. For $p \leq 4$, it turns out, the minimizing value of m_0 is proportional to $(\Delta t)^{-1/4}$; hence, the final estimates are

$$(1.29) \quad u^n - u = \begin{cases} O[(\Delta t)^{p/4}] = O[(\Delta x)^{p/2}] & \text{for } p \leq 3 \\ O(\Delta t |\ln \Delta t|) = O[(\Delta x)^2 |\ln \Delta x|] & \text{for } p = 4 \\ O(\Delta t) = O[(\Delta x)^2] & \text{for } p > 4. \end{cases}$$

If the initial function $\varphi(x)$ is a broken linear function, as in the example illustrated in Figures 1.1 and 1.2, then p has the value 1, and the error goes to zero as $(\Delta x)^{1/2}$, as $\Delta x \to 0$; at the other extreme, if $\varphi(x)$ is analytic, the error vanishes as $(\Delta x)^2$.

If, as discussed above, we make Δt and Δx go to zero in such a way that $\sigma \Delta t/(\Delta x)^2 = 1/6$, the argument can be modified to give the estimates:

$$(1.30) \quad u^n - u = \begin{cases} O[(\Delta t)^{p/3}] & \text{for } p \leq 5 \\ O[(\Delta t)^2 |\ln \Delta t|] & \text{for } p = 6 \\ O[(\Delta t)^2] & \text{for } p > 6. \end{cases}$$

All these estimates can be improved by a more complicated argument, but they illustrate the trends. Note that for smooth initial data ($p > 6$) the more accurate difference equation gives considerably more rapid convergence, whereas for rough data ($p = 1$) the gain is only from $(\Delta t)^{1/4}$ to $(\Delta t)^{1/3}$, and even this gain results from the peculiarity of the heat flow equation that for any initial data the solution is analytic for all $t > 0$. For hyperbolic problems, there is generally no gain at all, for rough data, from use of the more accurate difference equations.

If the exact solution $u(x, t)$ is assumed to have continuous derivatives up to the fourth order in x, then the simpler error estimate made in Section 1.6 shows that in any case the error goes to zero as Δt.

1.8. Comments on High-Order Formulas and Rounding Errors

For partial differential equations, it is a general experience that those difference formulas that claim a high order of accuracy, for example by having truncation error $O[(\Delta t)^4]$, are usually quite disappointing in practice. The situation is in contrast with that for ordinary differential equations, where methods like the Runge-Kutta method often achieve a phenomenal accuracy with a quite small amount of labor; the latter methods have no true counterpart for partial differential equations. The reason for the difference is a basic one: for an ordinary differential equation, with the independent variable t, the specification of a finite (and usually quite small) number of quantities corresponding to $t = t_0$ suffices (at least in principle) to determine the solution completely and exactly for all $t > t_0$; how accurately one then calculates the solution for $t = t_0 + \Delta t$ depends only on the skill with which one makes use of the information available; for a partial differential equation, one of whose independent variables is t, it would be necessary to specify the values of infinitely many quantities (or the equivalent) at $t = t_0$ to determine the solution for $t > t_0$; therefore, in calculating the solution for $t = t_0 + \Delta t$, one is limited not merely by a possible lack of skill, but also by a lack of needed information.

One can give more information about the initial functions than merely their values at the net points by specifying that they are smooth (say differentiable a certain number of times); hence, for smooth initial functions, a modest increase in accuracy can be obtained by use of improved difference formulas provided that all boundary conditions, singularities, jump conditions on internal boundaries, etc. are treated with comparable accuracy. This question will be taken up again in Chapter 13.

In any case, however, finite-difference methods for partial differential equations seldom achieve more than a rather modest accuracy. It is the authors' impression that, generally, rounding errors are consequently of not much importance in this field. When stable difference equations are used, the rounding errors are not amplified as time goes on; they merely accumulate roughly in proportion to the square root of the number of steps in the calculation, if rounding is unbiased and if care is used to prevent quantities from getting badly out of scale. To be sure, modern rapid automatic computing machines are capable of handling calculations with a very large number of steps, and are often called upon to do so, so that rounding errors can accumulate appreciably, but these machines usually also carry a fairly large number of decimal or binary places, and an increase by a factor of a million in speed can be compensated for, so far as the accumulation of unbiased rounding errors is concerned, by merely carrying three more places of decimals. Furthermore, significant-digit arithmetic in some form (see Ashenhurst and Metropolis, 1959) may eventually be incorporated in most machines and can monitor the effect of rounding errors more or less automatically, whereas truncation errors and other errors of approximation must be evaluated by analysis.

1.9. Outline of the Remainder of the Book

It has been seen that the basic considerations for difference methods of the kind discussed in this chapter are (1) convergence to the true solution as the mesh is refined (in the above example, this depended on stability and on vanishing of the truncation error as $\Delta x, \Delta t \to 0$), (2) the rate of convergence, (3) construction of difference systems to obtain stability and rapid convergence, and (4) methods of solving implicit systems. Additional considerations for more complicated problems are (5) influence of variable coefficients and non-linearities on stability, (6) special procedures for boundary conditions, external or internal, and their effect on stability, (7) practical stability criteria for cases in which the formal theory is either too difficult to apply or inadequate. A general theoretical discussion of the considerations (1) to (6) is given in Part I of this book; various applications to initial-value problems of mathematical physics are then discussed in Part II from a practical point of view.

Chapters 2 and 3 give the general theory for linear problems. Various concepts are defined in terms of operators in a Banach space, leading to the Lax theorem on the equivalence of stability and convergence: the notion of a properly posed initial-value problem is defined, and finite difference approximations to it are considered; a consistency condition is formulated, which ensures that the difference system does in fact approximate the given problem rather than, for example, some other problem; stability of the approximation is then defined as the uniform boundedness of a certain set of operators. The Lax theorem states that if one has a properly posed initial-value problem and the consistency condition is satisfied, then stability is necessary and sufficient for convergence to the true solution for arbitrary initial data. Certain alternative definitions of stability are mentioned briefly in Section 5.2 (see also Section 3.9).

In Chapter 4, the theory is specialized to pure (linear) initial-value problems with constant coefficients. As observed many years ago by von Neumann (see O'Brien, Hyman, and Kaplan, 1951), the considerations can in this case be put into a particularly simple and useful form by the use of Fourier transformations, which yield immediately the von Neumann (necessary) condition for stability. During the years since the first edition of this book was published, general sufficient conditions have been obtained through the work of Kato, Buchanan, and Kreiss; their results are given in detail, with simplified proofs.

Chapter 5 is devoted to problems with variable coefficients (mostly parabolic and hyperbolic) and to a brief discussion of certain non-linear (quasi-linear) problems in the last section. The important notion of dissipative difference schemes, due to Kreiss, is introduced. For such schemes, von Neumann's idea of stability as a local consideration is rigorously justified. Furthermore, there seems to be no difficulty in practice (and no loss of accuracy), in always choosing dissipative schemes for hyperbolic problems and other problems which are non-dissipative on the differential-equation level. Lastly, empirical evidence indicates, as shown in the last section, that dissipativity of the right amount and kind often provides a cure for non-linear instabilities.

Chapter 6 takes up the question of the influence of boundary conditions, mainly for problems with one space variable. There are two main methods for investigating stability in such problems: the so-called energy method and the normal-mode method of Godunov and Ryabenkii; these often complement each other.

Chapter 7 is a sort of appendix to Part I in which it is shown how the theory of Chapter 3 on linear problems can be extended to multi-level difference systems.

Part II deals with the practical details of methods that have been found useful in some applications and some methods that are still very much in the exploratory stage (for multidimensional fluid dynamics). In a number of cases it is found that the theoretical stability concept of Part I, which is concerned with the behavior of the calculation in the limit as Δt, $\Delta x, \ldots \to 0$, has to be augmented by a practical stability criterion, discussed in Section 10.5 and elsewhere, concerned with finite values of Δt, Δx, etc.

In conclusion, it should be mentioned that the developments described in this book have been strongly paralleled by much excellent work in this area by Soviet mathematicians; e.g. in Ladyzhenskaya's paper of 1952, she proved the stability and convergence of an implicit scheme of Crank-Nicholson type for general hyperbolic systems. For general discussion of finite-difference methods, see Ladyzhenskaya (1957a) and the book of Godunov and Ryabenkii (1964). Discontinuous solutions of hyperbolic equations, to which we refer again in Chapter 13, have been discussed by Ladyzhenskaya (1957b), by Oleinik and Vvedenskaya (1957), by Rozhdestvenskii (1960), and by Yanenko (1964).

CHAPTER 2

Linear Operators

2.1. The Function Space of an Initial-Value Problem

In the solution of an initial-value problem, the time variable t plays a special role. At any stage of the process one has at hand one or more functions of some other variables, which we shall call space variables, and these functions describe an instantaneous state of a physical system. When a computing machine is used, tabular values of these functions are stored in the storage or memory units of the machine. As time goes on, the state of the system changes according to the differential equations, and the function values become replaced by ones describing later states of the system. It is convenient to think of these functions, for fixed t, as represented by an element or point in a function space, and to denote them by a single symbol u. Each point in the function space represents a state of the system, and the succession of states of the system as time goes on is represented by a motion of the representative point in the function space.

For discussion of approximations and errors, one needs to define a measure of the difference between two states of the system, which can be interpreted as the distance between the two representative points.

If the two states are almost identical, the distance between the representative points should be very small, and conversely. For example, if the general point u represents a continuous function of certain space variables, like the temperature in the heat flow problem, one may choose the measure of the difference of two states to be the maximum magnitude of the difference of the corresponding functions $u_1(\mathbf{x})$ and $u_2(\mathbf{x})$, i.e., the maximum of the function $|u_1(\mathbf{x}) - u_2(\mathbf{x})|$. Another familiar choice of measure of difference is the root-mean-square difference, $[(1/V) \int |u_1(\mathbf{x}) - u_2(\mathbf{x})|^2 dv]^{1/2}$, where the integral is over some region of volume V.

For the remainder of this chapter, and in the next five, the discussion will be restricted to linear problems. In this case the measure of difference of two states u and v is *always* some property of the difference $w = u - v$ which may be interpreted as the magnitude of the element w. It is called the *norm* of w and written $\|w\|$. Clearly, the norm of w should be a positive number if u and v are different states and should be zero if they are the same, because the norm is to have the interpretation of the magnitude of the difference of the two states. The representation as a distance suggests as a further requirement the triangle inequality, i.e., that the norm of the sum of two elements, $u + v$, should not exceed the sum of the norms of u and v, corresponding to the fact that the length of one side of a triangle cannot exceed the sum of the lengths of the other two. It is perhaps not obvious whether this requirement is dictated by the physical interpretation, but in fact the triangle inequality is satisfied by any choice of norm that is likely to be used in practice, and it simplifies the operator theory to assume that it is always satisfied, and to make certain other assumptions which will appear formally in the next section.

If, as above, u represents a single continuous function and its norm is taken as the maximum of its magnitude (this is called the *maximum norm*), the convergence[1] of a sequence of points in the function space corresponds to uniform convergence of the corresponding functions. On the other hand, if u represents a measurable and quadratically integrable function and its norm is taken as its root-mean-square, convergence of the points corresponds to convergence-in-the-mean[2] of the functions.

By the sum or difference of two elements, u and v, is of course meant the function or set of functions obtained by adding or subtracting the functions or sets of functions represented by u and v. The result is supposed to be another element of the function space. In this way we may arrive at elements which have no *direct* interpretation as states of the physical system. For example, in the problem considered in the first chapter, as applied to the diffusion of the molecules of a gas through some permeable system, a function with negative values would have no direct meaning since there is no such thing as a negative number of molecules. But it is desirable to admit such functions to the function space, nevertheless, and to think of them as in some sense representing generalized states

---

[1] By this is meant that the distance between the variable point and the limit-point tends to zero.

[2] See Courant and Hilbert, Vol. I (1953), pages 51 to 54.

of the system; then the individual terms and partial sums of a Fourier series, for example, are represented by points in the space. It is convenient to generalize still further and admit complex-valued functions, in accordance with certain rules characterizing a linear space.

By considerations of this sort, it is seen that the axioms of Banach space theory are suitable for the characterization of the function spaces arising in linear initial-value problems. The relevant portions of the theory of linear operators in a Banach space will occupy the remainder of this chapter.

The function values obtained from a finite-difference calculation of course fail to give a complete description of the functions. But this set of function values can nevertheless be represented by a point in the function space if some linear rule or procedure is given whereby the function values at space points between the net points are somehow filled in.

For example, we may imagine the values at intermediate points to be defined by linear or other interpolation. Another way, useful for the problems considered in Chapter 4, is to regard the difference equations as holding for all values of the space variables. For the heat flow example, equation (1.9) would take the form

$$\frac{u^{n+1}(x) - u^n(x)}{\Delta t} = \sigma \frac{u^n(x + \Delta x) - 2u^n(x) + u^n(x - \Delta x)}{(\Delta x)^2},$$

and if the initial function $\varphi(x)$ is given for all x, this equation determines the functions $u^n(x)$ inductively ($n = 1, 2, 3, \ldots$) for all x; they are actually calculated, of course, only at a finite set of values of x.

Some authors prefer to consider both the exact and approximate functions as net-functions, i.e., as functions defined only at the net points, in which case the maximum norm is usually used. We think it slightly more in accord with the physical concepts to regard the calculated points as describing, by curve-fitting of some kind, approximations to the complete functions.

2.2. Banach Spaces

The function space, which will henceforth be called $\mathscr{B}$, of a linear initial-value problem is supposed to satisfy the following axioms of an abstract space (Banach, 1932).

SEC. 2.2] BANACH SPACES 31

A Banach space is first of all a linear vector space; that is, it consists of elements for which the operations of addition and of multiplication by a number (real or complex) are defined, subject to the following rules:

If u, v, w are elements (also called points) of $\mathscr{B}$ and a, b, c are numbers, then

$u + v$ is an element of $\mathscr{B}$,
$u + v = v + u$,
$u + (v + w) = (u + v) + w$,
au is an element of $\mathscr{B}$,
$a(bu) = (ab)u$,
$(a + b)u = au + bu$,
$a(u + v) = au + av$.

It is obvious that the function space of an initial-value problem, as discussed above, satisfies these axioms, provided that the addition of two elements is interpreted in the usual way as the addition of the corresponding function values, and multiplication by a is interpreted as multiplication of the function values by a.

Secondly, $\mathscr{B}$ is a normed space; that is, to each u of $\mathscr{B}$ there corresponds a finite non-negative real number, denoted by $\|u\|$, such that

$\|au\| = |a| \; \|u\|$,
$\|u + v\| \leq \|u\| + \|v\|$,
$\|u - v\| = 0$ if and only if $u = v$

("$u - v$" is an abbreviation for "$u + (-1)v$").

Finally $\mathscr{B}$ is a complete space; that is, if u_1, u_2, u_3, ... is a sequence of elements of $\mathscr{B}$ such that $\|u_n - u_m\| \to 0$ as n, $m \to \infty$ independently (such a sequence is called a *Cauchy sequence*), then $\mathscr{B}$ contains an element u, called the *limit* of the sequence, such that $\|u_n - u\| \to 0$ as $n \to \infty$.

A familiar example of a function space which is a Banach space is the following: let $\mathscr{B}$ be the class of continuous functions $f(x)$ for $a \leq x \leq b$, and let the norm of an element $f(x)$ be the maximum of $|f(x)|$ for $a \leq x \leq b$. The foregoing axioms are readily found to be true in this case. Convergence in $\mathscr{B}$ corresponds to uniform convergence of functions, and the completeness of $\mathscr{B}$ comes about as follows: a Cauchy sequence $f_n(x)$ converges for each x because of the Cauchy convergence theorem, and the limit function is an element of $\mathscr{B}$ because the limit of a uniformly convergent sequence of continuous functions is continuous. This function space and its generalizations are much used in numerical analysis, but for the purposes of this book it will be more convenient to use spaces based on the Hilbert norm, e.g.,

$$\left[\int_a^b |f(x)|^2 dx\right]^{1/2},$$

as will be discussed in Chapter 4.

We now state some simple consequences of the above axioms (anyone interested in the axiomatic method should derive them from the axioms, but for the purposes of this book it suffices to observe that they are also obviously true for a linear function space):

$$1u = u,$$
$$u - u = v - v = w - w, \text{ etc.},$$

(we therefore define $\phi = u - u$; ϕ is called the null element of $\mathscr{B}$),

$$\|\phi\| = 0,$$
$$0u = \phi,$$
$$a\phi = \phi,$$
$$\|u - v\| \geq |\|u\| - \|v\||.$$

If a sequence has a limit, the limit is unique.

Certain notions concerning sets of elements of $\mathscr{B}$ are defined as follows: An element u is called a *limit point* of a set $\mathscr{S}$ if there are points of $\mathscr{S}$ arbitrarily close to u, i.e., if u is the limit of some sequence contained in $\mathscr{S}$. A set is called *closed* if it contains all its limit points. The *closure* of a set is the set obtained by adding to the given set (if it is not already closed) all its limit points. A set $\mathscr{S}$ is *dense in* $\mathscr{B}$ if any element of $\mathscr{B}$ can be approximated arbitrarily closely by an element of $\mathscr{S}$; in other words, if every element of $\mathscr{B}$ is the limit of some sequence of $\mathscr{S}$, i.e., if the closure of $\mathscr{S}$ is $\mathscr{B}$. For example, if $\mathscr{B}$ is the class C of continuous functions with maximum norm mentioned above, the set of polynomials of x is dense in $\mathscr{B}$ because a continuous function on a finite closed interval can be uniformly approximated by a polynomial. Denseness of one set in another is similarly defined; any element of the second set is the limit of some sequence of elements of the first set. If u is a point in $\mathscr{B}$ and $\mathscr{S}$ is a set of points in

$\mathscr{B}$, the *distance* $(u, \mathscr{S})$ is the greatest lower bound of $\|u - v\|$ for v in $\mathscr{S}$. It may be shown that if $\mathscr{S}$ is a closed set and u is a point not in $\mathscr{S}$, this distance is necessarily greater than zero (assuming it to be zero leads to a contradiction). If u_0 is an element of $\mathscr{B}$ and r is a positive number, the set of points u satisfying $\|u - u_0\| \leq r$ is called a *sphere* with u_0 as *center* and r as *radius*; it is a closed set.

2.3. Linear Operators in a Banach Space

The operations that one performs on functions in connection with solving initial-value problems (like differentiating, integrating, differencing, multiplying by a given function) have the effect of replacing one function (or set of functions) by another. The result is a transformation of the Banach space into itself or of a part of the Banach space into another part of it. Generally, a *transformation* T in $\mathscr{B}$ is a many–one correspondence between elements of a set $\mathscr{D} = \mathscr{D}(T)$, called the *domain* of T and those of a set $\mathscr{R} = \mathscr{R}(T)$, called the *range* of T. Sometimes one considers transformations from one space to another, but in most cases of interest to us, $\mathscr{R}$ and $\mathscr{D}$ are both sets of elements in $\mathscr{B}$. Transformations T and T' are said to be *identical* or *equal* if and only if their domains are the same, and if for any u in the common domain, $Tu = T'u$.

An *operator* or *operation* is a transformation; these three words are used interchangeably, although we shall think of "transformation" as applying to an abstract Banach space and of "operator," to the function space it represents. Note that according to this usage specifying the domain is an essential part of specifying an operator. An operator d/dx that is permitted to apply to any differentiable function is not regarded as the same as an operator d/dx if the domain is restricted to, say, twice-differentiable functions or to functions that vanish at $x = 0$. (Needless to say, the symbol "d/dx" should not be used for one of these operators unless it is made clear which one is meant.) However, the first of these differential operators is a so-called extension of either of the others. A transformation T' is an *extension* of the transformation T if $\mathscr{D}(T)$ is contained in $\mathscr{D}(T')$ and $Tu = T'u$ for any u in $\mathscr{D}(T)$. The extension of an operator has the same effect as the original operator whenever the original one can be applied, but it can be applied to a wider class of functions.

A transformation T is called *linear* if and only if for any elements u and v in the domain and any complex numbers a and b, $au + bv$ is in $\mathscr{D}(T)$ and $T(au + bv) = aTu + bTv$. That is, not only must the effect of the operator depend linearly on the element operated on, but the domain must be a linear manifold.

A linear transformation T is called *bounded* if and only if there is a real number K such that $\|Tu\| \leq K\|u\|$ for all u in $\mathscr{D}(T)$. The smallest value of K for which this is true is called the *bound* of T and is denoted by $\|T\|$. (The reason for this notation is that under some circumstances a set of operators can also be regarded as a set of elements in another Banach space, in which the norm of an element is identified as the bound of the corresponding operator.) Note that the triangle inequality holds for the bounds of operators: $\|T_1 + T_2\| \leq \|T_1\| + \|T_2\|$.

2.4. The Extension Theorem

A bounded linear transformation T whose domain is dense in $\mathscr{B}$ has a unique bounded linear extension T' whose domain is the entire space and such that $\|T'\| = \|T\|$.

To see that this is so, let u be any element of $\mathscr{B}$. By hypothesis there exists a sequence of elements $u_1, u_2, u_3, \ldots$ all belonging to the domain of T and converging to u. Therefore,

$$\|Tu_n - Tu_m\| = \|T(u_n - u_m)\| \leq \|T\| \cdot \|u_n - u_m\| \to 0 \text{ as } n, m \to \infty.$$

Therefore, the sequence of elements $Tu_1, Tu_2, Tu_3, \ldots$ is a Cauchy sequence and has a limit, by the axiom of completeness. We shall identify the limit with $T'u$. The reader can readily verify that (1) the limit is just Tu if u is in the domain of T, (2) the limit is unique (independent of the choice of the sequence $u_1, u_2, u_3, \ldots$ for given u), (3) the limit depends linearly on u, and (4) $\|T'u\| \leq \|T\| \cdot \|u\|$, thus proving the theorem.

2.5. The Principle of Uniform Boundedness

A collection or class C of linear transformations, defined on the whole space, is called *uniformly bounded* if there is a real number K such that

THE PRINCIPLE OF UNIFORM BOUNDEDNESS

$\|Tu\| \leq K\|u\|$ for all u in $\mathscr{B}$ and all T in C. Clearly, in this case all the transformations of C are individually bounded. But a class C of bounded transformations is not necessarily uniformly bounded (unless it is a finite class) unless some additional condition is imposed. Such a condition is contained in the following theorem: *If C is a class of bounded linear transformations defined in all $\mathscr{B}$, and if, for each element u of $\mathscr{B}$, there exists a real number $K_1(u)$ such that $\|Tu\| \leq K_1(u)$ for all T in C, then the class C is uniformly bounded.*

Proof: Let $K_0(T)$ denote the bound $\|T\|$ of T. What has to be proved is that the inequalities

$$\|Tu\| \leq K_0(T)\|u\|$$

and

$$\|Tu\| \leq K_1(u)$$

yield the existence of a constant K_2 not depending on either u or T such that

$$\|Tu\| \leq K_2\|u\| ;$$

each inequality holding for all u in $\mathscr{B}$ and all T in C. We consider a sequence of sets $M_1, M_2, M_3, \ldots$ of which M_p is defined as the set of all elements u of $\mathscr{B}$ such that $\|Tu\| \leq p\|u\|$ for all T in C. Our aim is to show that one of these sets fills the entire space. The steps are:

1. Each u in $\mathscr{B}$ belongs to at least one of the sets M_p, because ϕ belongs to all M_p and any u not equal to ϕ belongs to all M_p for which $p \geq K_1(u)/\|u\|$.

2. Each M_p is closed; for, if $u_1, u_2, \ldots$ is a sequence of elements of M_p having u as its limit and if ε is any positive number, we can choose n so large that $\|u - u_n\| < \varepsilon$, hence so that $\|Tu - Tu_n\| \leq K_0(T)\varepsilon$. Therefore, by the triangle inequality,

$$\begin{aligned}\|Tu\| &\leq \|Tu_n\| + K_0(T)\varepsilon \\ &\leq p\|u_n\| + K_0(T)\varepsilon \\ &\leq p\|u\| + p\varepsilon + K_0(T)\varepsilon.\end{aligned}$$

Since this can be done for any positive ε, it follows that $\|Tu\| \leq p\|u\|$, so that u is in M_p and M_p is closed.

3. At least one M_p contains a sphere. To prove this, first assume the contrary. Let S_1 be the sphere consisting of all u for which $\|u\| \leq 1$. By assumption, M_1 does not contain S_1, hence there is a point in S_1 that is not in M_1. In fact, there is a point in the interior of S_1 that is not in M_1, for if M_1 contained the entire interior of S_1, it would have to contain the surface also, since M_1 is closed. About such a point construct a sphere S_2, with radius not greater than $1/2$ and in any case so small that S_2 is entirely in S_1 but entirely outside of M_1—this is possible because the point in question is at a positive distance from the closed set M_1 and also from the surface of S_1 (which is also a closed set), according to the last paragraph of Section 2.2. Similarly, construct a sphere S_3 with radius not greater than $1/4$ and lying entirely in S_2 and entirely outside of M_2. Similarly, construct for each n a sphere S_n with radius not greater than 2^{-n} and lying entirely in S_{n-1} and outside of M_{n-1}. The centers of these spheres form a Cauchy sequence, and it is easy to see that the limit of this sequence lies in all the spheres but not in any of the M_n. This contradicts 1. above, and shows our assumption to be false.

4. Let p be such that M_p contains a sphere, and let u_0 and r be the center and radius of the sphere. If $u \neq \phi$ is any point in $\mathscr{B}$, the point $g = u_0 + (r/\|u\|) u$ lies on the surface of the sphere and hence in M_p. Therefore, $\|Tg\| \leq p\|g\| \leq p(\|u_0\| + r)$. But also $u = (\|u\|/r)(g - u_0)$, so that $\|Tu\| \leq (\|u\|/r) [p(\|u_0\|+r) + \|Tu_0\|]$. This is the result claimed by the theorem (it clearly holds also for $u = \phi$), if we identify the constant K_2 as

$$K_2 = \frac{1}{r}[p(\|u_0\|+r) + K_1(u_0)],$$

and proves the principle of uniform boundedness.

The following may be noted: At the beginning of the proof, it was stated that our aim was to show that one of the M_p fills the whole Banach space. Instead of doing this directly, we first showed that one of the M_p contains an entire sphere, although perhaps a very small one. This was the difficult part of the proof, and after that it was easy to show, without further use of the axiom of completeness, that some later member of the sequence of the M_p fills the entire space. This is in fact true for any M_p for which $p \geq K_2$.

2.6. A Fundamental Convergence Theorem

The following theorem, which is due to Kantorowitch (1948), serves as an abstract prototype of all theorems stating the equivalence of stability and convergence.

Let $\mathscr{B}'$ and $\mathscr{B}''$ be two Banach spaces with norms $\|\ \|'$ and $\|\ \|''$, and let T and T_m ($m = 1, 2, 3, \ldots$) be linear operators mapping $\mathscr{B}'$ into $\mathscr{B}''$. Suppose that (i) for every f in $\mathscr{B}''$, the equation $Tu = f$ has a unique solution u in $\mathscr{B}'$; (ii) each operator T_m has a bounded inverse T_m^{-1} and approximates T in the sense that, for every $u \in \mathscr{B}'$, $\|Tu - T_m u\|'' \to 0$ as $m \to \infty$.

Then the solution u_m of $T_m u_m = f$ converges to the solution of u of $Tu = f$ for every f in $\mathscr{B}''$ if and only if the approximations are stable, i.e., the inverse operators T_m^{-1} are uniformly bounded.

Proof: If $\|T_m^{-1}\|' \leq M$ where M is independent of m, then $\|u - u_m\|' = \|T_m^{-1} T_m(u - u_m)\|' \leq M\|T_m(u - u_m)\|''$. But $\|T_m(u - u_m)\|'' = \|(T_m - T)u + Tu - T_m u_m\|'' = \|(T_m - T)u\|'' \to 0$ as $m \to \infty$, by (ii). On the other hand, if for any given f, $\|u - u_m\|' \to 0$ as $m \to \infty$, then the sequence $\|u_m\|' = \|T_m^{-1} f\|'$ is bounded. Hence, by the principle of uniform boundedness, the norms $\|T_m^{-1}\|'$ are uniformly bounded.

If the above theorem were applied directly to initial-value problems, the elements of the two spaces $\mathscr{B}'$ and $\mathscr{B}''$ would be functions of both space and time. It is preferable in this application that they be functions of the space variables only, so some modification of the basic theorem is necessary. It is also desirable to weaken somewhat the hypothesis (i).

2.7. Closed Operators

Although the operators that will appear are not generally bounded and, in particular, differential operators are not, they often have a property called *closure*. An operator T with domain $\mathscr{D}(T)$ is called *closed* if it has the property that whenever a sequence $\{u_j\}$ of elements, all in $\mathscr{D}(T)$, converges, say to u^*, and is such that $\{Tu_j\}$ also converges, say to w^*, then u^* is in $\mathscr{D}(T)$ and $Tu^* = w^*$.

All the operators that will appear can be taken as closed; this will be proved in Section 3.6 of the next chapter. Roughly speaking, only closed operators need appear if one is sufficiently careful about choosing their domains adequately.

As an example, consider the operator $T = d/dx$, acting on functions $u(x)$ in a Banach space $\mathscr{B} = L_2(0, 1)$, i.e., the space of functions that are square integrable over the interval $(0, 1)$. If the domain $\mathscr{D}(T)$ is taken as the set of absolutely continuous functions $u(x)$, defined for $0 \leq x \leq 1$ (an absolutely continuous function has a derivative almost everywhere and is equal to the Lebesgue integral of its derivative—see Titchmarsh, 1939, page 364), then T is closed, for it can be proved that if the functions of a sequence $\{u_j(x)\}$ are absolutely continuous, if $Tu_j(x)$, $\varphi(x)$, and $\psi(x)$ are all in $L_2(0, 1)$ and are such that

$$\|u_j - \varphi\| \to 0,$$

$$\left\|\frac{d}{dx} u_j - \psi\right\| \to 0,$$

then $\varphi(x)$ is absolutely continuous, and $d\varphi/dx = \psi$, wherever the derivative exists.

CHAPTER 3

Linear Difference Equations

In this chapter we present a slightly simplified version of a general theory, due to Peter Lax, of the approximation of linear initial-value problems by finite-difference equations.

3.1. Properly Posed Initial-Value Problems

The problems we consider can be represented abstractly, using Banach space terminology. One wishes to find a one-parameter family $u(t)$ of elements of $\mathscr{B}$, where t is a real parameter, such that

(3.1) $$\frac{d}{dt} u(t) = A u(t), \quad 0 \leq t \leq T$$

and

(3.2) $$u(0) = u_0,$$

where A is a linear operator and u_0 is a given element of $\mathscr{B}$ describing the initial state of the physical system. The derivative $du(t)/dt$ will be defined below.

Boundary conditions, if any, are assumed to be linear and homogeneous and are taken care of by assuming that the domain of A is restricted to those functions satisfying the boundary conditions.

The application of (3.1) is not restricted to first-order equations, since higher order systems can be reduced to first-order ones with a correspondingly increased number of dependent variables and differential equations. However, the manner in which this is done is not immaterial; it will be seen that a problem may be properly posed, in the sense defined below, for one choice of function space as $\mathscr{B}$ but not for another.

The application of (3.1) is not even restricted to differential equation systems. There are, for example, applications of the theory presented here to initial-value problems of integro-differential equations.

In Lax's original presentation (seminar, New York University, 1953) the operator A was permitted to depend explicitly on the parameter t, as is the case, for example, if the coefficients of a differential equation are time-dependent. The theory is almost identical to the one presented here, but requires a slightly more cumbersome notation.

To be more specific about the interpretation of (3.1), we define a *genuine solution* of the initial-value problem as a one-parameter family $u(t)$ such that $u(t)$ is in the domain of A for each t in the interval $0 \leq t \leq T$ and

(3.3) $\quad \left\| \dfrac{u(t + \Delta t) - u(t)}{\Delta t} - Au(t) \right\| \to 0, \quad \text{as } \Delta t \to 0, \, 0 \leq t \leq T.$

A problem is characterized as properly posed if the family of genuine solutions is sufficiently large and if the solutions depend uniquely and continuously on the initial data, in a certain sense. (The task of determining whether a given operator A is such as to lead to a properly posed initial-value problem according to our characterization will not be considered part of the general theory but will be discussed in connection with specific examples.)

Let $\mathscr{D}$ be the set of elements u_0 of $\mathscr{B}$ for each of which there is a unique genuine solution $u(t)$ such that $u(0) = u_0$ and such that the convergence in (3.3) is uniform in t. (It will be shown below that this convergence is automatically uniform for any genuine solution of a properly posed initial-value problem of the kind considered in this chapter. However, to show that a particular problem is properly posed, one must show the uniqueness of $u(t)$ for each u_0 of a sufficiently large set $\mathscr{D}$, and this task is greatly simplified if one can assume the uniformity of convergence in advance.) For any fixed t, the correspondence between u_0 and $u(t)$ defines a transformation in $\mathscr{B}$ with $\mathscr{D}$ as domain. It is easy to verify that this transformation is linear, and we denote it by $E_0(t)$, so that the equation

$$u(t) = E_0(t)u_0 \quad \text{for } u_0 \text{ in } \mathscr{D},$$

gives the solution of the initial-value problem for those initial states for which a genuine solution exists. In a few cases an explicit formula for $E_0(t)$ can be given (the integral representation of solutions of the linear heat flow problem is an example), but this is not necessary for the present arguments.

The initial-value problem determined by the linear operator A will be called *properly posed* (this notion is due to Hadamard—several slightly different versions of it have appeared subsequently) if the following two conditions are satisfied:

1. the domain $\mathscr{D}$ of $E_0(t)$ is dense in $\mathscr{B}$;
2. the family $E_0(t)$ of operators is uniformly bounded; i.e., there is a K such that $\|E_0(t)\| < K$ for $0 \leq t \leq T$.

The second of these conditions says in effect that the solution depends continuously on the initial data; for, if $u(t)$ and $v(t)$ are genuine solutions corresponding to initial elements u_0 and v_0, then $\|u(t) - v(t)\| \leq K\|u_0 - v_0\|$, so that if the initial states are very nearly equal, the later states are very nearly equal also, at corresponding times. The first condition says that even though for some choice of initial element u_0 a genuine solution may not exist, we can approximate this initial element as closely as we wish by one for which a genuine solution does exist. An example is provided again by the linear heat flow problem: if the initial temperature distribution is discontinuous, we can approximate it by a sequence of continuous (and in fact twice-differentiable) ones that approach it in the limit; the sequence of solutions obtained then approaches a function which is interpreted as the solution of the heat flow problem for the discontinuous initial temperature distribution.

According to the extension theorem of Section 2.4, since $E_0(t)$ is a bounded linear transformation with domain dense in $\mathscr{B}$, it has an extension $E(t)$, called the *generalized solution operator*, defined for all $\mathscr{B}$ and having the same bound K as $E_0(t)$. The equation $u(t) = E(t)u_0$ is interpreted as giving the generalized solution for an arbitrary initial element u_0.

The solution operator has the *semi-group property*:

$$E(s + t) = E(s)E(t)$$

for $s \geq 0, t \geq 0$. Although this does not hold in the more general case in which A depends explicitly on t, and is furthermore not essential to the theory, we shall make use of it whenever convenient to simplify the argument.

In order for the initial-value problem of a partial differential equation in a bounded domain to be properly posed, the boundary conditions must be appropriate, for they influence the domain of A and hence that of

$E_0(t)$. Consider, for example, the simple heat flow equation for $a \leq x \leq b$. If we have too many boundary conditions, say $u(x, t) = 0$ and $(\partial/\partial x) u(x, t) = 0$ for $x = a$ and $x = b$, then there are no genuine solutions at all, unless $u_0(x) \equiv 0$; hence, $\mathscr{D}$ is not dense in $\mathscr{B}$. On the other hand, if there are too few boundary conditions, say none, then there is no *unique* genuine solution for any $u_0(x)$, hence $\mathscr{D}$ is again not dense in $\mathscr{B}$.

It follows from (3.3) that a genuine solution is continuous, in the sense of the norm of $\mathscr{B}$: $\|u(t + \Delta t) - u(t)\| \to 0$ as $\Delta t \to 0$, for $0 \leq t \leq T$, and it then follows by a simple use of the triangle inequality that any generalized solution is also continuous, for $0 \leq t \leq T$.

For u_0 in $\mathscr{D}$, $(1/\delta)[E(\delta) - I]u_0$ approximates Au_0, and since $E(t)$ commutes with $E(\delta)$, one can show, by taking limits, that

$$E(t)Au_0 = AE(t)u_0.$$

From this it follows that the quantity appearing in the definition (3.3) of a genuine solution can be written as

$$\left\| E(t)\left\{ \frac{E(\Delta t) - I}{\Delta t} - A \right\} u_0 \right\|;$$

since $\|E(t)\| < K$ for $0 \leq t \leq T$, we see that the convergence in (3.3) is uniform in t for a genuine solution of a properly posed initial-value problem.

When A depends explicitly on the time, the solution operator becomes a function of two variables $E(s, t)$, and if the initial state u_0 is given at time t_0, then $u(t) = E(t, t_0)u_0$; corresponding to the semi-group property, we have the equation $E(t_2, t_0) = E(t_2, t_1)E(t_1, t_0)$.

3.2. Finite-Difference Approximations

According to the remarks at the end of Section 2.1, the approximations obtained from finite-difference equations can also be thought of as represented by points in the space $\mathscr{B}$. In place of the one-parameter family $u(t)$ we have a sequence of points $u^0, u^1, \ldots$, of which u^n is supposed to approximate $u(n\Delta t)$, Δt being a small increment.

The finite difference equations are

(3.4) $$B_1 u^{n+1} = B_0 u^n$$

where $B_0 = B_0(\Delta t, \Delta x_1, \Delta x_2, \ldots)$ and $B_1 = B_1(\Delta t, \Delta x_1, \Delta x_2, \ldots)$ denote linear finite-difference operators depending, as indicated, on the sizes of the finite increments $\Delta t, \Delta x_1, \Delta x_2, \ldots$ of the variables as well as the space variables themselves; as with the operator A, we suppose that B_0 and B_1 are independent of t. At each point of the space, each side of equation (3.4) will consist of a linear sum of the function values at some set of neighboring points. We assume that, by some procedure, unique function values u^{n+1} can be calculated from any u^n, on which they should depend continuously. Thus, we suppose B_1^{-1} exists and $B_1^{-1}B_0$ is a bounded linear transformation in $\mathscr{B}$, its domain being the whole Banach space. These assumptions will clearly be satisfied for any reasonable difference scheme.

A formula of type (3.4) is called a *two-level formula*, since only two "time levels" t^n and t^{n+1} are involved. It should be noted however that use of (3.4) does not restrict us to differential equations of the first order in time. As noted earlier, higher order systems can be reduced to first-order systems by the introduction of new dependent variables. The introduction of these new variables will reduce the corresponding finite-difference system to a two-level formula unless the original number of time levels was greater than one plus the order, in t, of the original differential equation system. (A number of three-level formulas for the diffusion equation are discussed in Chapter 8—see examples 7, 8, 9, 10, 11, 13 of Table 8.1) Multi-level formulas will be considered in Chapter 7. The terms "one-step" and "multi-step" are also used.

The concepts of stability and convergence with which we deal here suppose an infinite sequence of calculations with increasingly finer meshes. We assume relations $\Delta x_i = g_i(\Delta t)$, $i = 1, 2, \ldots, d$, where d is the number of space dimensions, which tell how the space increments approach zero as the time increment goes to zero along the sequence. Then we set
$$B_1^{-1}(\Delta t, g_1(\Delta t), g_2(\Delta t), \ldots)B_0(\Delta t, g_1(\Delta t), g_2(\Delta t), \ldots) = C(\Delta t),$$
so that

(3.5) $$u^{n+1} = C(\Delta t)u^n.$$

We come now to the consistency condition. Since
$$\frac{u^{n+1} - u^n}{\Delta t}$$

is to be an approximation to the time derivative, the ratio

$$\frac{C(\Delta t)u - u}{\Delta t}$$

must be an approximation, in some sense, to Au. We cannot expect this to be true for all u in $\mathscr{B}$, because in general A is not even defined for all u in $\mathscr{B}$. But we want it to be true for nearly all u that can appear in a genuine solution of the initial-value problem. We shall say that the family $C(\Delta t)$ of operators provides a *consistent approximation* for the initial-value problem if, for every $u(t)$ in some class $\mathscr{U}$ of genuine solutions whose initial elements $u(0)$ are dense in $\mathscr{B}$,

(3.6a) $\quad \left\| \left\{ \dfrac{C(\Delta t) - I}{\Delta t} - A \right\} u(t) \right\| \to 0, \quad \text{as } \Delta t \to 0, 0 \leq t \leq T.$

Here, I stands for the identity operator. (3.6a) is called the *consistency condition*.

Since this condition is applied only to genuine solutions, an alternative form for it can be obtained by combining it with (3.3) to give

(3.6b) $\quad \left\| \dfrac{u(t + \Delta t) - C(\Delta t)u(t)}{\Delta t} \right\| \to 0, \quad \text{as } \Delta t \to 0, 0 \leq t \leq T.$

The quantity under the norm here is called the *truncation error* and measures how closely a genuine solution of the initial-value problem satisfies the finite-difference equations. It will be assumed that the limits indicated in (3.6a) and (3.6b) are uniform in t; then, finally, for every positive ε there is a positive δ such that

(3.6c) $\quad \|\{C(\Delta t) - E(\Delta t)\}u(t)\| < \varepsilon \Delta t, \quad \text{for } \begin{matrix} 0 \leq t \leq T \\ 0 < \Delta t < \delta \end{matrix}$

3.3. Convergence

Operating on u_0 n times with $C(\Delta t)$ gives $u^n = C(\Delta t)^n u_0$ which, it is hoped, approximates $u(n\Delta t)$. Since $u(t) = E(t)u_0$, we therefore make the following definition: the family of operators $C(\Delta t)$ provides a *convergent* approximation for the initial-value problem if the following condition is satisfied: let $\Delta_1 t, \Delta_2 t, \ldots$ be a sequence of time increments tending to zero; let t be fixed ($0 \le t \le T$), and let n_j be an integer close to $t/\Delta_j t$ for each j, in the sense that $n_j \Delta_j t \to t$ as $j \to \infty$; it is then required that

(3.7) $\qquad \|C(\Delta_j t)^{n_j} u_0 - E(t) u_0\| \to 0, \quad \text{as } j \to \infty,$

for every u_0 in $\mathscr{B}$.

3.4. Stability

If in a sequence of calculations with $\Delta_j t \to 0$ each calculation is carried from $t = 0$ to $t = T$, then the operators used are those of the infinite set

(3.8) $\qquad C(\Delta_j t)^n, \quad \begin{array}{l} 0 \le n\Delta_j t \le T, \\ j = 1, 2, 3, \ldots, \end{array}$

all applied to u_0. The essence of stability is that there should be a limit to the extent to which any component of an initial function can be amplified in the numerical procedure. Therefore, the approximation $C(\Delta t)$ is said to be stable, if for some $\tau > 0$, the infinite set of operators

(3.9) $\qquad C(\Delta t)^n, \quad \begin{array}{l} 0 < \Delta t < \tau, \\ 0 \le n\Delta t \le T, \end{array}$

is uniformly bounded.

It will be assumed henceforth that the finite-difference operator $C(\Delta t)$ depends continuously on Δt, for Δt positive and less than some $\tau > 0$. That is, the coefficients of the difference operator, hence also its norm, are continuous functions of Δt. Then, the above set of operators would automatically be uniformly bounded, for all Δt in an interval $\tau' \le \Delta t \le \tau$ where $\tau' > 0$; the essence of the definition of stability is that the set remains uniformly bounded as $\tau' \to 0$.

The concept of stability makes no reference to the differential equations one wishes to solve, but is a property solely of the sequence of difference equations.

3.5. Lax's Equivalence Theorem

Given a properly posed initial-value problem and a finite-difference approximation to it that satisfies the consistency condition, stability is the necessary and sufficient condition for convergence.

According to the definition of Section 3.3, this means convergence for an arbitrary initial element u_0. As noted in the first chapter—see discussion following equation (1.15)—an unstable scheme may sometimes give convergence for special initial elements.

We prove first that a convergent scheme is necessarily stable. We assert that if the scheme is convergent, and if u_0 is any element of $\mathscr{B}$, the quantities

$$\|C(\Delta t)^n u_0\| \quad \begin{array}{l} 0 < \Delta t < \tau \\ 0 \leq n\Delta t \leq T \end{array}$$

are bounded, for some $\tau > 0$. If this were not so, we could choose a sequence $\Delta_1 t, \Delta_2 t, \ldots$ and a corresponding sequence of elements $C(\Delta_1 t)^{n_1} u_0$, $\ldots, C(\Delta_j t)^{n_j} u_0, \ldots$ whose norms tend to infinity (this shows that $\Delta_j t$ must $\to 0$, because of the continuous dependence of $C(\Delta t)$ on Δt for positive Δt); then we could choose a sub-sequence of these elements for which the quantities $n_j \Delta_j t$ tend to a limit t in the interval $0 \leq t \leq T$; this would contradict convergence, because for convergence the norms have to approach the finite value $\|E(t)u_0\|$. Hence, there exists a function $K_1(u)$ such that $\|C(\Delta t)^n u\| \leq K_1(u)$ for all members of the set (3.9) and all u in $\mathscr{B}$. Therefore, by the principle of uniform boundedness, the set (3.9) is uniformly bounded. That is, the approximation is stable by the definition of Section 3.4.

To prove that, conversely, stability implies convergence, let $u(t) = E(t)u_0$ be a genuine solution belonging to the set $\mathscr{U}$ referred to in the definition of consistency. Let ε and δ be as in the consistency condition in the form (3.6c); let n_j and $\Delta_j t$ be as in the definition of convergence, and call ψ_j the difference between the calculated and exact values of u at $n_j \Delta_j t$, namely,

$$\psi_j = [C(\Delta_j t)^{n_j} - E(n_j \Delta_j t)]u_0$$

(3.10)

$$= \sum_{0}^{n_j - 1}{}_{(k)} C(\Delta_j t)^k [C(\Delta_j t) - E(\Delta_j t)] E((n_j - 1 - k)\Delta_j t) u_0.$$

The third member of this equation, when written out in full, is seen to be the same as the second member by cancellation of all its terms except the

SEC. 3.5] LAX'S EQUIVALENCE THEOREM

first and last. The norm of this quantity may be estimated by (3.6c) via the triangle inequality:

$$(3.11) \qquad \|\psi_j\| < K_2 \sum_{0}^{n_j-1}{}_{(k)} \varepsilon \Delta_j t = K_2 \varepsilon n_j \Delta_j t \leq K_2 \varepsilon T,$$

for $0 < \Delta_j t < \delta$, where K_2 denotes the uniform bound of the set (3.9). Since ε was arbitrary, $\|\psi_j\| \to 0$ as $\Delta_j t \to 0$, $n_j \Delta_j t \to t$. Convergence will then follow, if it can be shown that the effect of replacing $E(n_j \Delta_j t)$ in (3.10) by $E(t)$ is negligible in the limit as $j \to \infty$. The semi-group property of $E(t)$ says that $E(s + t') = E(s)E(t')$ for $s \geq 0$, $t' \geq 0$, and we take $t' = \text{Min}\{t, n_j\Delta_j t\}$, $s = |t - n_j\Delta_j t|$; then $E(n_j\Delta_j t) - E(t) = \pm[E(s) - I]E(t')$, where the sign depends on whether $t > n_j\Delta_j t$ or conversely. In either case, $\|[E(n_j\Delta_j t) - E(t)]u_0\| \leq K_E\|(E(s) - I)u_0\|$, where K_E is the maximum bound of $E(t)$ for $0 \leq t \leq T$. The above expression goes to zero as s goes to zero, and therefore as j goes to infinity. Therefore, $\|[C(\Delta_j t)^{n_j} - E(t)]u_0\|$ can be made as small as one pleases by requiring $\Delta_j t$ and $|t - n_j\Delta_j t|$ to be sufficiently small. This is true for any u_0 which can be the initial element of a genuine solution of the class $\mathscr{U}$; but these initial elements are dense in $\mathscr{B}$, so that if u is any element, there is a sequence $u_1, u_2, \ldots$ of elements of $\mathscr{U}$ converging to u. Then

$$(3.12) \quad [C(\Delta_j t)^{n_j} - E(t)]u = [C(\Delta_j t)^{n_j} - E(t)]u_m \\ + C(\Delta_j t)^{n_j}(u - u_m) - E(t)(u - u_m).$$

The last two terms on the right can be made as small as one wishes by choosing m sufficiently large, by the uniform boundedness of the class (3.9) and of $E(t)$. The first term on the right can then be made as small as one wishes by choosing $\Delta_j t$ and $|t - n_j\Delta_j t|$ sufficiently small. Since u was arbitrary, convergence is now established, and the equivalence theorem is proved.

As a corollary of this theorem it may be noted that for given u_0 the convergence is uniform in t for $0 \leq t \leq T$ in the sense that the restriction which must be placed on $\Delta_j t$ and on $|t - n_j\Delta_j t|$ to make expression (3.12) arbitrarily small does not depend on t or on the particular sequence $\Delta_j t$ used. This uniformity is of practical importance because it is the aim in numerical work to find an increment Δt for which the approximate solution will be acceptably accurate for the entire interval $0 \leq t \leq T$. One often varies Δt during the course of a calculation, but there should be a positive limiting value for Δt below which it is not necessary to go.

The theorem can be applied in a simple way to the implicit difference equation (1.18) for diffusion in one dimension:

$$\frac{u_j^n - u_j^{n-1}}{\Delta t} = \sigma \frac{u_{j+1}^n - 2u_j^n + u_{j-1}^n}{(\Delta x)^2}.$$

We first show that the solution of this equation satisfies the following maximum principle[1]: We suppose that the equation is to be used in a rectangular region $0 \leq x \leq a$ and $0 \leq t \leq T$ and that we choose Δx equal to a/J, where J is an integer. Then the *maximum value M_1 attained by u_j^n in the interior of the rectangle cannot exceed the maximum value M_2 attained by the initial and boundary values* (i.e. the values on $t = 0$ and on $x = 0$ and $x = a$). To prove this, assume the contrary, namely, that $M_1 > M_2$ and let (n, j) be the first interior net point where $u_j^n = M_1$ (first in the sense of having the smallest value of n and of j). It is then immediately clear that at this net point the left member of the above equation would have to be positive and the right member negative, because u_j^n by assumption exceeds its neighbor u_{j-1}^n on the left and its neighbor u_j^{n-1} below, and is at least equal to its neighbor u_{j+1}^n on the right. Therefore, the assumption was false and the maximum principle is established. Obviously, we can apply the argument also to $-u_j^n$ and thereby conclude that $|u_j^n|$ is bounded, independent of the fineness of the net, in other words, that the difference equation is stable.

This argument can also be used to prove the stability of certain analogous difference equations for the more general problem

$$\frac{\partial u}{\partial t} = a(x) \frac{\partial}{\partial x} b(x) \frac{\partial u}{\partial x}, \quad a(x) > 0, b(x) > 0,$$

(details are left to the reader). The consistency condition is also satisfied in these cases, and convergence is therefore assured independently of the manner in which Δt and $\Delta x \to 0$.

---

[1] Joel Franklin, private communication (1954); Jim Douglas, Jr. (1955); P. Laasonen (1949).

3.6. The Closed Operator A'

In this section it will be proved that if the operator A appearing in (3.1) determines a properly posed initial-value problem, then there is a closed operator A', which also determines a properly posed initial-value problem such that every genuine solution of the first problem is also a genuine solution of the second, and the two problems have the same complete solution operator $E(t)$, so that for practical purposes they are the same problem. The operator A' will be important in the following section on inhomogeneous problems.

Roughly speaking, A' is defined as the t-derivative of $E(t)$, at $t = 0$. If an element u of $\mathscr{B}$ is such that $(1/\Delta t)[E(\Delta t) - I]u$ has a limit w in $\mathscr{B}$ as $\Delta t \to 0$, we say that u is an element of the domain $\mathscr{D}(A')$ and that w is the value of $A'u$; i.e.,

$$\lim_{\Delta t \to 0} \frac{1}{\Delta t}[E(\Delta t) - I]u = A'u$$

for all u such that the limit exists. It is evident that $\mathscr{D}(A')$ is a linear subspace of $\mathscr{B}$, that A' is a linear operator, and, by comparison of the above equation with (3.3), that $A'u = Au$ for any u that can appear in a genuine solution of the original problem.

In the terminology of the theory of operators (see Dunford and Schwartz, 1958, Vol. 1), A' is called the *infinitesimal generator* of $E(t)$, which is a strongly continuous semi-group.

Since $E(t)$ is a bounded operator and commutes with $E(\Delta t)$, we see that if $u_0 \in \mathscr{D}(A')$ then $E(t)(1/\Delta t)[E(\Delta t) - I]u_0$ or $(1/\Delta t)[E(\Delta t) - I]E(t)u_0$ has a limit, as $\Delta t \to 0$, namely, $E(t)A'u_0$; but this is equal to $A'E(t)u_0$ by definition of the operator A' applied to $E(t)u_0$, so that A' and $E(t)$ commute when applied to u_0. Therefore, $E(t)u_0 \in \mathscr{D}(A')$ and the function $u(t) = E(t)u_0$ satisfies

$$\left| \frac{u(t + \Delta t) - u(t)}{\Delta t} - A'u(t) \right| \to 0 \quad \text{as } \Delta t \to 0,$$

i.e., it is a solution of the initial-value problem

(3.13) $$\frac{d}{dt}u(t) = A'u(t), \quad u(0) = u_0.$$

Conversely, we shall now prove that the only solution of (3.13) is $u(t) = E(t)u_0$. Let $u(t)$ be a solution; for fixed $t > 0$, the function $g(s) = E(t - s)u(s)$ $(0 \leq s \leq t)$ will be shown to be a constant. In fact,

$$\left.\frac{d}{ds} g(s)\right|_{s=s_0} = \left.\frac{d}{ds} E(t - s)u(s_0)\right|_{s=s_0} + \left.\frac{d}{ds} E(t - s_0)u(s)\right|_{s=s_0}.$$

The first term on the right can be written as

$$\left.-\frac{d}{dw} E(w)u(s_0)\right|_{w=t-s_0} = -A'E(t - s_0)u(s_0)$$

because the function $E(t)u(s_0)$ satisfies the differential equation (3.13), and since $E(t - s_0)$ is a bounded operator, the second term can be written as

$$E(t - s_0) \frac{d}{ds} u(s_0) = E(t - s_0)A'u(s_0).$$

Since A' and $E(t - s_0)$ commute, $dg(s)/ds = 0$; hence $g(s)$ is a constant, and therefore $g(t) = g(0)$ or $u(t) = E(t)u(0)$, as was to be proved; therefore (3.13) is a properly posed initial-value problem.

It will now be proved that A' is a closed operator, as defined in Section 2.7 of the preceding chapter, after a parenthetical reminder on integration. If $w(s)$ is any continuous one-parameter family of elements of $\mathscr{B}$, one can define the Riemann integral

$$I = \int_{s_1}^{s_2} w(s)\,ds$$

as the limit of Riemann sums

$$\sum_{(j)} w(s'_j)(s_{j+1} - s_j) \quad (s_j \leq s'_j \leq s_{j+1}),$$

just as for an ordinary continuous function. It is left as an exercise for the reader to verify, by use of the triangle inequality, that the limit is unique, and that the integral has all the usual properties, such as

$$\frac{dI}{ds_2} = w(s_2),$$

$$\|I\| \leq \int_{s_1}^{s_2} \|w(s)\|\,ds \leq |s_2 - s_1| \underset{s_1 \leq s \leq s_2}{\text{Max}} \|w(s)\|.$$

It will be shown that for u_0 in $\mathscr{D}(A')$

(3.14) $$u(t) - u_0 = \int_0^t A'u(s)ds,$$

where $u(t) = E(t)u_0$; this is in agreement with the equation $du/dt = A'u$. Equation (3.13) shows that, given $\varepsilon > 0$, there is a $\delta > 0$ such that

$$\left\|\frac{E(\delta) - I}{\delta} u_0 - A'u_0\right\| < \varepsilon$$

or, applying the operator $E(j\delta)$, $\|E((j+1)\delta)u_0 - E(j\delta)u_0 - E(j\delta)A'u_0\delta\|$ $< \varepsilon\delta K$, where $K = \text{Max} \|E(t)\|$; summing on j, using the triangle inequality, and calling $n\delta = t$ gives

$$\left\|E(t)u_0 - u_0 - \sum_{j=0}^{n-1} E(j\delta)A'u_0\delta\right\| < \varepsilon tK.$$

Since $E(s)A'u_0$ is continuous, the sum approximates the integral, and, since ε was arbitrary, (3.14) follows.

Now suppose that $u_1, u_2, \ldots, u_n, \ldots$ is a sequence in $\mathscr{D}(A')$ converging to some u^* in $\mathscr{B}$ and that $A'u_n$ converges to some w^* in $\mathscr{B}$. It will be shown that $u^* \in \mathscr{D}(A')$ and that $w^* = A'u^*$, i.e., that A' is a closed operator. In fact,

$$E(\delta)u^* - u^* = \lim_{n \to \infty} (E(\delta)u_n - u_n) = \lim_{n \to \infty} \int_0^\delta A'E(s)u_n ds$$

$$= \lim_{n \to \infty} \int_0^\delta E(s)A'u_n ds$$

$$= \int_0^\delta E(s) \lim_{n \to \infty} A'u_n ds.$$

(The convergence is uniform because $E(s)$ is bounded uniformly in s.) Therefore,

$$\frac{E(\delta)u^* - u^*}{\delta} = \frac{1}{\delta}\int_0^\delta E(s)w^* ds$$

which approaches $E(0)w^* = w^*$ as $\delta \to 0$, by continuity of the integrand, thus proving that A' is closed.

The domains of A' and of all its powers can be shown to be dense in $\mathscr{B}$, by a method due to Gel'fand and Gårding (J. L. Lions, private communication). Let $\varphi(t)$ be a C^∞ function which vanishes outside an interval $0 < t < t_0$. (In particular, φ and all its derivative vanish at $t = 0$.) If u is any element of $\mathscr{B}$, we call

$$T(\varphi)u = \int_0^\infty E(t)u\varphi(t)dt;$$

this defines, for each $\varphi(\cdot)$, an operator $T(\varphi)$ with domain equal to $\mathscr{B}$; we shall show that $T(\varphi)u$ is always in $\mathscr{D}(A')$:

$$E(s)T(\varphi)u = \int_0^\infty E(s+t)u\varphi(t)dt$$

$$= \int_0^\infty E(t)u\varphi(t-s)dt;$$

$$\frac{E(s)T(\varphi)u - T(\varphi)u}{s} = \int_0^\infty E(t)u \frac{\varphi(t-s) - \varphi(t)}{s} dt \to T(-\varphi')u,$$

where φ' denotes the function $d\varphi/dt$. Therefore, by the definition of A', $T(\varphi)u$ is in $\mathscr{D}(A')$ and $A'T(\varphi)u = T(-\varphi')u$. By repetition of this process, $(A')^q T(\varphi)u$ is also defined and is equal to $T((-)^q\varphi^{(q)})u$. Now let $\mathscr{B}_0$ be the set of all elements $T(\varphi)u$, for all choices of $\varphi(\cdot)$ and of u in $\mathscr{B}$. If u is any element of $\mathscr{B}$, it can be approximated arbitrarily closely by $T(\varphi_m)u$, where $\{\varphi_m\}$ is a sequence of functions of the kind specified which approaches the Dirac function (distribution) $\delta(t)$ (that $T(\varphi_m)u \to u$ follows from the continuity of $E(t)u$); hence, $\mathscr{B}_0$ is dense in $\mathscr{B}$, and we have shown that $\mathscr{D}((A')^q) \equiv \mathscr{B}_0$; hence, $\mathscr{D}((A')^q)$ is dense.

3.7. Inhomogeneous Problems

The first aim of this section is to solve the initial-value problem

(3.15)
$$\frac{d}{dt}u(t) - Au(t) = g(t),$$
$$u(0) = u_0,$$

where u_0 and $g(t)$ are given and $g(t)$ is uniformly continuous in t, for $0 \leq t \leq T$, with respect to the norm in $\mathscr{B}$. Piecewise uniform continuity of $g(t)$ would suffice, but the verification of this is left to the reader. A is assumed to be a closed operator all of whose powers have domains dense in $\mathscr{B}$ (this is no restriction, as shown in the preceding section).

We shall prove that under further restrictions (1) that u_0 and $g(t)$ are in $\mathscr{D}(A)$, (2) that $g(t)$ is in $\mathscr{D}(A^2)$, and (3) that $Ag(t)$ and $A^2g(t)$ are continuous, the solution of (3.15) is

$$(3.16) \qquad u(t) = E(t)u_0 + \int_0^t E(t-s)g(s)ds.$$

Such a solution is called a *genuine solution*. When u_0 and $g(t)$ are unrestricted, except for the continuity of $g(t)$—or more generally, when $g(t)$ is sufficiently well behaved that the integral in (3.16) exists—the function $u(t)$ given by (3.16) is defined to be the *generalized* solution of (3.15). These results were obtained by R. J. Thompson (1964); he discussed also the quasi-linear case, in which $g = g(t, u(t))$.

From the boundedness of $E(t)$ and the continuity of $g(t)$ and of $E(t)u$ for any u, it follows by a simple use of the triangle inequality that $E(t-s)g(s)$ is continuous in s and therefore can be integrated. To prove the result we shall show that

$$(3.17) \qquad \frac{d}{dt}\int_0^t E(t-s)g(s)ds = E(0)g(t) + \int_0^t \frac{d}{dt}E(t-s)g(s)ds.$$

The existence of the integral in the right member of this equation is assured because the integrand is equal to $AE(t-s)g(s) = E(t-s)Ag(s)$ and is continuous, by continuity of $Ag(s)$ and boundedness of $E(t-s)$. The differentiation under the integral sign is justified, as for ordinary functions, if the difference quotient converges uniformly in s to the derivative, that is, if

$$(3.18) \qquad \left\|\frac{E(\delta)-I}{\delta}E(t-s)g(s) - AE(t-s)g(s)\right\| \to 0 \quad \text{uniformly in } s,$$
$$\text{as } \delta \to 0.$$

Since $E(t-s)$ is bounded and commutes with A, the requirement is that

$$\left\|\frac{E(\delta)-I}{\delta}g(s) - Ag(s)\right\| \to 0 \quad \text{uniformly in } s, \text{ as } \delta \to 0.$$

For fixed s, the function $h(t) = E(t)g(s)$ satisfies not only $dh/dt = Ah$ but also $d^2h/dt^2 = A^2h$, under our assumptions; therefore, using (3.14) twice, we find

$$\frac{E(\delta) - I}{\delta} g(s) - Ag(s) = \frac{1}{\delta}\left(\int_0^\delta Ah(t)dt - Ah(0)\delta\right) =$$

$$\frac{1}{\delta}\int_0^\delta dt \int_0^t dw A^2 h(w) = \frac{1}{\delta}\int_0^\delta dt \int_0^t dw E(w) A^2 g(s);$$

therefore,

$$\left\|\frac{E(\delta) - I}{\delta} g(s) - Ag(s)\right\| \leq \frac{1}{2}\delta \operatorname{Max}\|E(w)A^2 g(s)\|;$$

the indicated maximum is finite by the continuity of $A^2 g(s)$ and the uniform boundedness of $E(w)$. This proves the uniformity of (3.18) and establishes (3.17).

Equation (3.17) takes the form

$$\frac{d}{dt}\int_0^t E(t - s)g(s)ds = g(t) + \int_0^t AE(t - s)g(s)ds.$$

The closure property of the operator A allows us to move A outside the integral, because any Riemann sum that approximates $\int AE(t - s)g(s)ds$ is equal to A times a Riemann sum that approximates $\int E(t - s)g(s)ds$; because of the closure of A we can go to the limit, and therefore

$$\frac{d}{dt}\int_0^t E(t - s)g(s)ds = g(t) + A\int_0^t E(t - s)g(s)ds,$$

from which it is seen that the function $u(t)$ given by (3.16) solves the initial-value problem (3.15).

The uniqueness of the solution follows from the uniqueness of the solution of the corresponding homogeneous problem, which has been tacitly assumed since the existence of the solution operator $E(t)$ presupposes that the original problem was properly posed.

For the discussion of the generalized solutions, we can take $u_0 = 0$ without loss of generality, because the first term of (3.16) is simply the solution of the corresponding homogeneous problem. Let $\mathscr{B}'$ be an auxiliary Banach space, each of whose elements is a continuous curve $v(t)$ ($0 \leq t \leq T$) in $\mathscr{B}$, and in which the norm is defined as

$$\|v(\cdot)\|_{\mathscr{B}'} = \operatorname{Max}_{(t)}\|v(t)\|_{\mathscr{B}}.$$

SEC. 3.7]   INHOMOGENEOUS PROBLEMS

The mapping

$$g(t) \to u(t) = \int_0^t E(t-s)g(s)ds$$

is a bounded linear transformation F defined in all $\mathscr{B}'$; we call F_0 the restriction of F to elements $g(\cdot)$ of $\mathscr{B}'$ for which a genuine solution of (3.15) exists, as defined above, and we shall show that $\mathscr{D}(F_0)$ is dense in $\mathscr{B}'$. Let $g(\cdot)$ be any element of $\mathscr{B}'$; we subdivide $[0, T]$ into N subintervals of length δ; at each point of subdivision we approximate $g(n\delta)$ by an element $\bar{g}^n$ of $\mathscr{D}(A^2)$, and then we define $\bar{g}(t)$ as a piecewise linear function that connects the values $\bar{g}^n$. Clearly, $\bar{g}(t) \in \mathscr{D}(A^2)$ for $0 \leq t \leq T$ while $\bar{g}(t)$, $A\bar{g}(t)$, and $A^2\bar{g}(t)$ are continuous; furthermore, if δ is small enough and the approximations $\bar{g}^n$ are close enough, clearly $\|g(t) - \bar{g}(t)\|$ can be made uniformly small for $0 \leq t \leq T$; this shows that $\mathscr{D}(F_0)$ is dense in $\mathscr{B}'$. Therefore, we define the *generalized solution* of (3.15) as given by (3.16) for any u_0 in $\mathscr{B}$ and any $g(\cdot)$ in $\mathscr{B}'$, with the same justification as for the homogeneous problems: it is the limit of approximating genuine solutions and is unique because F is the unique bounded extension of F_0.

The finite-difference equations are

(3.19) $$B_1 u^{n+1} - B_0 u^n = g^{n+1}, \quad u^0 = u_0,$$

where B_1 and B_0 are the difference operators discussed in Section 3.2, and g^n or $g^n_{\Delta t}$ is an approximation to $g(n\Delta t)$ or to $g((n - 1/2)\Delta t)$; we assume merely that $\|g^n_{\Delta t} - g(n\Delta t)\|$ can be made arbitrarily small, uniformly in n for $0 \leq n\Delta t \leq T$, by suitable choice of Δt. As in Section 3.2, the difference equations can be rewritten as

$$u^{n+1} - C(\Delta t)u^n = D(\Delta t)g^{n+1},$$

where the operators $C(\Delta t) = B_1^{-1}B_0$ and $D(\Delta t) = B_1^{-1}$ are bounded and are functions of Δt according to the assumed relations connecting Δx, Δy, etc. with Δt. For explicit equations, $D(\Delta t)$ is simply $\Delta t\, I$, where I is the identity. In any case, we assume that $D(\Delta t)$ is such that the elements $\bar{g}^n$, defined by $D(\Delta t)g^n = \Delta t \bar{g}^n$, approximate the g^n uniformly in n, for sufficiently small Δt.

Iteration of the difference equation gives

(3.20) $$u^n = C(\Delta t)^n u_0 + \Delta t \sum_{j=1}^n C(\Delta t)^{n-j} \bar{g}^j,$$

and we wish to show that the sum in (3.20) approximates the integral in (3.16), under the conditions of the Lax equivalence theorem, thereby establishing the convergence of the approximate solution to the exact one. From the uniform continuity of $g(s)$ and the uniform boundedness of the operators $E(t)$ and $C(\Delta t)$—see (3.9)—it follows that the convergence of

$$\|\{E(m\Delta t) - C(\Delta t)^m\}g(s)\|$$

to zero, when Δt and m assume values in sequence such that $\Delta t \to 0$ and $m\Delta t \to t'$ (for fixed t'), is uniform with respect to s, for $0 \leq s \leq T$. (We have already found in Section 3.5 that it is uniform with respect to t'.) Therefore, given $\varepsilon > 0$, we can choose Δt small enough so that (1) the integral in (3.16) is approximated to within ε by a Riemann sum $\Delta t \sum_{(j)} E(t - j\Delta t) g(j\Delta t)$, (2) this sum is approximated to within ε by $\Delta t \sum_{(j)} C(\Delta t)^{n-j} g(j\Delta t)$, where $t = n\Delta t$, and (3) this sum is approximated to within ε by the sum in equation (3.20). Convergence is thus established.

It follows that the Lax equivalence theorem is valid also for the inhomogeneous problems.

In the foregoing, the boundary conditions were still assumed homogeneous, since they were taken into account by restricting the domain of A. Let A_0 be the extension of A obtained by ignoring this restriction on the domain, that is, by ignoring the boundary conditions, and consider the problem

$$\frac{d}{dt} u(t) - A_0 u(t) = g(t),$$

$$A_1 u(t) = h(t),$$

$$u(0) = u_0.$$

In the second equation, which represents the boundary condition, A_1 is some linear operator and $h(t)$ is a given function; this operator takes us out of the Banach space $\mathscr{B}$ into some other space of functions defined just on the boundaries, and $h(t)$ is such a function.

Let $w(t)$ be any smooth function with values in $\mathscr{B}$ and in fact in the domains of A_0 and A_1 and such that $A_1 w(t) = h(t)$; $w(t)$ is highly arbitrary because it only has to satisfy the boundary conditions; then, the above problem can be solved by the following well-known procedure: Call $v(t) = u(t) - w(t)$, and require $v(t)$ to satisfy the equations

$$\frac{d}{dt}v(t) - A_0 v(t) = g(t) - \frac{d}{dt}w(t) - A_0 w(t),$$

$$A_1 v(t) = 0,$$

$$v(0) = u_0 - w(0),$$

whose right-hand sides are known. The boundary condition is now again homogeneous, so if A is the restriction of A_0 to elements v of $\mathscr{B}$ such that $A_1 v = 0$, then we have a problem of the type (3.15), and all the preceding considerations of this section apply.

R. J. Thompson (1964) has extended most of these results to the quasi-linear case in which $g(t)$ is replaced by $g(t, u(t))$, where $g(t, u)$ is a function defined for $0 \leq t \leq T$ and for $u \in \mathscr{B}$, continuous in t for each u, and uniformly Lipschitz continuous with respect to u, by which is meant that there exists a constant M such that $\|g(t, u) - g(t, v)\| \leq M \|u - v\|$ for all u, v, and t.

3.8. Change of Norm

Although the choice of a suitable norm for $\mathscr{B}$ is part of the formulation of a problem, and a problem may be properly posed in one norm but not in another, there is nevertheless much arbitrariness in the choice.

Suppose that, in addition to $\|u\|$, there is defined, for all u in $\mathscr{B}$, a second real-valued function $\|u\|'$ having all the properties of a norm, as given in Chapter 2. Suppose also that there are two positive constants K_1 and K_2 such that

(3.21) $\qquad K_1 \|u\| \leq \|u\|' \leq K_2 \|u\|, \quad \text{for all } u \text{ in } \mathscr{B}.$

Then, we say that the two norms are *equivalent*. All topological properties of $\mathscr{B}$ are the same in the one norm as in the other; in particular, convergence of a sequence in $\mathscr{B}$, boundedness of a set in $\mathscr{B}$, boundedness of a linear operator, and uniform boundedness of a family of operators are all invariant concepts under a change from the one norm to the other.

3.9. Stability and Perturbations

The following theorem (Kreiss, 1962; Strang, 1964) shows that stability is not destroyed by a small perturbation. *If the difference system*

$$(3.22) \qquad u^{n+1} = C(\Delta t)u^n$$

is stable, and $Q(\Delta t)$ is a bounded family of operators, then the difference system

$$(3.23) \qquad u^{n+1} = [C(\Delta t) + \Delta t Q(\Delta t)]u^n$$

is also stable.

By definition, the stability of (3.22) implies the existence of a constant K such that

$$(3.24) \qquad \|C(\Delta t)^n\| < K, \quad \text{for } \begin{cases} 0 < \Delta t < \tau \\ 0 \leq n\Delta t \leq T. \end{cases}$$

Suppose further that

$$3.25) \qquad \|Q(\Delta t)\| \leq H, \quad \text{for } 0 < \Delta t < \tau$$

as required in the statement of the theorem. Then, the expression $(C + \Delta t Q)^n$ consists of 2^n terms, of which $\binom{n}{j}$ consist of a product of j factors $\Delta t Q$ and $n - j$ factors C, in some order. In such a term, the C's are grouped into at most $j + 1$ sequences of consecutive factors (since there are only j factors $\Delta t Q$ to separate them), and the norm of each such sequence is bounded by K. Hence,

$$(3.26) \qquad \|(C + \Delta t Q)^n\| \leq \sum_{j=0}^{n} \binom{n}{j} K^{j+1}(\Delta t H)^j$$

$$= K(1 + \Delta t K H)^n \leq K e^{n\Delta t K H} \leq K e^{T K H},$$

showing that the perturbed system is stable.

It should be mentioned that certain alternative definitions of stability are not equally insensitive to perturbations. According to one definition, used by several authors, the operators (3.24) are not required to be uniformly bounded but are permitted to grow with some power of $1/\Delta t$ as

SEC. 3.9] STABILITY AND PERTURBATIONS 59

$\Delta t \to 0$. Then, according to various examples due to Kreiss, one of which will be described in Section 5.2, a fairly simple perturbation can cause the operators to grow, in bound, exponentially, with an exponent containing $(1/\Delta t)^{1-1/p}$, where p is the number of components in the solution vector u.

If the initial-value problem is improperly posed, in the sense that the domain of $E_0(t)$ can be chosen to be dense in $\mathscr{B}$ but that, whenever it is so chosen, $E_0(t)$ is not bounded in any interval $0 \leq t < \tau$, then no difference scheme that is consistent with this problem can be stable. The unboundedness of $E_0(t)$ implies that there is a sequence of genuine solutions $u_j(t) = E_0(t)u_j^0$ ($j = 1, 2, \ldots$), normalized, let us say, by $\|u_j^0\| = 1$ for each j, and, at the same time, a sequence $t_j > 0$ tending to zero and such that

(3.27) $$\|u_j(t_j)\| = \|E_0(t_j)u_j^0\| \to \infty, \quad \text{as } j \to \infty.$$

Let $\varepsilon > 0$; then, for any genuine solution $u_j(t)$, according to the consistency condition—see (3.6c)—there is a $\Delta_j t = t_j/n_j$ for some integer n_j such that $\|\{C(\Delta_j t) - E(\Delta_j t)\}u_j(t)\| < \varepsilon \Delta_j t$. The single further assumption of stability leads to a contradiction, for (3.10) and (3.11) then show that

$$\|\{C(\Delta_j t)^{n_j} - E(t_j)\}u_j^0\| < \varepsilon K_2 T,$$

where K_2 is the bound appearing in the definition of stability, so that, since $\|u_j^0\| = 1$, $\|C(\Delta_j t)^{n_j}\| \geq \|C(\Delta_j t)^{n_j}u_j^0\| \to \infty$, by use of equation (3.27) and the triangle inequality.

Therefore, for the initial-value problem of Laplace's equation $\partial^2 u/\partial t^2 + \partial^2 u/\partial x^2 = 0$ (usually called a "Cauchy problem" in this case), for the heat flow equation with the sign of t reversed, etc., it is useless to look for schemes that are stable.

CHAPTER 4

Pure Initial-Value Problems with Constant Coefficients

4.1. The Class of Problems

In this chapter we consider problems with constant coefficients and with simple boundary conditions so that the Fourier series or Fourier integral representation of the solution can be used. In this case the concepts introduced in the preceding chapter take simple forms, as will be seen below. In particular, the Fourier transformation reduces the stability condition to a form from which a number of practical stability criteria are obtained. The simplest and earliest, historically, is the von Neumann necessary condition, stated below. In the first edition of this book, a number of sufficient conditions were given for special cases. Since then, as a result of the work of T. Kato (1960), H. O. Kreiss (1962), and M. L. Buchanan (1963a, b), these have been successively improved and finally incorporated in a single algebraic stability condition, whose derivation is one of the main objectives of this chapter, and through whose use one can, in principle, determine the stability of any difference scheme with constant coefficients. A number of special cases are then given to make the investigation of stability more tractable in general practice. The work of Kreiss also lays the foundation for an alternative class of techniques for investigating stability, the so-called energy methods, which we shall discuss in greater detail in Chapter 6.

In order to apply the Fourier transform method, we shall either suppose that the boundary conditions can be replaced by periodicity conditions, as was done in the simple heat flow problem in Chapter 1, or that the functions involved are quadratically integrable over all space, as in quantum mechanics. The Fourier representation theorems that will be used are those based on an L_2 norm, namely, the Riesz-Fischer theorem for Fourier series and Plancherel's theorem for the integrals. This fixes the choice of the Banach space $\mathscr{B}$, once the function elements u have been chosen. The set of Fourier coefficients or the Fourier transform determines a point in a second Banach space $\mathscr{B}'$ and the Fourier transformation provides a one-to-one norm-preserving mapping or isomorphism between $\mathscr{B}$ and $\mathscr{B}'$. In consequence, all the considerations of the preceding chapter apply in $\mathscr{B}'$ as well as in $\mathscr{B}$. This is the advantage gained, for the problems with constant coefficients, by use of the L_2 norm rather than, for example, the maximum norm, which is sometimes regarded as the natural norm for work in numerical analysis.

The Banach spaces $\mathscr{B}$ and $\mathscr{B}'$ are also Hilbert spaces. With every pair of elements u and v of $\mathscr{B}$ is associated a complex number, denoted by (u, v) and called the *inner product* of u and v. If u, v, w are any elements of $\mathscr{B}$ and α, β are complex numbers,

$$(u, \alpha v + \beta w) = \alpha(u, v) + \beta(u, w)$$
$$(v, u) = \overline{(u, v)}$$
$$(u, u) \geq 0$$
$$(u, u) = 0 \text{ if and only if } u = 0.$$

The first shows that the function (u, v) is linear in the second argument; this, together with the second, shows that it is anti-linear in the first argument: $(\alpha u + \beta v, w) = \bar{\alpha}(u, w) + \bar{\beta}(v, w)$, where the bar denotes complex conjugate. The inner product is related to the norm by the equation $\|u\|^2 = (u, u)$.

4.2. Fourier Series and Integrals

If there are p dependent variables in d space variables, the functions corresponding to a point in the Banach space are denoted, in vector notation, by $\mathbf{u}(\mathbf{x})$, where $\mathbf{u}$ is a vector of p components and $\mathbf{x}$ a vector of d components. The solution of an initial-value problem is a function $\mathbf{u}(\mathbf{x}, t)$.

If $\mathbf{u}(\mathbf{x})$ is periodic in each component x_i of $\mathbf{x}$ and has period L_i in x_i, then its Fourier series is a summation

(4.1a) $$\mathbf{u}(\mathbf{x}) = \left(\frac{1}{L_1 L_2 \ldots L_d}\right)^{1/2} \sum_{\mathscr{L}}{}_{(\mathbf{k})} \hat{\mathbf{u}}(\mathbf{k}) e^{i\mathbf{k} \cdot \mathbf{x}}.$$

Here, **k** is a d-component vector which ranges, in the summation, over a lattice $\mathscr{L}$ of points such that each component k_i of **k** runs over the values $2\pi r_i/L_i$, r_i being any integer: $\hat{\mathbf{u}}(\mathbf{k})$ is the Fourier coefficient vector; as a function of **k** defined on $\mathscr{L}$, it represents a point in the Banach space $\mathscr{B}'$. The relation inverse to (4.1a) is:

$$(4.2a) \qquad \hat{\mathbf{u}}(\mathbf{k}) = \left(\frac{1}{L_1 L_2 \cdots L_d}\right)^{1/2} \int_0^L d\mathbf{x}\, \mathbf{u}(\mathbf{x}) e^{-i\mathbf{k}\cdot\mathbf{x}},$$

where the expression

$$\int_0^L d\mathbf{x}$$

is an abbreviation for

$$\int_0^{L_1} dx_1 \cdots \int_0^{L_d} dx_d.$$

The so-called Parseval relation is

$$(4.3a) \qquad \int_0^L d\mathbf{x}\,|\mathbf{u}(\mathbf{x})|^2 = \sum_{\mathscr{L}(\mathbf{k})} |\hat{\mathbf{u}}(\mathbf{k})|^2;$$

for a function $\mathbf{u}(\mathbf{x})$ in $\mathscr{B}$, the norm $\|\mathbf{u}\|$ is taken as the square root of the left member and for a function $\hat{\mathbf{u}}(\mathbf{k})$ in $\mathscr{B}'$, the norm $\|\hat{\mathbf{u}}\|$ is taken as the square root of the right member. Hence, (4.1a) and (4.2a) establish a one-to-one norm-preserving mapping from a set in $\mathscr{B}$ onto a set in $\mathscr{B}'$.

If, on the other hand, $\mathbf{u}(\mathbf{x})$ is quadratically integrable over all space, it is represented by a Fourier integral

$$(4.1b) \qquad \mathbf{u}(\mathbf{x}) = \frac{1}{(2\pi)^{d/2}} \int d\mathbf{k}\, \hat{\mathbf{u}}(\mathbf{k}) e^{i\mathbf{k}\cdot\mathbf{x}}$$

where now the integration is over all **k**-space, and $\hat{\mathbf{u}}(\mathbf{k})$ is given in this space by

$$(4.2b) \qquad \hat{\mathbf{u}}(\mathbf{k}) = \frac{1}{(2\pi)^{d/2}} \int d\mathbf{x}\, \mathbf{u}(\mathbf{x}) e^{-i\mathbf{k}\cdot\mathbf{x}}.$$

In this case the Parseval relation is

$$(4.3b) \qquad \int d\mathbf{x}\,|\mathbf{u}(\mathbf{x})|^2 = \int d\mathbf{k}\,|\hat{\mathbf{u}}(\mathbf{k})|^2.$$

For well-behaved functions, the above equations can be taken literally, but if $\mathscr{B}$ and $\mathscr{B}'$ were restricted to such functions, they would fail to be complete. Therefore, to make the foregoing rigorous, one should adhere to the mathematical refinements of the L_2 theory of regarding two functions as identical if they differ only on a set of measure zero, of interpreting the integrals in the Lebesgue sense, and of regarding the limits implied in (4.1a), (4.1b), and (4.2b) as limits in the mean, that is, in the norm of the Banach space. Thus, if $\mathbf{u}_n(\mathbf{x})$ is a sequence of partial sums of the series (4.1a) in which, say, each r_i runs from $-n$ to n, then the limit is understood in the sense

$$\|\mathbf{u} - \mathbf{u}_n\| \to 0, \quad \text{as } n \to \infty.$$

Similarly, if $\mathbf{u}_K(\mathbf{x})$ is the integral as in (4.1b) but with each component of $\mathbf{k}$ restricted by $-K \leq k_i \leq K$, then

$$\|\mathbf{u} - \mathbf{u}_K\| \to 0, \quad \text{as } K \to \infty.$$

The Banach spaces are then complete, (4.2a) maps all of $\mathscr{B}$ into $\mathscr{B}'$, the Fourier theorem says that (4.1a) and (4.2a) are inverse mappings, and the Riesz-Fischer theorem says that (4.1a) maps all of $\mathscr{B}'$ into $\mathscr{B}$; that is, any set of numbers $\hat{\mathbf{u}}(\mathbf{k})$ defined on $\mathscr{L}$, such that the right member of (4.3a) is a convergent sum, are the coefficients of some function $\mathbf{u}(\mathbf{x})$ in $\mathscr{B}$. Similarly, the Plancherel theorem says that (4.1b) is a one-to-one mapping of all $\mathscr{B}'$ onto all $\mathscr{B}$ and that (4.2b) is its inverse.

In the following, the foregoing refinements will be implicitly assumed, whenever equations like the above are written. We shall usually use the notation of (4.1b) to (4.3b) to cover both cases and let $\mathscr{L}$ denote either the whole $\mathbf{k}$-space or the lattice defined above, as appropriate.

4.3. Properly Posed Initial-Value Problems

The general linear differential operator with constant coefficients for functions in $\mathscr{B}$ can be obtained formally by taking a function $P(\mathbf{q})$, or $P(q_1, \ldots, q_d)$, which is a $p \times p$ matrix whose elements are polynomials in $q_1, \ldots, q_d$, and substituting $\partial/\partial x_1$ for q_1, $\partial/\partial x_2$ for q_2, etc. If A is such an operator and we apply it to the element $\hat{\mathbf{u}} e^{i\mathbf{k}\cdot\mathbf{x}}$, where $\hat{\mathbf{u}}$ is a constant vector, the result is simply to multiply this element by the matrix $P(i\mathbf{k})$.

64   PURE INITIAL-VALUE PROBLEMS   [CHAP. 4

Therefore if the solution of the initial-value problem is expressed as a Fourier integral

$$\mathbf{u}(\mathbf{x}, t) = (2\pi)^{-d/2} \int d\mathbf{k}\, \hat{\mathbf{u}}(\mathbf{k}, t) e^{i\mathbf{k}\cdot\mathbf{x}},$$

the Fourier transform $\hat{\mathbf{u}}(\mathbf{k}, t)$ must satisfy the equation

$$\frac{\partial}{\partial t} \hat{\mathbf{u}}(\mathbf{k}, t) = P(i\mathbf{k})\hat{\mathbf{u}}(\mathbf{k}, t),$$

which is the equivalent, for the space $\mathscr{B}'$, of the original differential equation

$$\frac{\partial}{\partial t} \mathbf{u}(\mathbf{x}, t) = A\mathbf{u}(\mathbf{x}, t).$$

Hence, the solution is

$$\hat{\mathbf{u}}(\mathbf{k}, t) = e^{tP(i\mathbf{k})}\hat{\mathbf{u}}(\mathbf{k}, 0),$$

where $\hat{\mathbf{u}}(\mathbf{k}, 0)$ is the Fourier transform of the initial element $\mathbf{u}(\mathbf{x}, 0)$. In terms of $\mathscr{B}$ the solution is therefore

(4.4) $$\hat{\mathbf{u}}(\mathbf{x}, t) = (2\pi)^{-d/2} \int d\mathbf{k}\, e^{tP(i\mathbf{k})}\hat{\mathbf{u}}(\mathbf{k}, 0) e^{i\mathbf{k}\cdot\mathbf{x}},$$

where

$$\hat{\mathbf{u}}(\mathbf{k}, 0) = (2\pi)^{-d/2} \int d\mathbf{x}\, \mathbf{u}(\mathbf{x}, 0) e^{-i\mathbf{k}\cdot\mathbf{x}}.$$

In the above it is understood that the function e^M of a $p \times p$ matrix M is defined by the power series $e^M = 1 + M + \frac{1}{2!}M^2 + \frac{1}{3!}M^3 + \cdots$ which converges absolutely, element by element, and has the property that $d/dt\, e^{tM} = M e^{tM}$.

The first requirement for a properly posed initial-value problem, namely that the domain of the solution operator $E(t)$ be dense in $\mathscr{B}$ (see Chapter 3), is automatically satisfied for the problem considered in this chapter. The above solution is certainly valid whenever the summation or integration in (4.4) is over a finite range, i.e., when the initial element $\mathbf{u}(\mathbf{x}, 0)$ is a trigonometric polynomial or has a Fourier transform with compact support,[1] and such elements are dense in $\mathscr{B}$. The second

---

[1] The *support* of a function is the closure of the point set on which the function is non-zero. A function then has *compact support* if its support forms a compact set.

requirement for a properly posed initial-value problem takes the form that $\|e^{tP(i\mathbf{k})}\|$ should be bounded uniformly in $\mathbf{k}$; and then we have

(4.5) $$\|E(t)\| = \text{Max}_{(\mathbf{k})}\|e^{tP(i\mathbf{k})}\|.$$

Whether this condition is satisfied must be investigated separately in each case. As a simple example consider the equation of heat flow in which A is $\sigma\partial^2/\partial x^2$. Then $\|e^{tP(i\mathbf{k})}\| = e^{-\sigma t k^2}$ which is bounded by unity if $\sigma t \geqq 0$. But if $\sigma t < 0$ there is no bound which is uniform in k: thus, the problem of solving the heat flow equation backwards in time is not well posed.

However, the purpose of this book is to discusss the solution of problems, not their formulation, and we shall include the proper posing of an initial-value problem as part of its formulation—in other words, we shall assume that only properly posed problems are given to us to solve. But it should be pointed out that the choice of Banach space is then to some extent a part of the problem formulation: as we have seen above, the choice of the L_2 norm is indicated by the desire to use Fourier analysis, but the vector $\mathbf{u}$ whose norm is to be taken is still often open to choice. This is particularly so when a higher order differential equation has to be reduced to a first order system before the theory can be applied. To illustrate this, we show in Chapter 10 that the initial-value problem of wave motion is properly posed for one simple choice of Banach space but not for another equally simple choice. In the simple cases, the proper formulation is usually a rather trivial matter; in other cases, like problems of multi-dimensional fluid dynamics, it is not at present clear whether a proper formulation is possible without some new physical principles.

4.4. The Finite-Difference Equations

We suppose that the difference equations are given by a two-level formula. Multi-level formulas will be discussed in a later chapter.

Let $\mathbf{u}^n = \mathbf{u}^n(\mathbf{x})$ denote the approximation to $\mathbf{u}(\mathbf{x}, n\Delta t)$ obtained from the difference equations. And let T^β denote the translation operator which replaces the value of a function at a point $\mathbf{x} = (x_1, x_2, \ldots, x_d)$ by its value at a neighboring net point $(x_1 + \beta_1\Delta x_1, x_2 + \beta_2\Delta x_2, \ldots, x_d + \beta_d\Delta x_d)$ of the lattice in $\mathbf{x}$-space, where the components $\beta_1, \beta_2, \ldots, \beta_d$ of the multi-index $\boldsymbol{\beta}$ are integers. Each difference equation at a point $\mathbf{x}$ equates to zero a certain linear combination of the components of $\mathbf{u}^n$ and $\mathbf{u}^{n+1}$ at some group of such neighbors. Thus, the difference equations $B_1\mathbf{u}^{n+1} - B_0\mathbf{u}^n = 0$ can be written in the form

$$(4.6) \qquad \sum_{\mathcal{N}_1} B_1^{\boldsymbol{\beta}} T^{\boldsymbol{\beta}} \mathbf{u}^{n+1} - \sum_{\mathcal{N}_0} B_0^{\boldsymbol{\beta}} T^{\boldsymbol{\beta}} \mathbf{u}^n = 0,$$

where the sums are to be extended over the finite sets $\mathcal{N}_0$ and $\mathcal{N}_1$ of the neighbors[2] of the point $\mathbf{x}$. For each value of $\boldsymbol{\beta}$, $B_0^{\boldsymbol{\beta}}$ and $B_1^{\boldsymbol{\beta}}$ are $p \times p$ matrices with constant elements; that is, the elements may depend on Δt, Δx_1, etc., but not on $\mathbf{x}$ or t. For an explicit system of difference equations the set $\mathcal{N}_1$ reduces to the single point $\boldsymbol{\beta} = 0$; then the matrix B_1^0 is usually diagonal, and in any case non-singular. If the equations are implicit, we assume that they are always solvable for periodic $\mathbf{u}^n$ and that the solution $\mathbf{u}^{n+1}$ is uniquely determined by the equations and the periodicity condition.

For numerical work, the equations (4.6) are used only at the lattice points, but they will be taken to apply equally well at other points, so that if $\mathbf{u}^n(\mathbf{x})$ is given for all $\mathbf{x} = (x_1, \ldots, x_d)$, $\mathbf{u}^{n+1}(\mathbf{x})$ is determined for all $\mathbf{x}$ by the equations (4.6) and the periodicity condition.

If Fourier series are substituted for $\mathbf{u}^n(\mathbf{x})$ and $\mathbf{u}^{n+1}(\mathbf{x})$ in (4.6), the typical terms contain a factor

$$\exp\{i[k_1(x_1 + \beta_1 \Delta x_1) + \cdots + k_d(x_d + \beta_d \Delta x_d)]\},$$

and we may cancel out the common part, $\exp\{i\mathbf{k} \cdot \mathbf{x}\}$ from all the terms of the equation. With $\mathbf{v}(\mathbf{k})$ instead of $\hat{\mathbf{u}}(\mathbf{k})$ denoting the Fourier coefficients, what is left can be written as

$$(4.7) \qquad H_1 \mathbf{v}^{n+1}(\mathbf{k}) - H_0 \mathbf{v}^n(\mathbf{k}) = 0,$$

where H_1 is an abbreviation for the matrix

$$(4.8) \qquad H_1 = \sum_{\mathcal{N}_1} B_1^{\boldsymbol{\beta}} \exp\{i[k_1 \beta_1 \Delta x_1 + \cdots + k_d \beta_d \Delta x_d]\}$$

and similarly for H_0. As in the discussion of the last chapter, we assume that the manner of refinement of the mesh has been set in advance, so

---

[2] The point $(x_1, \ldots, x_d)$ itself is of course also generally included in the set $\mathcal{N}_1$.

SEC. 4.5]   ORDER OF ACCURACY AND THE CONSISTENCY CONDITION   67

that the manner in which $\Delta x_1, \ldots, \Delta x_d$ go to zero as Δt goes to zero is given by the functions

(4.9) $$\Delta x_1 = g_1(\Delta t), \text{ etc.}$$

The matrices H_1 and H_0 then depend only on Δt and on $\mathbf{k}$. The solvability assumption made in the preceding paragraph is tantamount to the assumption that H_1 has an inverse. Therefore, we can write

(4.10) $$\mathbf{v}^{n+1}(\mathbf{k}) = G(\Delta t, \mathbf{k})\mathbf{v}^n(\mathbf{k}),$$

where

(4.11) $$G(\Delta t, \mathbf{k}) = (H_1)^{-1} H_0.$$

Clearly, $G(\Delta t, \mathbf{k})$ is the representation, in the space $\mathscr{B}'$, of the difference operator $C(\Delta t)$ of the Lax theory, and (4.10) is the form taken by the equation (3.5) when transformed to the space $\mathscr{B}'$ of the Fourier coefficients. $G(\Delta t, \mathbf{k})$ is called the amplification matrix.[3]

4.5. Order of Accuracy and the Consistency Condition

There is an obvious ambiguity in the formula (4.6) in that the equation could be multiplied through by any non-singular matrix, and an exactly equivalent equation would be obtained. This freedom is sometimes made use of, when the equations are explicit, to put the equations in a form in which they are already solved for the unknown $\mathbf{u}^{n+1}$ (e.g., in the simple heat flow equation, one multiplies through by Δt to replace $(u_j^{n+1} - u_j^n)/\Delta t$ by $u_j^{n+1} - u_j^n$). For the purpose of the present section, however, we must avoid this ambiguity, and we shall assume that the left member of (4.6) has been obtained *directly* from the expression $\partial \mathbf{u}/\partial t - A\mathbf{u}$ by such steps as replacement of derivatives by difference quotients, replacement of quantities by interpolated or extrapolated values, and the like. That is, the difference operator on the left of (4.6), when applied to any smooth function, must have approximately the same effect as the operator $\partial/\partial t - A$. As before, the manner of refinement of the mesh is supposed to be fixed in advance by relations like (4.9). Then we define the *order of*

---

[3] When $p = 1$ so that the amplification matrix has a single element, we also call $G(\Delta t, \mathbf{k})$ the *amplification factor*.

accuracy of the difference equations as the largest[4] real number ρ for which

(4.12) $\qquad \|B_1 \mathbf{u}(t + \Delta t) - B_0 \mathbf{u}(t)\| = O[(\Delta t)^\rho], \quad \text{as } \Delta t \to 0,$

for all sufficiently smooth genuine solutions of the differential equation $\partial \mathbf{u}/\partial t = A\mathbf{u}$. By our assumption of the solvability of the difference equations for $\mathbf{u}^{n+1}$, B_1 must have an inverse $(B_1)^{-1}$. Moreover, from the way in which we have normalized the equations, an expansion of B_1 in ascending (possibly fractional) powers of Δt will have a leading term proportional to $(\Delta t)^{-1}$. Thus, we will have $\|(B_1)^{-1}\| = O(\Delta t)$ (see Section 7.4, where the general multi-level difference scheme will be treated in more detail), so that (4.12) may be written

(4.12a) $\qquad \|\mathbf{u}(t + \Delta t) - (B_1)^{-1} B_0 \mathbf{u}(t)\| = O[(\Delta t)^{\rho+1}], \quad \text{as } \Delta t \to 0.$

Comparing this with the consistency condition (3.6a) in which we have put $C = (B_1)^{-1} B_0$, we see that the requirement of consistency is equivalent to demanding that the order of accuracy of the difference equations be positive. Very often ρ will be an integer, in which case consistency requires that the equations be at least "first order accurate."

An alternative definition of the order of accuracy, involving the amplification matrix G, can be obtained by taking the Fourier transform of the expression under the norm in (4.12a). From (4.4), (4.7), and (4.8) this becomes

(4.13) $\qquad \|e^{\Delta t P(i\mathbf{k})} - G(\Delta t, \mathbf{k})\| = O[(\Delta t)^{\rho+1}], \quad \text{as } \Delta t \to 0.$

The expression in (4.12) is what was called the truncation error in Chapter 1. It can be readily estimated from a Taylor series expansion of $\mathbf{u}(\mathbf{x}, t)$ with remainder. The smoothness of $\mathbf{u}$ required above, in fact, is determined by the number of derivatives needed in this expansion. We shall assume that solutions with these properties are dense in the class of all genuine solutions of the differential equation.

4.6. Stability

We pointed out in Section 4.5 that equation (4.10) corresponds to the equation $u^{n+1} = C(\Delta t) u^n$ of the general theory of Chapter 3 and that this

---

[4] In practice, the relations (4.9) will set the Δx_i proportional to rational powers of Δt; ρ will then be a rational number and there will be no difficulty about this maximum actually being attained.

equation therefore defines an operator in $\mathscr{B}'$ equivalent to the operator $C(\Delta t)$. The operation consists of multiplying the vector $\mathbf{v}^n$ by the matrix $G(\Delta t, \mathbf{k})$ ($\mathbf{k}$ appears only as a parameter). Clearly, the stability condition therefore will involve the matrices

$$G(\Delta t, \mathbf{k})^n \quad \text{for} \quad \begin{cases} 0 < \Delta t < \tau \\ 0 \leq n\Delta t \leq T \end{cases}$$

and will in fact state that the operators corresponding to these matrices should be uniformly bounded.

If L is an operator in $\mathscr{B}'$ determined by a matrix $F(\mathbf{k})$ according to an equation $\mathbf{w}(\mathbf{k}) = F(\mathbf{k})\mathbf{v}(\mathbf{k})$, the bound of L is

(4.14) $$\|L\| = \operatorname*{Max}_{\mathscr{L}}{}_{(\mathbf{k})} \|F(\mathbf{k})\|,$$

where, for given $\mathbf{k}$, the bound of the matrix $F(\mathbf{k})$ is given by

(4.15) $$\|F(\mathbf{k})\| = \operatorname*{Max}_{|\mathbf{v}|=1} |F(\mathbf{k})\mathbf{v}| = \operatorname*{Max}_{\mathbf{v} \neq 0} \frac{|F(\mathbf{k})\mathbf{v}|}{|\mathbf{v}|}.$$

Here, as in Section 4.2, the magnitude $|\mathbf{v}|$ of a p-component vector $\mathbf{v}$ means $(|v_1|^2 + \cdots + |v_p|^2)^{1/2}$. Thus, $\|F(\mathbf{k})\|$ is the norm of the matrix $F(\mathbf{k})$ regarded as a transformation of the p-dimensional Euclidean space of the vectors $\mathbf{v}, \mathbf{w}, \ldots$.

The stability condition requires therefore that for some positive τ the matrices

(4.16) $$G(\Delta t, \mathbf{k})^n \quad \text{for} \begin{cases} 0 < \Delta t < \tau \\ 0 \leq n\Delta t \leq T \\ \text{all } \mathbf{k} \text{ in } \mathscr{L} \end{cases}$$

be uniformly bounded.

If F is a $p \times p$ matrix and $\lambda_1, \ldots, \lambda_p$ are its eigenvalues, the maximum of $|\lambda_i|$ $(i = 1, \ldots, p)$ is called the spectral radius. If R denotes the spectral radius of F, clearly $\|F\| \geq R$, because the maximum value of $|F\mathbf{v}|/|\mathbf{v}|$ is not less than the value obtained by substituting for $\mathbf{v}$ an eigenvector of F corresponding to the largest eigenvalue. The spectral radius of F^n is R^n since the eigenvalues of F^n are the nth powers of the eigenvalues of F. Furthermore,

$$\|F^2\| = \operatorname*{Max}_{\mathbf{v} \neq 0} \frac{|F(F\mathbf{v})|}{|F\mathbf{v}|} \frac{|F\mathbf{v}|}{|\mathbf{v}|} \leq \operatorname*{Max}_{\substack{\mathbf{v} \neq 0 \\ \mathbf{w} \neq 0}} \frac{|F\mathbf{w}|}{|\mathbf{w}|} \frac{|F\mathbf{v}|}{|\mathbf{v}|} = \|F\|^2,$$

and similarly $\|F^n\| \leq \|F\|^n$.

In what follows, we shall denote by $R(\Delta t, \mathbf{k})$ the spectral radius of $G(\Delta t, \mathbf{k})$. We have generally[5]

(4.17) $$R(\Delta t, \mathbf{k})^n \leq \|G(\Delta t, \mathbf{k})^n\| \leq \|G(\Delta t, \mathbf{k})\|^n.$$

4.7. The von Neumann Condition

A necessary condition for stability is that there exist a constant C_1 such that

$$R(\Delta t, \mathbf{k})^n \leq C_1 \quad \text{for} \quad \begin{cases} 0 < \Delta t < \tau \\ 0 \leq n\Delta t \leq T \\ \text{all } \mathbf{k} \text{ in } \mathscr{L}. \end{cases}$$

Without loss of generality, we can assume that $C_1 \geq 1$. Therefore,

$$R(\Delta t, \mathbf{k}) \leq C_1^{1/n}, \quad 0 \leq n \leq \frac{T}{\Delta t};$$

thus, in particular,

$$R(\Delta t, \mathbf{k}) \leq C_1^{\Delta t/T}$$

For Δt in the interval $0 < \Delta t < \tau$, the expression $C_1^{\Delta t/T}$ is bounded by a linear expression of the form $1 + C_2\Delta t$ (see upper line in the schematic graph, Figure 4.1). Remembering the definition of spectral radius, we have the condition

(4.18) $$|\lambda_i| \leq 1 + O(\Delta t) \quad \text{for} \quad \begin{cases} 0 < \Delta t < \tau \\ \text{all } \mathbf{k} \text{ in } \mathscr{L} \\ i = 1, \ldots, p \end{cases}$$

where $\lambda_1, \ldots, \lambda_p$ are the eigenvalues of the amplification matrix $G(\Delta t, \mathbf{k})$: (4.18) *is the von Neumann necessary condition for stability.*

In the simple heat flow examples of Chapter 1 we used a stability condition of the form

(4.19) $$|\lambda_i| \leq 1.$$

---

[5] If $G(\Delta t, \mathbf{k})$ is a normal matrix, all three members of (4.17) are equal. A normal matrix is one that commutes with its adjoint (i.e., with its hermitian conjugate). Necessary and sufficient for a matrix to have a complete orthogonal set of eigenvectors is that it be normal. A normal matrix can be written as $A + iB$ where A and B are commuting hermitian matrices. The amplification matrix $G(\Delta t, \mathbf{k})$ is very often not normal.

SEC. 4.7] THE VON NEUMANN CONDITION 71

But in some problems it is possible for a component of the exact solution to grow exponentially with increasing time, and the condition (4.19) would not permit such a growth (and in fact could not be satisfied at all without violating consistency). Then one must use, instead, condition

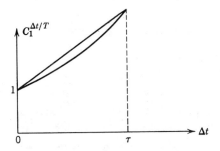

Fig. 4.1. A bound for a portion of an exponential curve.

(4.18) which is more generous than (4.19) by just enough to permit a "legitimate" exponential growth. An example is provided by the modified heat flow problem

$$\frac{\partial u}{\partial t} = \sigma \frac{\partial^2 u}{\partial x^2} + bu, \quad \sigma > 0, \, b > 0,$$

in which a source term has been added. If we write the difference equations as

$$\frac{u_j^{n+1} - u_j^n}{\Delta t} = \sigma \frac{u_{j+1}^n - 2u_j^n + u_{j-1}^n}{(\Delta x)^2} + bu_j^n,$$

the amplification factor is

$$G(\Delta t, k) = 1 - 4 \frac{\sigma \Delta t}{(\Delta x)^2} \sin^2 \frac{k \Delta x}{2} + b \Delta t,$$

and it is seen that the von Neumann condition (4.18) is satisfied if

$$\frac{\sigma \Delta t}{(\Delta x)^2} = \text{constant} \leq \tfrac{1}{2},$$

as was found for the unmodified equation.

It is perhaps worth noting that one can obtain the amplification factors (the eigenvalues of G) by simply substituting $\mathbf{u}_0 \exp\{i\mathbf{k}\cdot\mathbf{x}\}$ and

$\lambda \mathbf{u}_0 \exp\{i\mathbf{k}\cdot\mathbf{x}\}$ for $\mathbf{u}^n$ and $\mathbf{u}^{n+1}$ directly into the difference equations (4.6); this gives p linear equations for the p components of $\mathbf{u}_0$, and the vanishing of their determinant determines λ. This procedure saves some labor if the difference scheme is implicit.

4.8. A Simple Sufficient Condition

It was noted in Section 4.6 that for a normal matrix the spectral radius and the bound are equal. Therefore, *if $G(\Delta t, \mathbf{k})$ is a normal matrix, the von Neumann condition is sufficient as well as necessary for stability.*

This applies in particular when $p = 1$, because all 1×1 matrices commute. We therefore have as a corollary: *For two-level difference equations with only one dependent variable, the von Neumann condition is sufficient as well as necessary for stability.* This applies to two-level difference equations for heat flow in any number of independent variables.

By way of further example we refer to the difference equations discussed in Chapter 10 for sound waves in one space variable. The wave equation is of the second order in time, so normally $p = 2$. For the simplest explicit equation the amplification matrix is

$$(4.20) \qquad G = \begin{bmatrix} 1 & ia \\ ia & 1 - a^2 \end{bmatrix}$$

where a is an abbreviation for $(2c\Delta t/\Delta x) \sin(k\Delta x/2)$; it is readily seen that G is not normal, because $G^*G - GG^* \neq 0$. For the simplest *implicit* equation, the amplification matrix is

$$G = \begin{bmatrix} \dfrac{1 - \frac{1}{4}a^2}{1 + \frac{1}{4}a^2} & \dfrac{ia}{1 + \frac{1}{4}a^2} \\ \dfrac{ia}{1 + \frac{1}{4}a^2} & \dfrac{1 - \frac{1}{4}a^2}{1 + \frac{1}{4}a^2} \end{bmatrix}$$

which is obviously a unitary matrix and therefore normal. Consequently, the von Neumann condition is sufficient for stability of the implicit equation.

A slightly more general sufficient condition can be obtained from the right-hand inequality of (4.17): *If, for some M and some $\tau > 0$,*

$$\|G(\Delta t, \mathbf{k})\| \leq 1 + M\Delta t, \qquad \text{for } 0 < \Delta t < \tau,$$

then the difference equations are stable, for then $\|G\|^n \le \exp(MT)$, for $0 \le n\Delta t \le T$.

In many practical problems, neither of these conditions is applicable; hence, more powerful criteria are needed.

4.9. The Kreiss Matrix Theorem

Rather than searching for further conditions which are sufficient for stability under certain special circumstances, it is preferable at this point to develop a theory of stable families of matrices. This will yield necessary and sufficient stability criteria which, although rather complicated, lead easily to further sufficient conditions. Moreover, we shall need later a new definition of stability, namely, that there should exist a new but equivalent norm to be denoted by $\| \ \|_H$ for which

$$\|\mathbf{u}^{n+1}\|_H \le (1 + O(\Delta t))\|\mathbf{u}^n\|_H,$$

i.e., $$\|C(\Delta t)\|_H \le 1 + O(\Delta t).$$

One of the results of this section will be that, for the present constant coefficient problems, this new definition is equivalent to that already adopted.

It is convenient, first of all, to eliminate the restriction on n which occurs in (4.16). We therefore shall say that a family $\mathscr{F}$ of $p \times p$ matrices A is stable if the following statement is true:

(A) *There exists a constant C such that for all $A \in \mathscr{F}$ and all positive integers ν*

(4.21) $$\|A^\nu\| \le C.$$

It is clear that, if the powers of the amplification matrices G of (4.16) have the uniform bound K, then the family of matrices $\{e^{-\alpha \Delta t} G(\Delta t, \mathbf{k})\}$, where $\alpha = T^{-1} \log K$, is stable in the above sense. For any positive integer ν can be written in the form[6]

$$\nu = m(T/\Delta t) + n, \quad \text{where } 0 \le n\Delta t < T,$$

so that

$$\|(e^{-\alpha \Delta t} G)^\nu\| \le \|(e^{-\alpha \Delta t} G)^{T/\Delta t}\|^m \|(e^{-\alpha \Delta t} G)^n\| \le K.$$

---

[6] There is clearly no loss in generality in assuming T an integral multiple of Δt.

Conversely, if for some constant α, the family $\{e^{-\alpha\Delta t}G(\Delta t, \mathbf{k})\}$ satisfies (4.21), then when $0 \leq n\Delta t \leq T$,

$$\|G^n\| \leq Ce^{\alpha T} = \text{constant}.$$

Hence, this usage of the term "stable" is consistent with that in our present theory. Moreover, a necessary condition for statement (A) to hold is that all the eigenvalues $\varkappa_i$ of A lie inside or on the unit circle. This means that it is necessary for the eigenvalues λ_i of G to satisfy

$$|\lambda_i| \leq e^{\alpha\Delta t}$$

for some constant α, which is equivalent to the von Neumann condition (4.18).

Three important characterizations of the stability of $\mathscr{F}$ are contained in a theorem due to Kreiss (1962).

The stability of the family $\mathscr{F}$ of matrices A is equivalent to each of the following statements:

(R) *There exists a constant C_R such that for all $A \in \mathscr{F}$ and all complex numbers z with $|z| > 1$, $(A - zI)^{-1}$ exists and*

(4.22) $$\|(A - zI)^{-1}\| \leq \frac{C_R}{(|z| - 1)}.$$

(S) *There exist constants C_S and C_B and, to each $A \in \mathscr{F}$, a non-singular matrix S such that (i) $\|S\|, \|S^{-1}\| \leq C_S$; and (ii) $B = SAS^{-1}$ is upper triangular and its off-diagonal elements satisfy*

(4.23) $$|B_{ij}| \leq C_B \, \text{Min}\,(1 - |\varkappa_i|, 1 - |\varkappa_j|),$$

where $\varkappa_i$ are the diagonal elements of B, i.e., the eigenvalues of A and B (note that this implies $|\varkappa_i| \leq 1$):

$$B = \begin{bmatrix} \varkappa_1 & B_{12} & B_{13} & \cdots & B_{1p} \\ 0 & \varkappa_2 & B_{23} & \cdots & B_{2p} \\ 0 & 0 & \varkappa_3 & & \cdot \\ \vdots & \vdots & \vdots & & \vdots \\ 0 & 0 & 0 & \cdots & \varkappa_p \end{bmatrix}.$$

(H) *There is a constant* $C_H > 0$ *and, for each* $A \in \mathscr{F}$, *a positive definite hermitian matrix* H *with the properties*[7]

(4.24) $$C_H^{-1} I \leq H \leq C_H I,$$

(4.25) $$A^* H A \leq H.$$

For general matrices, all known conditions imposed on the elements of A to ensure stability depend upon putting A into triangular form. This displays the eigenvalues as the diagonal elements and the von Neumann condition on these has to be supplemented by inequalities on the off-diagonal elements. Statement (S) is an example of this type of condition and others will be given below. The first statement provides a characterization which is extremely useful in theoretical work and, in particular, will be our main tool in proving (S). It gives a limit on how fast the resolvent $(A - zI)^{-1}$ can grow as the unit circle is approached and is therefore called the *resolvent condition*. Finally, statement (H) provides for the existence of the new norm $\| \ \|_H$: in the Fourier space $\mathscr{B}'$ we define

$$\|\mathbf{v}\|_H^2 = \mathbf{v}^* H \mathbf{v}$$

so that, if $G = e^{\alpha \Delta t} A$, (4.25) gives

$$\|\mathbf{v}^{n+1}\|_H^2 = e^{2\alpha \Delta t}(\mathbf{v}^n)^* A^* H A \mathbf{v}^n \leq [1 + O(\Delta t)] \|\mathbf{v}^n\|_H^2.$$

This inequality can then be taken over to the original space $\mathscr{B}$.

An alternative form for this statement is that the vector space in which A operates can be so transformed that the norm of a transformed A is bounded by one. For, from each H, one can obtain a matrix T with $T^* T = H$ and $\|T\|, \|T^{-1}\| \leq C_H^{1/2}$: then, in the space of vectors $\mathbf{w} = T\mathbf{v}$, A is replaced by $T A T^{-1}$ and

$$\|T A T^{-1}\|^2 = \underset{|\mathbf{u}|=1}{\text{Max}}\ \mathbf{u}^* T^{-1*} A^* T^* T A T^{-1} \mathbf{u}$$

$$\leq \underset{|\mathbf{u}|=1}{\text{Max}}\ \mathbf{u}^* T^{-1*} T^* T T^{-1} \mathbf{u}^* = 1.$$

Kreiss' theorem is proved by showing that each of the statements (A), (R), (S), and (H) implies the succeeding one, with the loop closed by proving that (H) implies (A). The only difficult step is to show that (R) implies (S), so we shall leave this one till last.

---

[7] In saying of two hermitian matrices A and B that $A \leq B$, we mean that $\mathbf{u}^* A \mathbf{u} \leq \mathbf{u}^* B \mathbf{u}$ for any vector $\mathbf{u}$.

As we have already remarked, if (A) is true, the eigenvalues $\varkappa_i$ of A lie within the closed unit disk and therefore $(A - zI)^{-1}$ exists for $|z| > 1$. Moreover,

$$\|(A - zI)^{-1}\| = \|\sum_{\nu=0}^{\infty} A^\nu z^{-\nu-1}\| \leq C \sum_{\nu=0}^{\infty} |z|^{-\nu-1} = C(|z| - 1)^{-1},$$

so that (R) is satisfied with $C_R \leq C$.

To deduce (H) from (S), we introduce the real diagonal matrix

$$D = \begin{bmatrix} \Delta & & & 0 \\ & \Delta^2 & & \\ & & \ddots & \\ 0 & & & \Delta^p \end{bmatrix} \quad \text{with } \Delta > 1.$$

Then Δ can be chosen so large that

(4.26) $$D - B^*DB \geq 0,$$

i.e., $M = I - (D^{-1/2}B^*D^{1/2})(D^{1/2}BD^{-1/2}) \geq 0$. To show this we use Gerschgorin's theorem[8] which states that, if a circle of radius

$$\sum_{\substack{j \neq i}}^{(j)} |M_{ij}|$$

is drawn in the complex plane about each point M_{ii} ($i = 1, 2, \ldots, p$), then every eigenvalue of M lies in the point set consisting of the union of these circles. Now, the $(i, j)^{\text{th}}$ element of $D^{1/2}BD^{-1/2}$ is $\Delta^{1/2(i-j)}B_{ij}$, so that the diagonal elements of M are

$$M_{ii} = 1 - \sum_{(k)} |B_{ki}|^2 \Delta^{k-i} = 1 - |\varkappa_i|^2 + \varepsilon_i,$$

where, from (4.23), $|\varepsilon_i| \leq C_B^2(1 - |\varkappa_i|)^2/(\Delta - 1)$; the sum of the off-diagonal elements equals

$$\delta_i = \sum_{\substack{j \neq i}}^{(j)} |M_{ij}| \leq \sum_{\substack{j \neq i}}^{(j)} \sum_{(k)} |B_{ki}|\,|B_{kj}|\Delta^{(2k-i-j)/2}.$$

---

[8] Let λ be an eigenvalue and $\mathbf{v} = (v_1 \ldots, v_p)$ a corresponding eigenvector, and let the largest component of $\mathbf{v}$ be v_i. Then

$$(\lambda - M_{ii})v_i = \sum_{\substack{j \neq i}}^{(j)} M_{ij}v_j,$$

$$|\lambda - M_{ii}||v_i| \leq \sum_{\substack{j \neq i}}^{(j)} |M_{ij}||v_j| \leq |v_i| \sum_{\substack{j \neq i}}^{(j)} |M_{ij}|.$$

Therefore, this eigenvalue λ lies in the ith circle, Q.E.D. Compare Gershgorin (1931).

THE KREISS MATRIX THEOREM

Because of the triangularity of B, the non-zero terms in the second sum are obtained for $k \leq i$ and $k \leq j$. Thus, as $i \neq j$, $2k - i - j \leq -1$ and one obtains

$$\delta_i \leq C_B^2(1 - |\varkappa_i|)2\Delta/(\Delta - 1)(\Delta^{1/2} - 1).$$

Hence, by choosing $\Delta > 9(1 + C_B^2)^2$, we make $2\Delta/(\Delta - 1) < 5/2$, $C_B^2/(\Delta^{1/2} - 1) < 1/3$, $C_B^2/(\Delta - 1) < 1/18$ and thus $\delta_i + |\varepsilon_i| \leq 1 - |\varkappa_i|^2$, so that Gerschgorin's theorem can be applied to show that the eigenvalues of M, which is hermitian, are all non-negative. It follows that (4.26) is satisfied and therefore that

$$S^{-1*}A^*S^*DSAS^{-1} - D \leq 0,$$

i.e.,
$$A^*HA - H \leq 0,$$

with $H = S^*DS$ clearly satisfying the requirements of statement (H); in particular, $C_H \leq \Delta^p C_S^2$.

Suppose, now, that (H) is satisfied and consider the iteration $\mathbf{w}_\nu = A\mathbf{w}_{\nu-1} = A^\nu \mathbf{w}_0$. Then, $\mathbf{w}_\nu^* H \mathbf{w}_\nu = \mathbf{w}_{\nu-1}^* A^* HA \mathbf{w}_{\nu-1} \leq \mathbf{w}_{\nu-1}^* H \mathbf{w}_{\nu-1} \leq \ldots \leq \mathbf{w}_0^* H \mathbf{w}_0$ and so, from (4.24), $|\mathbf{w}_\nu|^2 \leq C_H^2 |\mathbf{w}_0|^2$, i.e., $\|A^\nu\| \leq C_H$.

This completes the circuit apart from the proof that (R) implies (S), the core of which is contained in the lemma below. We begin by noting[9] that any matrix A can be put into upper triangular form by a unitary transformation U^*AU. This clearly leaves the resolvent condition unaffected so we shall assume from now on that each A is of this form. Then we shall obtain the similarity transformation which yields the required result by working on each upper diagonal of A successively, that is, (4.23) will be obtained for $j - i = 1, 2, 3, \ldots$ in turn.

The first step is to note that each corner (upper left or lower right principal sub-matrix) of A satisfies (R) with the same constant C_R.[10]

---

[9] This is Schur's Theorem (see MacDuffie, 1946, page 75). We shall need a stronger statement later, namely, that the triangularization can be accomplished with an arbitrary ordering of the eigenvalues on the diagonal. To see this, let $J = S^{-1}AS$ be the Jordan canonical form of A, obtain the desired ordering of J by the permutation $P^{-1}JP = (SP)^{-1}A(SP)$, and then perform a Gram-Schmidt orthonormalization process on the columns of SP. The result of the last operation is a unitary matrix $U = SPT$, where T is upper triangular, so we finally have $U^*AU = T^{-1}P^{-1}JPT$ in the required form.

[10] This follows from the triangularity of A:

$$\left\| \begin{bmatrix} A_1 & A_2 \\ 0 & A_3 \end{bmatrix}^{-1} \begin{bmatrix} \mathbf{v}_1 \\ \mathbf{v}_2 \end{bmatrix} \right\| \leq \alpha \left\| \begin{bmatrix} \mathbf{v}_1 \\ \mathbf{v}_2 \end{bmatrix} \right\|$$

for all $\mathbf{v}_1, \mathbf{v}_2$ implies $|A_1^{-1}\mathbf{v}_1| \leq \alpha|\mathbf{v}_1|$ for all $\mathbf{v}_1$, by putting $\mathbf{v}_2 = 0$.

Further, each element of the first upper diagonal of A lies in a (2×2) corner of such a corner, so we can apply the inequality (4.22) to such a (2×2) matrix. But the off-diagonal element

$$\begin{bmatrix} \varkappa_i - z & A_{i,i+1} \\ 0 & \varkappa_{i+1} - z \end{bmatrix}^{-}$$

is $-A_{i,i+1}/(\varkappa_i - z)(\varkappa_{i+1} - z)$ so one obtains

(4.27) $$|A_{i,i+1}| \leq C_R \frac{|z - \varkappa_i| |z - \varkappa_{i+1}|}{|z| - 1}.$$

To obtain the more useful form,

(4.28) $$|A_{i,i+1}| \leq 16 C_R \operatorname{Max} \{\gamma_{i,i+1}, |\varkappa_i - \varkappa_{i+1}|\},$$

where

$$\gamma_{ij} = \operatorname{Min} (1 - |\varkappa_i|, 1 - |\varkappa_j|),$$

we first prove

(4.28a) $$|A_{i,i+1}| \leq 16 C_R \operatorname{Max} \{1 - |\varkappa_{i+1}|, |\varkappa_i - \varkappa_{i+1}|\}$$

by making use of the freedom in the choice of z. Putting $z = 3$ gives $|A_{i,i+1}| \leq 8 C_R$, so (4.28a) is satisfied if $|\varkappa_{i+1}| \leq \frac{1}{2}$. On the other hand, if $|\varkappa_{i+1}| > \frac{1}{2}$ we put $z = t/\bar{\varkappa}_{i+1}$, where $t > 1$, and obtain from (4.27)

$$|A_{i,i+1}| \leq C_R \frac{(t - |\varkappa_{i+1}|^2)|t - \varkappa_i \bar{\varkappa}_{i+1}|}{|\varkappa_{i+1}|(t - |\varkappa_{i+1}|)}.$$

On letting $t \to 1$, there results $|A_{i,i+1}| \leq 3 C_R |1 - \varkappa_i \bar{\varkappa}_{i+1}|$. But

$$|1 - \varkappa_i \bar{\varkappa}_{i+1}| = |1 - |\varkappa_{i+1}|^2 + \bar{\varkappa}_{i+1}(\varkappa_{i+1} - \varkappa_i)|$$
$$\leq 3 \operatorname{Max} \{1 - |\varkappa_{i+1}|, |\varkappa_i - \varkappa_{i+1}|\},$$

so that (4.28a) holds for this case also. Inequality (4.28) then follows by symmetry.

We can now prove that (4.23) can be established for each element $B_{i,i+1}$ of the first upper diagonal, with $C_B \leq 16 C_R$. If $\gamma_{i,i+1} \geq |\varkappa_i - \varkappa_{i+1}|$, it is already true for $A_{i,i+1}$, and no transformation is required: otherwise, $A_{i,i+1}$ can be annihilated by a bounded similarity transformation. For in general, the element A_{ij} with $j > i$ is annihilated by the transformation $S_{ij} A S_{ij}^{-1}$, where $S_{ij} = I + T_{ij}$ and T_{ij} is a matrix all of whose elements are zero except the $(i, j)^{\text{th}}$ which has the value $A_{ij}/(\varkappa_i - \varkappa_j)$. Thus, when the transformation is needed, (4.28) provides a bound for it and its inverse $S_{i,i+1}^{-1} = I - T_{i,i+1}$, the bound being the same for all A in $\mathscr{F}$. By composing at most $(p - 1)$ such transformations we fulfill the requirements of (S) for the first upper diagonal with $C_S \leq 1 + 16C_R$.

To continue this process we need the following key lemma: *If the family of $(m \times m)$ upper triangular matrices A satisfy the resolvent condition with a constant C_1, and if all their off-diagonal elements except A_{1m} satisfy the inequality $|A_{ij}| \leq C_2 \gamma_{ij}$, then*

(4.29) $\quad |A_{1m}| \leq 16C_1(1 + (m - 2)C_2^2)^{1/2} \operatorname{Max}(\gamma_{1m}, |\varkappa_1 - \varkappa_m|).$

The (2×2) case, in which A_{12} is the only off-diagonal element, has already been proved and the general case is proved by reducing it to this case. To do this we permute the 2nd and the mth rows and columns of $(A - zI)$, which clearly leaves the resolvent condition unchanged. Then we partition the result E in the following way:

$$E = \begin{bmatrix} E_1 & E_2 \\ E_3 & E_4 \end{bmatrix} \equiv \begin{bmatrix} \zeta_1 & A_{1m} & A_{13} & \cdots & A_{1,m-1} & A_{12} \\ 0 & \zeta_m & 0 & \cdots & 0 & 0 \\ \hline 0 & A_{3m} & & & & \\ \vdots & \vdots & & & E_4 & \\ 0 & A_{m-1,m} & & & & \\ 0 & A_{2m} & & & & \end{bmatrix}$$

where $\zeta_i = \varkappa_i - z$ and the detailed form of E_4 is not required. Now it is clear that $E_3 E_2 = 0$ and, indeed, $E_3 E_1^{-1} E_2 = 0$. Hence, if we perform a triangular decomposition of E into $E = LU$, we have

$$L = \begin{bmatrix} I & 0 \\ E_3 E_1^{-1} & I \end{bmatrix}, \quad U = \begin{bmatrix} E_1 & E_2 \\ 0 & E_4 \end{bmatrix}.$$

The resolvent condition therefore gives, for any vector $\mathbf{u}$,

$$|E^{-1}\mathbf{u}|^2 = \mathbf{u}^*(L^{-1})^*(U^{-1})^* U^{-1} L^{-1} \mathbf{u} \leq (C_1/\zeta)^2 |\mathbf{u}|^2,$$

where $\zeta = |z| - 1$. Putting $\mathbf{u} = L\mathbf{v}$, where all except the first two elements of $\mathbf{v}$ (which subvector we denote by $\mathbf{v}_1$) are zero, gives $|E_1^{-1}\mathbf{v}_1|^2 \leq (C_1/\zeta)^2 |L\mathbf{v}|^2$. But

$$|L\mathbf{v}|^2 \leq |\mathbf{v}_1|^2 (1 + \sum_{i=2}^{m-1} A_{im}^2 / \zeta_m^2) \leq [1 + (m - 2)C_2^2] |\mathbf{v}_1|^2$$

so that E_1 satisfies (4.23) with C_R replaced by $C_1[1 + (m - 2)C_2^2]^{1/2}$, and the reduction to the (2 × 2) case has been accomplished.

To complete the proof of statement (S) is now straightforward. When the kth upper diagonal has been made to satisfy (4.23), each element of the $(k + 1)^{th}$ is the top right element of a matrix to which the lemma can be applied. Then by use of the similarity transformations $S_{i,i+k+1}$ as defined above, (4.23) is extended to this upper diagonal. The success of this process, of course, depends on the fact that these transformations, apart from effecting the desired annihilations, change only elements in diagonals yet to be considered. Note also that they have the effect of multiplying the constant in the resolvent condition by the square of their norm.

This completes the proof of the Kreiss matrix theorem. The heart of the proof, that (R) implies (S), is based on the treatment of Morton and Schechter (1965). Further discussion of the theorem, particularly of the resolvent condition (R), is to be found in Morton (1964) and in Miller and Strang (1965).

4.10. The Buchanan Stability Criterion

The theorem in the last section still leaves one some distance from the goal of a practical stability criterion. Statement (S) merely states the existence of a similarity transformation putting A into a form which can be tested for stability and, although in the proof an explicit construction for this transformation was given, it is clearly not a practical proposition to do this in an actual case of any complexity. The next step is to consider whether stable families of matrices can be recognized merely by carrying out the unitary transformations which triangularize them. The best result in this direction is due to Buchanan (1963b) who showed that, if the eigenvalues of the matrices are ordered in a suitable way in the triangularization, then inequalities of the form (4.28) or (4.29) are both necessary and sufficient for stability.

We define a sequence of complex numbers $\varkappa_1, \varkappa_2, \ldots, \varkappa_p$ to be *nested* with nesting constant K if,

$$|\varkappa_r - \varkappa_s| \leq K|\varkappa_l - \varkappa_m| \quad \text{whenever } l \leq r \leq s \leq m.$$

SEC. 4.10] THE BUCHANAN STABILITY CRITERION

It is easy to see that any set of p numbers can be ordered so that it is nested with $K \leq 2^p$. For by taking any one as the first and selecting the second as the nearest to this, the third the nearest of the remainder to the second, and so on, we have

$$|x_{i-1} - x_i| \leq |x_{i-1} - x_j| \quad \text{for any } j \geq i.$$

Thus if $l \leq r \leq s \leq m$,

$$|x_r - x_s| \leq |x_r - x_m| + |x_s - x_m|,$$

and

$$|x_r - x_m| \leq |x_{r-1} - x_r| + |x_{r-1} - x_m|, \quad \text{if } r > 1,$$
$$\leq 2|x_{r-1} - x_m| \leq \cdots \leq 2^{r-l}|x_l - x_m|.$$

Similarly, $|x_s - x_m| \leq 2^{s-l}|x_l - x_m|$, so

$$|x_r - x_s| \leq (2^{r-l} + 2^{s-l})|x_l - x_m|$$
$$\leq 2^p|x_l - x_m|.$$

Then we have the following theorem:

Let $\mathscr{F}$ be a family of matrices A which are in upper triangular form with their eigenvalues nested with constant K_1. Then $\mathscr{F}$ is stable if and only if the von Neumann condition, $|x_i| \leq 1$, on the eigenvalues is satisfied and the off-diagonal elements satisfy

(4.30) $$|A_{ij}| \leq K_2 \text{ Max } (1 - |x_i|, 1 - |x_j|, |x_i - x_j|)$$

for some constant K_2 independent of A.

The proof of this theorem follows almost immediately from that of Kreiss' theorem. All we have to show is that, under the hypothesis of nesting, the similarity transformations $S_{ij}AS_{ij}^{-1}(j > i)$ introduced there, together with their inverses $S_{ij}^{-1}AS_{ij}$, leave the inequalities (4.30) unchanged apart from the constant. For if the family is stable, applying the transformations yields (4.23), and thus (4.30), for the transformed family. Hence, if we "undo" the transformations and retain (4.30) we have necessity. On the other hand, we have $1 - |x_i| \leq 1 - |x_j| + |x_i - x_j|$ and the same thing with i and j interchanged; so, if we denote by Γ_{ij} the quantity in parenthesis on the right-hand side of (4.30), we have

$$\Gamma_{ij} \leq 2\text{Max } \{\gamma_{ij}, |x_i - x_j|\} \leq 2\Gamma_{ij}.$$

Therefore, if (4.30) is satisfied, $|A_{ij}| \leq 2K_2 \text{Max}(\gamma_{ij}, |x_i - x_j|)$ and the successive application of these transformations to the upper diagonals of A will yield (4.23) and hence imply stability.

Thus, suppose the elements of A satisfy (4.30). Then the elements of $S_{ij} A S_{ij}^{-1}$ differ from them only on the ith row to the right of the $(i, j)^{\text{th}}$ element, where they are $\hat{A}_{i\nu} = A_{i\nu} + t_{ij} A_{j\nu}$, and on the jth column above this element, where they are $\hat{A}_{\mu j} = A_{\mu j} - t_{ij} A_{\mu i}$, and where $|t_{ij}|$ is bounded by a bound independent of A. For these we have $\mu \leq i \leq j \leq \nu$, and so

$$|A_{j\nu}| \leq K_2 \Gamma_{j\nu} \leq 2K_2 \text{Max} \{\gamma_{j\nu}, |x_j - x_\nu|\}$$
$$\leq 2K_2 \text{Max} \{1 - |x_\nu|, K_1 |x_i - x_\nu|\}$$
$$\leq 2K_2(1 + K_1)\Gamma_{i\nu},$$

so that

$$|\hat{A}_{i\nu}| \leq K_2[1 + 2|t_{ij}|(1 + K_1)]\Gamma_{i\nu}.$$

The same argument can be carried through for $A_{\mu i}$ and $\hat{A}_{\mu j}$. Hence, (4.30) is still satisfied with K_2 multiplied by the factor in square brackets above. For the inverse transformation, only the sign of t_{ij} is changed, so the theorem is proved.

By the remarks above regarding nesting, and footnote 9 any matrix can be put in the nested triangular form required in the theorem by means of a unitary transformation. Thus, in principle, any family of matrices may be tested for stability by checking (4.30) and the von Neumann condition after this preliminary transformation has been effected. This will be illustrated at the end of the next section, where a difference approximation to the wave equation will be treated.

It might be conjectured that the inequalities (4.30) are perhaps either necessary or sufficient for stability without any side conditions. That neither is the case can be shown by means of two simple counter examples.

For the first, consider the single matrix

$$\begin{bmatrix} 1 & 1 & 0 \\ 0 & 0 & 1 \\ 0 & 0 & 1 \end{bmatrix},$$

which satisfies (4.30) but whose n^{th} power is as above except that the top right element is $(n - 1)$ and therefore unbounded.

SEC. 4.11] FURTHER SUFFICIENT CONDITIONS FOR STABILITY

As the second example, consider the one parameter family

$$\begin{bmatrix} \alpha & 1 & -(2\alpha)^{-1} \\ 0 & -\alpha & 1 \\ 0 & 0 & \alpha \end{bmatrix}, \quad \tfrac{1}{2} < \alpha < 1.$$

The top right element does not satisfy (4.30) as $\alpha \to 1$, yet one can easil check that the resolvent condition holds so that the family is stable.

4.11. Further Sufficient Conditions for Stability

Application of the Buchanan criterion still requires a rather awkward initial triangularization, and we consider in this section some weaker sufficient conditions which are simpler to apply in practice. In doing so, we shall return to the use of the amplification matrices G defined by (4.11), remembering the connection between the stability of G and that of the family $\{e^{-\alpha \Delta t} G\}$, discussed at the beginning of Section 4.9.

One general observation should be made first. Suppose $G(\Delta t, \mathbf{k})$ is uniformly Lipschitz continuous at $\Delta t = 0$, in the sense that $G(\Delta t, \mathbf{k}) = G(0, \mathbf{k}) + O(\Delta t)$, as $\Delta t \to 0$, where the constant implied by the expression $O(\Delta t)$ does not depend on $\mathbf{k}$. Then the stability of the difference scheme may be determined by considering $G(0, \mathbf{k})$. This follows from the theorem on bounded perturbations given in Section 3.9; but note that it does not follow merely by comparing $G(0, \mathbf{k})$ and $e^{-\alpha \Delta t} G(\Delta t, \mathbf{k})$ as in the discussion above, for $G(\Delta t, \mathbf{k})$ is not necessarily bounded by $e^{\alpha \Delta t} G(0, \mathbf{k})$ for any α.

If, in addition, one can show that $\|G(0, \mathbf{k})\| \leq 1$, then stability follows immediately. In this connection it is useful to note that, even for non-normal G, $\|G\|$ equals the square root of the spectral radius of G^*G. For

$$\|G\|^2 = \operatorname*{Max}_{|\mathbf{v}|=1} |G\mathbf{v}|^2 = \operatorname*{Max}_{|\mathbf{v}|=1} \mathbf{v}^* G^* G \mathbf{v}$$

and G^*G is normal, so that, if $\mathbf{v}$ is expressed in terms of its eigenvectors, it is clear that the above maximum is attained when $\mathbf{v}$ is a normalized eigenvector corresponding to the largest eigenvalue of G^*G.

Applications of these observations will be found in Chapter 9 on the transport equation, and in particular to its approximation by the so-called

84   PURE INITIAL-VALUE PROBLEMS   [CHAP. 4

"spherical harmonic" method. For example, for one difference equation, due to Friedrichs, the amplification matrix is of the form:

$$G(\Delta t, \mathbf{k}) = (I + A\Delta t) \cos k\Delta z + iB \sin k\Delta z,$$

where A and B are fixed hermitian matrices of order equal to the order of the spherical harmonic approximation being used. This example also illustrates the value of sometimes replacing the variable $\mathbf{k}$ by $\boldsymbol{\xi} = (k_1 \Delta x_1, k_2 \Delta x_2, \ldots, k_d \Delta x_d)$. For, in this case with $\Delta z = \text{const.} \Delta t$, $G(\Delta t, \mathbf{k})$ is not uniformly Lipschitz continuous at $\Delta t = 0$; but regarded as a function $\tilde{G}(\Delta t, \boldsymbol{\xi})$ of Δt and $\boldsymbol{\xi}$ it is. Moreover, $\tilde{G}(\Delta t, \boldsymbol{\xi})$ is not normal for $\Delta t \neq 0$ because A and B do not commute, but $\tilde{G}(0, \boldsymbol{\xi})$ is normal, so the von Neumann condition is both necessary and sufficient for stability in this case. In other situations $G(\Delta t, \mathbf{k})$ may not itself be Lipschitz continuous at $\Delta t = 0$ while $G^*(\Delta t, \mathbf{k})G(\Delta t, \mathbf{k})$ is. Such a case arises in Chapter 9, where $G(\Delta t, \mathbf{k})$ is a polynomial in $\sqrt{\Delta t}$ and again stability can be decided by considering only $G(0, \mathbf{k})$.

Condition 1. After normal G, the next simplest case is that in which G is "uniformly diagonalizable," that is, for each G, there exists a matrix T such that $T^{-1}GT = \Lambda$ is diagonal and T and T^{-1} are bounded independently of $\mathbf{k}$ and (sufficiently small) Δt; thus,

$$GT = T \begin{bmatrix} \lambda_1 & & & 0 \\ & \lambda_2 & & \\ & & \ddots & \\ 0 & & & \lambda_p \end{bmatrix} = T\Lambda.$$

The von Neumann condition is then clearly sufficient for stability, since $G^n = T\Lambda^n T^{-1}$. But to check that G has the required property, we have to calculate a complete set of its eigenvectors which, as we see from the above, form the columns of the matrix T. We may suppose that they are normalized and thus $\|T\|$ is bounded by p, because the norm of any $p \times p$ matrix does not exceed p times the absolute value of its largest element.[11]

---

[11] If M is the absolute value of the largest element of the $p \times p$ matrix A, and $\mathbf{v}$ is any vector, then, by the Schwarz inequality,

$$|\textstyle\sum_{(j)} A_{ij} v_j|^2 \leq (\textstyle\sum_{(j)} |A_{ij}|^2)(\textstyle\sum_{(j)} |v_j|^2) \leq pM^2 |\mathbf{v}|^2.$$

Therefore,

$$|A\mathbf{v}|^2 \leq p^2 M^2 |\mathbf{v}|^2 \quad \text{and} \quad \|A\| \leq pM.$$

We could, of course, use here the rather stronger result that the norm of any $p \times p$ matrix is bounded by $p^{1/2}$ times the maximum of the norms of its columns (or rows); the proof is similar.

SEC. 4.11] FURTHER SUFFICIENT CONDITIONS FOR STABILITY

Further, if Δ^2 is the determinant of T^*T, i.e., the Gram determinant of the normalized eigenvectors, then

$$|(T^{-1})_{ij}| \leq \frac{1}{\Delta}.$$

For, by Cramer's rule, $(T^{-1})_{ij}$ is the ratio of the determinant of the cofactor of T_{ji} to the determinant of T, and the absolute value of any determinant is bounded by the product of the lengths of the vectors making up its columns (or rows). It follows that $\|T^{-1}\| \leq p/\Delta$, so we conclude:

If G has a complete set of eigenvectors and there exists a constant δ such that $\Delta \geq \delta > 0$, where Δ^2 is the Gram determinant of the normalized eigenvectors, then the von Neumann condition is sufficient as well as necessary for stability.

To make a comparison with the Buchanan criterion, we note that if a family of matrices A are in upper triangular form with their eigenvalues nested, then a necessary and sufficient condition that they be uniformly diagonalizable is that

$$|A_{ij}| \leq \text{const.} \, |A_{ii} - A_{jj}|.$$

That this condition is sufficient follows from our construction in the proof of Kreiss' theorem of the similarity transformations $S_{ij}AS_{ij}^{-1}$ annihilating A_{ij}, and from the fact that the above inequalities are also invariant under these transformations if the eigenvalues are nested. The necessity of this condition is a little more difficult to show and we shall not pause for it here.

This theorem can be applied to the explicit difference equation for sound waves. We mentioned in Section 4.8 that the amplification matrix G given by (4.20) for this problem is not normal; but if $\mathbf{v}^{(1)}$ and $\mathbf{v}^{(2)}$ denote the normalized eigenvectors of G (which are easily calculated) we find that

$$|\det(\mathbf{v}^{(1)}, \mathbf{v}^{(2)})|^2 = 1 - \left(\frac{c\Delta t}{\Delta x}\right)^2 \sin^2 \frac{k\Delta x}{2}$$

(this expression is valid when the von Neumann condition, $c\Delta t/\Delta x \leq 1$, is satisfied). This determinant is bounded away from zero if

(4.31) $$c\Delta t/\Delta x = \text{const.} < 1$$

as $\Delta t, \Delta x \to 0$, and we reach the well-known conclusion that the equations are stable if (4.31) is satisfied.

Condition 2. It is often very troublesome to obtain a set of eigenvectors for G as is required in the previous condition. The easiest conditions to apply are those requiring only the eigenvalues. There are several of these, the simplest being the following:

If the elements of $G(\Delta t, \mathbf{k})$ are bounded for $0 < \Delta t < \tau$ and all $\mathbf{k}$ in $\mathscr{L}$, and if all the eigenvalues λ_i of G, with the possible exception of one, lie in a circle inside the unit circle:

$$|\lambda_i| \leq \gamma < 1 \quad \text{for} \begin{cases} 0 < \Delta t < \tau \\ \text{all } \mathbf{k} \text{ in } \mathscr{L} \\ i = 2, \ldots, p, \end{cases}$$

the von Neumann condition is sufficient as well as necessary for stability.

The proof of this theorem follows immediately from the Buchanan criterion. For when G has been put into nested triangular form, the off-diagonal elements will still be bounded, by M, say. Further, the exceptional eigenvalue may clearly be taken as the first one without impairing the nesting; hence, (4.30) is satisfied with $K_2 \leq M/(1 - \gamma)$, and the von Neumann condition is simply a condition on this eigenvalue: $|\lambda_1| \leq 1 + O(\Delta t)$.

An example of the use of this theorem is supplied by the five-point difference equation

$$(4.32) \quad \frac{3}{2} \frac{u_j^{n+1} - u_j^n}{\Delta t} - \frac{1}{2} \frac{u_j^n - u_j^{n-1}}{\Delta t} = \sigma \frac{u_{j+1}^{n+1} - 2u_j^{n+1} + u_{j-1}^{n+1}}{(\Delta x)^2}$$

(item number 9 in the table in Chapter 8) for heat flow in one dimension. This is a three-level formula, but can be reduced to a two-level system[12] by calling $u_j^{n-1} = v_j^n$. The system is

$$(4.33) \quad \frac{3}{2} \frac{u_j^{n+1} - u_j^n}{\Delta t} - \frac{1}{2} \frac{u_j^n - v_j^n}{\Delta t} = \sigma \frac{u_{j+1}^{n+1} - 2u_j^{n+1} + u_{j-1}^{n+1}}{(\Delta x)^2},$$

$$v_j^{n+1} = u_j^n.$$

---

[12] This example is mentioned again at the beginning of Chapter 7, where general multi-level formulas are discussed. We shall see that when the equations have been reduced to an equivalent two-level system, the stability considerations are precisely the same as in this chapter.

SEC. 4.11]   FURTHER SUFFICIENT CONDITIONS FOR STABILITY

The amplification matrix is

(4.34) $$G(\Delta t, k) = \begin{bmatrix} \dfrac{4}{3 + 8\alpha \sin^2 \beta} & \dfrac{-1}{3 + 8\alpha \sin^2 \beta} \\ 1 & 0 \end{bmatrix},$$

where α and β are abbreviations:

$$\alpha = \frac{\sigma \Delta t}{(\Delta x)^2}, \qquad \beta = \tfrac{1}{2} k \Delta x.$$

The eigenvalues are

$$\lambda_1, \lambda_2 = \frac{2 \pm \sqrt{1 - 8\alpha \sin^2 \beta}}{3 + 8\alpha \sin^2 \beta},$$

and it is readily seen that the von Neumann condition is satisfied for any value of α. Figure 4.2 shows graphically the values of $|\lambda_1|$ and $|\lambda_2|$ plotted against $\alpha \sin^2 \beta$. The eigenvectors of G are

$$\varphi^{(1)}, \varphi^{(2)} = A \begin{pmatrix} 1 \\ 2 \mp \sqrt{1 - 8\alpha \sin^2 \beta} \end{pmatrix},$$

where A is a normalization factor. The determinant of the normalized eigenvectors is $\Delta = 2A^2 \sqrt{1 - 8\alpha \sin^2 \beta}$. If $\alpha > \tfrac{1}{8}$, Δ can vanish for certain values of β; therefore, we cannot use Condition 1 above for stability. However, Condition 2 applies, because $|\lambda_2| \leq \tfrac{1}{2}$ for all positive α and all β (this is seen in Figure 4.2). Hence, the system (4.34) is always stable.

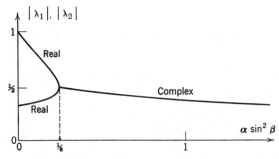

Fig. 4.2. Eigenvalues of the amplification matrix of the difference-equation system (4.33).

Kato's condition. Both the above conditions are special cases of the following condition due to Kato (1960):

Suppose $\|G(\Delta t, \mathbf{k})\|$ is uniformly bounded and the von Neumann condition is satisfied. And suppose that, for each G, a closed rectifiable curve Γ is drawn inside the circle $|\lambda| \leq R$, where R is the spectral radius of G, such that its length is uniformly bounded and its distance from the spectrum of G is uniformly bounded away from zero. Then the difference scheme is stable if there exists a constant $\delta > 0$, independent of Δt and k, such that each of the distinct eigenvalues λ_i, $i = 1, 2, \ldots, q$, which are outside Γ, has index one[13] and

either (i) the distance of λ_i from all other eigenvalues is greater than δ,

or (ii) for the set of λ_i not satisfying (i), the complete set of corresponding eigenvectors has Gram determinant Δ^2 satisfying $\Delta > \delta$.

In most practical cases Γ will be a circle; for the discussion of a single matrix, this will in fact always be sufficient, but in treating a family of matrices sharper results can sometimes be obtained with more general curves.

The main lines of the proof are clear. We put each G in nested triangular form taking the set of λ_i satisfying (ii) first, which is made possible by the hypotheses. Then it is easy to see that all elements in the columns above the remaining eigenvalues satisfy the Buchanan inequality (4.30). This leaves a square submatrix which can be diagonalized by a bounded similarity transformation as in Condition 1.

One further sufficient condition that can be very useful concerns the field of values or the numerical range of the matrix G:

The Lax-Wendroff condition.[14] *If $G(\Delta t, \mathbf{k})$ is such that*

$$|\mathbf{v}^*G\mathbf{v}| \leq [1 + O(\Delta t)]|\mathbf{v}|^2 \text{ for any vector } \mathbf{v},$$

then the corresponding difference scheme is stable.

The proof is a simple application of Kreiss' resolvent condition of

---

[13] For an eigenvalue λ of a matrix A, the set of vectors $\mathbf{v}$ for which $(A - \lambda)^n \mathbf{v} = 0$ for some positive integer n forms the *algebraic eigenspace* for λ. The smallest value of n which suffices for all elements in the eigenspace is called the *index* of λ. If the index is one, there is a complete set of eigenvectors spanning the eigenspace.

[14] Lax and Wendroff (1964).

SEC. 4.11] FURTHER SUFFICIENT CONDITIONS FOR STABILITY

(4.22). Choosing α so that $A = e^{-\alpha \Delta t} G$ satisfies $|\mathbf{v}^* A \mathbf{v}| \leq |\mathbf{v}|^2$, we see that all eigenvalues of A lie in the closed unit disk. Hence, $(zI - A)^{-1}$ exists for $|z| > 1$ so that, if $\mathbf{w}$ is any vector and $\mathbf{v} = (zI - A)^{-1}\mathbf{w}$,

$$|\mathbf{v}||\mathbf{w}| \geq |\mathbf{v}^*\mathbf{w}| = |\mathbf{v}^*(zI - A)\mathbf{v}| \geq (|z| - 1)|\mathbf{v}|^2,$$

i.e.,
$$|\mathbf{v}| = |(zI - A)^{-1}\mathbf{w}| \leq |\mathbf{w}|(|z| - 1)^{-1}.$$

The family of A therefore satisfies the resolvent condition and stability follows from Kreiss' matrix theorem.

If we write $\rho(A) = \underset{\mathbf{v}}{\text{Max}} \{|\mathbf{v}^* A \mathbf{v}| \,\big|\, |\mathbf{v}| = 1\}$, it is clear that $\rho(A) \leq \|A\|$. Equality holds if A is normal and, hence, by splitting any general matrix into a sum of its hermitian and anti-hermitian parts, it also follows that $\|A\| \leq 2\rho(A)$. Further, one can show—see Pearcy (1966)— that $\rho(A) \leq 1$ implies $\rho(A^n) \leq 1$ for any positive integer in n. Thus the Lax-Wendroff condition implies that $\|A^n\|$ is not only bounded, but is bounded by 2; this may be taken as some measure of its restrictiveness.

When all such sufficient conditions as the above have been exhausted and the question of stability remains unresolved, we fall back on the Buchanan criterion. Thus, let us consider the limiting case $c\Delta t = \Delta x$ for the explicit equations approximating the wave equation. We have given the amplification matrix G in (4.20), from which it can be seen that the eigenvalues are

$$1 - \tfrac{1}{2}a^2 \pm \tfrac{1}{2}ia\sqrt{4 - a^2}, \quad \text{with } a = (2c\Delta t/\Delta x) \sin \tfrac{1}{2}k\Delta x,$$

and the normalized eigenvectors are

$$\frac{1}{\sqrt{2}} \begin{bmatrix} 1 \\ \tfrac{1}{2}(\sqrt{4 - a^2} + ia) \end{bmatrix} \quad \text{and} \quad \frac{1}{\sqrt{2}} \begin{bmatrix} -\tfrac{1}{2}(\sqrt{4 - a^2} + ia) \\ 1 \end{bmatrix},$$

corresponding to the positive and negative signs, respectively. To triangularize G, we form $U^* G U$, where U is the unitary matrix

$$U = \frac{1}{\sqrt{2}} \begin{bmatrix} 1 & -\tfrac{1}{2}(\sqrt{4 - a^2} - ia) \\ \tfrac{1}{2}(\sqrt{4 - a^2} + ia) & 1 \end{bmatrix}.$$

We obtain

$$U^* G U = \begin{bmatrix} 1 - \tfrac{1}{2}a^2 + \tfrac{1}{2}ia\sqrt{4 - a^2} & \tfrac{1}{2}a^2(ia - \sqrt{4 - a^2}) \\ 0 & 1 - \tfrac{1}{2}a^2 - \tfrac{1}{2}ia\sqrt{4 - a^2} \end{bmatrix}.$$

The absolute value of the eigenvalues is one and their separation $a\sqrt{4-a^2}$ tends to zero as $a^2 \to 4$. But the off-diagonal element then has absolute value 4, so the system is unstable in this case.

It is worth noting that Courant, Friedrichs, and Lewy (1928) treated this case as stable but in doing so they allowed only smooth initial data. In Chapter 10 an unbounded solution of the difference equations is demonstrated (usually a quicker way of showing instability than using the Buchanan criterion), but the initial values are violently oscillatory as $\Delta x \to 0$.

This completes our study of difference schemes for general differential equations with constant coefficients and periodic boundary conditions. The theory we have described lies at the root of all the rest of the book and we shall constantly refer to it. In particular, the Kreiss matrix theorem forms the basis of our treatment of variable coefficient problems in the next chapter, while on the practical side the von Neumann condition will be found to be an essential tool in judging the difference schemes considered in Part II of the book. When a difference scheme is found to be conditionally stable (i.e., stable for some values of $\Delta t/\Delta x$ or the like), the von Neumann condition nearly always gives the correct stability range: it is usually only at the limits of the range, as with the explicit scheme for the wave equation, that it needs to be supplemented. However, there are some cases where our definition of stability is itself found to be wanting from the practical point of view. One such case occurs in the treatment of coupled sound and heat flow in Section 10.4, and others can be found in Sections 9.7, 11.6, and 12.16.

CHAPTER 5

Linear Problems with Variable Coefficients; Non-Linear Problems

5.1. Introduction

In the previous chapter we discussed in great detail the stability of finite-difference approximations to linear differential equations with constant coefficients. But in practical problems the coefficients are seldom constant and the equations are very often non-linear. Indeed, it is in checking the "local" stability of linearized equations obtained from truly non-linear equations that the constant coefficient theory is mainly of use. What this amounts to is the following. From non-linear difference equations for the approximation $\mathbf{u}^n(\mathbf{x})$ to the solution $\mathbf{u}(\mathbf{x}, n\Delta t)$ of the differential problem, one obtains the equations of first variation by substituting $\mathbf{u}^n(\mathbf{x}) = \mathbf{u}(\mathbf{x}, n\Delta t) + \mathbf{v}^n(\mathbf{x})$, where $\mathbf{v}^n(\mathbf{x})$ is regarded as a small quantity of first order, and dropping quantities of second and higher order. The resulting equations for $\mathbf{v}^n(\mathbf{x})$ are therefore linear with coefficients depending on the zero order quantity $\mathbf{u}(\mathbf{x}, t)$. Then, at each point $(\mathbf{x}, t)$, one considers the system of equations obtained by taking the coefficients as constant, with the values obtaining at that point, and tests for stability by the constant coefficient theory. The usefulness of the latter theory then depends on the extent to which this local stability at every point of the $\mathbf{x}, t$-space corresponds to the convergence of $\mathbf{u}^n(\mathbf{x})$ to the solution $\mathbf{u}(\mathbf{x}, t)$. Practical experience, which shows that instabilities usually start as local phenomena, suggests that the correspondence is very close; and it was a conjecture along these lines which led von Neumann to develop the treatment of stability theory on the basis of Fourier analysis.

It is clear that the first problem is to treat linear equations with variable coefficients, and we shall concentrate on this in the present chapter. As a further restriction we shall consider only those cases in which the coefficients depend only on the space variable $\mathbf{x}$. The main aim will be to show that, in several important classes of equations, a mild strengthening of the local stability conditions is sufficient to ensure overall stability. However, these local conditions are not always even necessary and a little time must first be spent in clarifying this situation.

The problem arises in the first instance in considerations of the well-posedness of differential equation problems. Kreiss (1962) gives an example[1] of an equation where the problem with variable coefficients is properly posed yet all corresponding constant coefficient problems are improperly posed; on the other hand, he gives an example in which the constant coefficient problems are all properly posed, yet the variable coefficient problem is not. As we have already observed in Section 3.9 an improperly posed problem cannot be approximated by a difference scheme which is both consistent and stable. So Kreiss' examples show that, in general, local stability is neither necessary nor sufficient for the overall stability of a variable coefficient problem. However, as Strang (1966) points out, if $\partial \mathbf{u}/\partial t = A\mathbf{u}$ leads to a properly posed problem, then if A is replaced by its principal part (i.e., all the derivatives of highest order) and then the coefficients are taken as constant, the resulting problem is properly posed. Thus, in particular, first-order systems must be properly posed locally.

Strang also proves a similar result for explicit difference schemes. Suppose the differential equation $\partial \mathbf{u}/\partial t = A\mathbf{u}$, where A is an mth order differential operator, is approximated by an explicit scheme $\mathbf{u}^{n+1} = C(\Delta t)\mathbf{u}^n$. Then, the highest powers of $(1/\Delta x)$ occurring in $C(\Delta t)$ will be the $(1/\Delta x)^m$ arising from the principal part of A. These will be multiplied by Δt and if a relation of the form $\Delta t = \text{const.} (\Delta x)^m$ holds, which will normally be necessary for stability, taking $\Delta t \to 0$ will result in $C(\Delta t)$ tending to a limit $C(0)$ corresponding to an approximation to only the principal part of A; $C(0)$ is therefore called the principal part of $C(\Delta t)$. We can then obtain the following theorem.

---

[1] The following similar but somewhat simpler example is given by Strang (1966):

$$\frac{\partial u}{\partial t} = Au, \quad \text{where } A = i\frac{\partial}{\partial x}\sin x \frac{\partial}{\partial x} = i\sin x \frac{\partial^2}{\partial x^2} + i\cos x \frac{\partial}{\partial x}.$$

At $x = 0$, one obtains $\partial u/\partial t = i\partial u/\partial x$ which is certainly improperly posed. On the other hand, for any solution u of the full problem, an integration by parts shows that $\frac{\partial}{\partial t}(u, u) = (u, Au) + (Au, u) = 0$, so this is properly posed.

SEC. 5.1] INTRODUCTION

Suppose that the explicit difference scheme,

(5.1) $\quad \mathbf{u}^{n+1}(\mathbf{x}) = (C(\Delta t)\mathbf{u}^n)(\mathbf{x}) \equiv \sum_{(\beta)} c^\beta(\mathbf{x}, \Delta t)\mathbf{u}^n(\mathbf{x} + \beta\Delta x),$

has a common mesh interval Δx in all coordinate directions, which is related to Δt by an equation of the form (4.9). Suppose also that it is stable and that, as $\Delta t \to 0$, each $c^\beta(\mathbf{x}, \Delta t)$ converges to a bounded function $c^\beta(\mathbf{x}, 0)$. Then, for any interior point $\mathbf{a}$ of a set in which this convergence is uniform and $c^\beta(\mathbf{x}, 0)$ is continuous, the scheme

(5.2) $\quad \mathbf{U}^{n+1}(\mathbf{x}) = (C_\mathbf{a}(0)\mathbf{U}^n)(\mathbf{x}) \equiv \sum_{(\beta)} c^\beta(\mathbf{a}, 0)\mathbf{U}^n(\mathbf{x} + \beta\Delta x)$

is stable.

To prove this result, we suppose it is false at some point $\mathbf{a}$ and that continuity and uniform convergence hold in a sphere S_0 centered at $\mathbf{a}$. Then, for any constant M, however large, there is some vector $\mathbf{V}$, frequency $\mathbf{k}$, mesh Δx and number N, such that,[2] for $\mathbf{x} = \mathbf{a}$ and $N\Delta t < 1$,

(5.3) $\quad |\Gamma(\mathbf{x})^N \mathbf{V}| > M|\mathbf{V}|,$

where

$$\Gamma(\mathbf{x}) = \sum_{(\beta)} c^\beta(\mathbf{x}, 0)e^{i\mathbf{k} \cdot \beta\Delta x}.$$

By the continuity, this will therefore be true for all $\mathbf{x}$ in some sphere S_1 centered at $\mathbf{a}$, which we suppose has radius strictly less than that of S_0: and, if we fix $\xi = \mathbf{k}\Delta x$, it remains true as Δt and $\Delta x \to 0$ with fixed N. Now, let $\rho(\mathbf{x})$ be a smooth function, zero outside S_1 but not identically zero inside, and set $\mathbf{v}(\mathbf{x}) = \mathbf{V}\rho(\mathbf{x})e^{i\mathbf{k}\cdot\mathbf{x}}$. Then,

$$(C\mathbf{v})(\mathbf{x}) = \sum_{(\beta)} c^\beta(\mathbf{x}, \Delta t)\mathbf{V}\rho(\mathbf{x} + \beta\Delta x)e^{i\mathbf{k}\cdot(\mathbf{x} + \beta\Delta x)}$$

$$= \Gamma(\mathbf{x})\mathbf{v}(\mathbf{x}) + \varepsilon_1(\mathbf{x}, \Delta t),$$

where ε_1 is zero outside some set, which lies wholly inside S_0 for sufficiently small Δt, and inside this set it tends to zero uniformly as $\Delta t \to 0$. Applying the operator C successively we obtain

$$(C^N\mathbf{v})(\mathbf{x}) = \Gamma(\mathbf{x})^N \mathbf{v}(\mathbf{x}) + \varepsilon_N(\mathbf{x}, \Delta t),$$

---

[2] Except where specifically stated otherwise, we shall hereafter use $|\ |$ to denote the Euclidean vector norm and $\|\ \|$ the L_2 norm, as in the previous chapter.

where, because N is fixed independent of Δt, ε_N has the same properties as ε_1. Thus, from (5.3) and taking Δt sufficiently small, it is clear that $\|C^N\| \geq M$ and therefore (5.1) is unstable, contrary to the hypothesis.

Because of considerations such as these, we shall limit our treatment in this chapter to two classes of equations for which the situation is relatively straightforward, namely, symmetric hyperbolic systems and parabolic equations. These embrace a large fraction of the physically important initial-value problems; and in both cases the linear, variable coefficient problems are well known to be properly posed under criteria which are purely local in character. In addition, we shall deal only with explicit difference schemes, although it is not too difficult to see how the results can be extended to implicit schemes. As we shall see below and in Chapter 8, stability criteria for parabolic equations depend to a large extent only on the principal part of the operator. Hence Strang's theorem shows that, in both the cases treated, local stability is always necessary.

These necessary criteria have to be mildly strengthened to yield conditions sufficient for the stability of the variable coefficient problems. One can see immediately that when $a(x)$ is variable even the simple difference $\Delta[a(x)\mathbf{u}(x)]$ yields an extra term $(\Delta a)\mathbf{u}$ that can cause instability, unless $a(x)$ is Lipschitz continuous and $\Delta x = O(\Delta t)$ so that the whole term is bounded by $O(\Delta t) \|\mathbf{u}\|$. For higher differences, e.g., $\Delta(a\Delta \mathbf{u})$, one obtains extra terms $(\Delta a)(\Delta \mathbf{u})$ that can cause trouble even when $\Delta x = O(\Delta t)$. Such terms as these cause a mixing of the Fourier components of $\mathbf{u}$ similar to but milder than that described for a non-linear problem in Phillips (1959); a good example of the phenomenon is described by Roberts and Weiss (1966). The consequent growth of high frequency components can be obviated by introducing some "dissipation" into the difference schemes. That is, one designs the scheme so that the eigenvalues λ_ν of the amplification matrix satisfy an inequality of the form

(5.4) $$|\lambda_\nu| \leq 1 - \delta|k\Delta x|^{2r}, \quad \text{when } |k\Delta x| \leq \pi,$$

for some real constant $\delta > 0$ and positive integer r. Such a requirement was first introduced by Fritz John (1952) for a single parabolic equation, where it arises quite naturally in explicit schemes—see Section 5.3; he showed that it was sufficient for stability under suitable smoothness conditions on the coefficients. Then, Lax (1961) and Lax and Wendroff (1962) found a sufficient stability condition for symmetric hyperbolic systems which essentially involved an inequality of the form (5.4) on the amplification matrix itself. This has been relaxed to a condition on the eigenvalues alone in a recent paper by Kreiss (1964), and it is this treatment which forms the basis of Section 5.4.

But before embarking on this, we first consider how else we might have defined stability. This is particularly relevant here since it is only when variable coefficient problems are treated that the differences in the definitions that have been adopted by various authors become important.

5.2. Alternative Definitions of Stability

Apart from differences in the norm used, there are two commonly used definitions of stability other than that on which we have based our treatment—one a weaker definition and one stronger. The former allows the solution operator $C(\Delta t)$ of a stable difference scheme to satisfy the inequality

(5.5) $\quad \|C(\Delta t)^n\| \leq \text{const.} \, (\Delta t)^{-\alpha} \quad$ for $\begin{cases} 0 < \Delta t < \tau \\ 0 \leq n\Delta t \leq T \end{cases}$
$\qquad\qquad\quad \leq \text{const.} \, n^\alpha,$

for some fixed, finite $\alpha \geq 0$, whereas stability in our sense requires this to be satisfied for $\alpha = 0$. This definition agrees with the empirical observation that instability is usually distinguished from stability by an exponential rather than a polynomial growth of the error, and it has been used by numerous writers including Strang (1960), Forsythe and Wasow (1960), and Riabenkii and Fillipov (1956).

At a more fundamental level, this "weak" stability springs from a more general definition of well-posedness. For, supposing that $\mathbf{u}(\mathbf{x}, t)$ possesses continuous derivatives in $\mathbf{x}$ up to order ρ, we define a norm $\|\mathbf{u}\|_\rho$ by

(5.6) $\qquad\qquad \|\mathbf{u}\|_\rho^2 = \sum_{|\mathbf{v}| \leq \rho} \|\partial_{x_1}^{v_1} \partial_{x_2}^{v_2} \ldots \partial_{x_d}^{v_d} \mathbf{u}\|^2,$

where $|\mathbf{v}| = v_1 + v_2 + \cdots + v_d$ denotes the order of the derivative whose L_2 norm appears under the summation sign. Then, we could call a problem properly posed or well posed if the following relation holds between its solution and the initial data:

(5.7) $$\|\mathbf{u}(\mathbf{x}, t)\| \leq \text{const.} \|\mathbf{u}(\mathbf{x}, 0)\|_\rho, \quad 0 \leq t \leq T,$$

for some fixed, finite ρ and for a dense set of initial data—this corresponds to the notion of ρ-continuous dependence on initial data introduced by Hadamard (1923). With this definition of well-posedness and the definition of weak stability, a theory similar to that given in Chapter 3 can be developed leading again to a theorem asserting the equivalence of convergence and stability. The arguments are a little more complicated because of the differing norms involved and the reader is referred to the works cited above for details. But the relationship between the α and ρ occurring in (5.5) and (5.7), respectively, can be easily indicated as follows. From an argument based on the principle of uniform boundedness, it is clearly necessary for convergence that

$$\|C(\Delta t)^n \mathbf{u}(\mathbf{x}, 0)\| \leq \text{const.} \|\mathbf{u}(\mathbf{x}, 0)\|_\rho$$

for any $\mathbf{u}(\mathbf{x}, 0)$ with ρ continuous derivatives and giving rise to a problem properly posed in the above sense. But the left-hand side depends only on the values of $\mathbf{u}(\mathbf{x}, 0)$ at the mesh points. For any $\mathbf{u}(\mathbf{x}, 0)$, we can always construct smooth initial data $\tilde{\mathbf{u}}(\mathbf{x}, 0)$ having the same values at these points, to which we can therefore apply the above inequality. Moreover, it is easy to show that $\|\tilde{\mathbf{u}}(\mathbf{x}, 0)\|_\rho \leq \text{const.} (\Delta x)^{-q} \|\mathbf{u}(\mathbf{x}, 0)\|$ where q depends on ρ and the dimension d. Hence, a stability relation of the form (5.5) must follow.

When one applies this theory to the constant coefficient problems treated in the last chapter, one finds a beautifully simple result due to Kreiss (1962), namely, *if the amplification matrix G is uniformly bounded, the von Neumann condition* (4.18) *is necessary and sufficient for weak stability*. In this case, it is the sufficiency that is easy to prove. Transforming G by a unitary matrix U into upper triangular form we have $UGU^* = D + S$, where

$$D = \begin{bmatrix} \lambda_1 & & 0 \\ & \ddots & \\ 0 & & \lambda_p \end{bmatrix}, \quad S = \begin{bmatrix} 0 & S_{12} & \cdots & S_{1p} \\ & \ddots & & \vdots \\ & & \ddots & S_{p-1,p} \\ 0 & & & 0 \end{bmatrix}$$

Hence

$$\|G^n\| = \|(UGU^*)^n\| = \|(D+S)^n\|$$
$$= \|D^n + \sum_{j=1}^{n} D^{n-j}SD^{j-1} + \sum_{j=1}^{n-1}\sum_{k=1}^{n-j} D^{n-j-k}SD^{j-1}SD^{k-1} + \cdots \|.$$

But the pth power of S is the null matrix, and this fact is unaffected by the interposition of diagonal matrices between the factors. So the above sum has a finite number of non-zero terms. Also, $\|S\|$ is bounded and, since $|\lambda_j| \leq 1 + \text{const.}\,\Delta t$, $\|D\|^n$ is bounded for $n\Delta t \leq T$. Thus, we have for $n \geq p - 1$,

$$(5.8) \qquad \|G^n\| \leq \sum_{j=0}^{p-1} \binom{n}{j} \|D\|^{n-j} \|S\|^j \leq \text{const.}\, n^{p-1}.$$

The proof that the von Neumann condition is necessary is considerably more difficult, and we refer the reader to Kreiss' paper.

With this simple algebraic stability condition it is an easy matter to consider how "local" and "global" stability are related. The result is disconcerting. For in contrast to what was proved for our definition of stability in Section 3.9, the addition of an operator bounded by $O(\Delta t)$ can cause a weakly stable operator to become unstable. As a simple example, suppose one uses the difference scheme

$$\mathbf{u}^{n+1} = (I + S\Delta_+^2)\mathbf{u}^n$$

to approximate $\partial \mathbf{u}/\partial t = 0$, where $\mathbf{u}$ is a two-component vector and S is a constant matrix whose only non-zero element is a one in the top right corner. Then, the corresponding amplification matrix $G = I + (e^{ik\Delta x} - 1)^2 S$ has unit eigenvalues, so the scheme is weakly stable. But, if S is perturbed by an element Δt in the bottom left corner, it is easily verified that the eigenvalues of the corresponding amplification matrix become

$$\lambda = 1 \pm (\Delta t)^{1/2}(e^{ik\Delta x} - 1)^2.$$

Hence, the scheme is no longer stable.

The way such a perturbation may come about in a variable coefficient problem is illustrated by the following example due to Kreiss (1962).

98   VARIABLE COEFFICIENTS AND NON-LINEAR PROBLEMS   [CHAP. 5

The differential equation is $\partial \mathbf{u}/\partial t = \partial \mathbf{u}/\partial x$; $\mathbf{u}$ is a vector of two components and, with $\Delta x = \Delta t$, the difference approximation is

$$\mathbf{u}^{n+1}(x) = [I + \Delta_+ + V(x)SV^{-1}(x)\Delta_+^2]\mathbf{u}^n(x),$$

where Δ_+ is the forward difference operator and

$$V(x) = \begin{bmatrix} \cos x & \sin x \\ -\sin x & \cos x \end{bmatrix}, \quad S = \begin{bmatrix} 0 & 1 \\ 0 & 0 \end{bmatrix}.$$

When $V(x)$ is replaced by its value $V(x_0)$ at some fixed point x_0, the amplification matrix of the above scheme is

$$V(x_0)(e^{ik\Delta x}I + (e^{ik\Delta x} - 1)^2 S)V^{-1}(x_0)$$

and so clearly satisfies the von Neumann condition for every x_0. To treat the variable coefficient case, we change the dependent variable to $\mathbf{v}(x, t) = V^{-1}(x)\mathbf{u}(x, t)$ to get

$$\mathbf{v}^{n+1}(x) = \mathbf{v}^n(x) + [V^{-1}(x)\Delta_+ + SV^{-1}(x)\Delta_+^2]V(x)\mathbf{v}^n(x).$$

But

$$\Delta_+[V(x)\mathbf{v}(x)] = V(x + \Delta x)\Delta_+\mathbf{v}(x) + [\Delta_+ V(x)]\mathbf{v}(x)$$

and

$$V^{-1}(x)V(x + \nu\Delta x) = \begin{bmatrix} \cos \nu\Delta x & \sin \nu\Delta x \\ -\sin \nu\Delta x & \cos \nu\Delta x \end{bmatrix}.$$

Hence, by separating off terms of order $(\Delta x)^2$ and higher, we obtain an equation for $\mathbf{v}$ with constant coefficients in the form

$$\mathbf{v}^{n+1}(x) = (A + (\Delta x)^2 B)\mathbf{v}^n(x),$$

where B is a uniformly bounded operator and

$$A = \begin{bmatrix} 1 & \Delta x \\ -\Delta x & 1 \end{bmatrix}(I + \Delta_+) + S\begin{bmatrix} 1 & 2\Delta x \\ -2\Delta x & 1 \end{bmatrix}\Delta_+^2 + 2S\begin{bmatrix} 0 & \Delta x \\ -\Delta x & 0 \end{bmatrix}\Delta_+.$$

Denoting the Fourier transform of B by $\hat{B}$, we obtain the amplification matrix

$$G = e^{ik\Delta x}\begin{bmatrix} 1 - 2\alpha\Delta x - 2\beta\Delta x & \Delta x + \alpha \\ -\Delta x & 1 \end{bmatrix} + (\Delta x)^2 \hat{B}$$

ALTERNATIVE DEFINITIONS OF STABILITY

where
$$\alpha = -4\sin^2 \tfrac{1}{2}k\Delta x, \qquad \beta = 2ie^{-\frac{1}{2}ik\Delta x}\sin \tfrac{1}{2}k\Delta x.$$
The characteristic equation of the displayed matrix is
$$(1 - \mu)(1 - \mu - 2(\alpha + \beta)\Delta x) + \Delta x(\Delta x + \alpha) = 0,$$
yielding eigenvalues
$$\mu = 1 - (\alpha + \beta)\Delta x \pm [-\alpha\Delta x + O[(\Delta x)^2]]^{\frac{1}{2}}.$$
And the eigenvalues of G are $\lambda = e^{ik\Delta x}\mu + O(\Delta x)$. Remembering that $\Delta t = \Delta x$ we see that the von Neumann condition is not satisfied and thus the difference equations for $\mathbf{v}(x)$, are not even weakly stable.

The conclusion to be drawn from these examples is that, although weak stability may be sufficient for constant coefficient problems or for those involving a scalar variable u, it is not a powerful enough criterion to apply to systems of difference equations with variable coefficients. Even our definition of stability is not always adequate when applied locally, as is shown by an example, similar to but more complicated than the above, given by Widlund (1966). In the search for a stronger definition, one may then turn to the idea of "strong" stability defined as follows:

A difference scheme is strongly stable if (i) for every fixed Δt, an operator $H(\Delta t)$ is everywhere defined and is such that if $(\mathbf{u}, H\mathbf{u})$ is denoted by $\|\mathbf{u}\|_H^2$ then $K_1^{-1}\|\mathbf{u}\|^2 \leq \|\mathbf{u}\|_H^2 \leq K_1\|\mathbf{u}\|^2$; and (ii) the solution of the difference scheme satisfies

$$(5.9) \qquad \|\mathbf{u}^{n+1}\|_H \leq (1 + K_2\Delta t)\|\mathbf{u}^n\|_H$$

where K_1 and K_2 are some fixed positive constants. We have seen in Section 4.9 that this definition of stability is equivalent to ours in the constant coefficient case. It is an easy matter to see that in general it is at least as strong. If it is satisfied,

$$\|\mathbf{u}^n\| \leq K_1^{\frac{1}{2}}\|\mathbf{u}^n\|_H \leq (1 + K_2\Delta t)^n K_1^{\frac{1}{2}}\|\mathbf{u}^0\|_H \leq K_1 e^{K_2 T}\|\mathbf{u}^0\|,$$

for $n\Delta t \leq T$.

This definition of stability is still very closely linked with the form in which the well-posedness condition is applied to the differential equation problem. For if a norm is constructed in which the solution of this problem satisfies (5.9) above, then under certain conditions of smoothness it can be used for the difference problem. It is these smoothness conditions, however, which are difficult to ascertain, and it is one of the great merits of Kreiss' treatment of the difference problem given in Section 5.4, that it circumvents these difficulties by a direct attack on the difference problem.

Finally, mention must be made of the possibility of using a basic norm other than the L_2 norm. In particular, the maximum norm is very attractive for numerical work. In Chapter 1 we gave an example showing how this may be used in a simple case, and it might be expected to be particularly appropriate for the parabolic equations treated in the next section. However, a condition on the amplification factor is given there which can also be applied to an explicit difference scheme for any equation of the form $\partial u/\partial t = \partial^m u/\partial x^m$. This condition was shown to be sufficient for stability in the maximum norm by Strang (1962), and the necessity has subsequently been proved by Thomée (1965a), who also generalized the results to implicit equations and to any L_p norm with $p \neq 2$, Thomée (1965b). [The L_p norm $\|\mathbf{u}\|_p$ is defined by $\|\mathbf{u}\|_p^p = \int d\mathbf{x}|\mathbf{u}|^p$.] From the condition, Thomée shows that for $m = 1$, the hyperbolic case, any stable scheme must have an odd order of accuracy. Hence, all the very useful second-order schemes are unstable in all but the L_2 norm. The growth rate is quite mild, for if $\|C^n\| = 1$, he shows that $\|C^n\|_p \leq (Kn)^{|\frac{1}{2} - 1/p|}$ for some constant K; but this is sufficient to render these norms inappropriate for these cases.

5.3. Parabolic Equations

For the parabolic equation in one space dimension

$$(5.10) \qquad \frac{\partial u}{\partial t} = a_0 \frac{\partial^2 u}{\partial x^2} + a_1 \frac{\partial u}{\partial x} + a_2 u, \quad a_0 > 0,$$

the general explicit difference scheme has the form:

$$(5.11) \qquad u^{n+1}(x) = (Cu^n)(x) \equiv \sum_{(\beta)} c^\beta u^n(x + \beta \Delta x),$$

where the sum extends over some fixed finite number of β. Here the a_i and c^β may be functions of x, and in addition, the c^β will depend on Δt and Δx.

From a Taylor series expansion of $u^n(x + \beta\Delta x)$, the consistency condition then leads to the relations

$$\lim_{\Delta t \to 0} \frac{1}{\Delta t} \sum_{(\beta)} (c^\beta - \delta_0^\beta) = a_2$$

(5.12)
$$\lim_{\Delta t \to 0} \frac{\Delta x}{\Delta t} \sum_{(\beta)} \beta c^\beta = a_1$$

$$\lim_{\Delta t \to 0} \frac{(\Delta x)^2}{\Delta t} \sum_{(\beta)} \tfrac{1}{2}\beta^2 c^\beta = a_0,$$

where the Kronecker delta $\delta_i^j = 0$ for $i \neq j$ and $= 1$ for $i = j$.

When the coefficients a_i are constants, the stability of (5.11) is determined by the amplification factor

(5.13) $$G = \sum_{(\beta)} c^\beta e^{i\beta\xi},$$

which is here a finite trigonometric polynomial in $\xi = k\Delta x$. It is therefore clear that, for stability, the coefficients c^β must be bounded as $\Delta t \to 0$. Since $a_0 > 0$, the last equation of (5.12) then implies that $\Delta t = O((\Delta x)^2)$ as $\Delta t \to 0$. We shall take $\Delta t/(\Delta x)^2 = $ const. and, with no loss of generality, we may take the constant as unity. With this relation, we assume that the coefficients c^β can be expanded in the form

(5.14) $$c^\beta(x, \Delta t) = c_0^\beta(x) + (\Delta t)^{1/2} c_1^\beta(x) + \Delta t\, c_2^\beta(x, \Delta t),$$

where each of the c_i^β are uniformly bounded and $c_2^\beta(x, \Delta t) \to c_2^\beta(x, 0)$ as $\Delta t \to 0$, uniformly in x. The consistency conditions may then be rewritten in the form

(5.15)
$$\sum_{(\beta)} c_0^\beta = 1, \quad \sum_{(\beta)} \beta c_0^\beta = 0, \quad \sum_{(\beta)} c_1^\beta = 0,$$
$$\sum_{(\beta)} \tfrac{1}{2}\beta^2 c_0^\beta = a_0, \quad \sum_{(\beta)} \beta c_1^\beta = a_1, \quad \sum_{(\beta)} c_2^\beta = a_2.$$

We first substantiate our earlier assertion that in most cases of constant coefficients, stability depends only on the principal part of the difference operator, i.e., the value of G at $\Delta t = 0$. Substituting the expansion (5.14) into (5.13), we write

(5.16) $$G(\Delta t, \xi) = G_0(\xi) + (\Delta t)^{1/2} G_1(\xi) + \Delta t\, G_2(\Delta t, \xi),$$

where

$$G_j = \sum_{(\beta)} c_j\beta e^{i\beta\xi}, \quad j = 0, 1, 2;$$

thus, $G(0, \xi) = G_0(\xi)$ and arises from an approximation to a differential operator which, by (5.15), has $a_1 = a_2 = 0$. Now, suppose that the principal part is stable, i.e., $|G_0(\xi)| \leq 1$. To show that (5.11) is stable, G_2 is clearly irrelevant and, using (5.15), we have

(5.17) $$G_0(\xi) = 1 - a_0\xi^2 + O(\xi^3)$$

(5.18) $$G_1(\xi) = ia_1\xi + O(\xi^2).$$

In the most obvious approximation, G_0 is wholly real and G_1 is imaginary; this holds in many other cases, too, and from it follows

$$|G_0 + (\Delta t)^{1/2}G_1| = (|G_0|^2 + \Delta t|G_1|^2)^{1/2} \leq 1 + O(\Delta t),$$

implying stability. In general, however, in order to prove our assertion we need to assume that $|G_0(\xi)| < 1$ in the period $(-\pi, \pi)$ except at $\xi = 0$. Then, it follows from (5.17) that

(5.19) $$|G_0(\xi)| \leq 1 - \delta\xi^2, \quad \text{for } |\xi| \leq \pi,$$

for some $\delta > 0$. Denoting the real and imaginary parts of G_j by E_j and F_j, respectively, we obtain

$$|G_0 + (\Delta t)^{1/2}G_1|^2 = [E_0 + (\Delta t)^{1/2}E_1]^2 + [F_0 + (\Delta t)^{1/2}F_1]^2$$
$$= |G_0|^2 + 2(\Delta t)^{1/2}(E_0E_1 + F_0F_1) + O(\Delta t).$$

But $E_0E_1 + F_0F_1 = O(\xi^2)$; hence, for small enough Δt, the second term will be smaller than $\delta\xi^2$ and so stability follows. A more detailed discussion of particular cases of this result can be found in Section 8.4.

It is also worth noting here that for any difference scheme approximating equation (5.10) with $a_1 = a_2 = 0$ and a_0 constant, Strang (1962) has shown that stability in the maximum norm is conditional on the amplification factor satisfying: either (i) $G(\xi) = ce^{i\beta\xi}$, with $|c| = 1$ and β real; or (ii) $|G(\xi)| < 1$, except for at most a finite number of points ξ_j in $|\xi| \leq \pi$, where $|G(\xi)| = 1$; near each such point there are constants $\alpha_j, \gamma_j, \nu_j$, where α_j is real, Re $\gamma_j > 0$, and ν_j is an even positive integer, such that

$$G(\xi_j + \xi) = G(\xi_j) \exp\left[i\alpha_j\xi - \gamma_j\xi^{\nu_j}[1 + o(1)]\right] \text{ as } \xi \to 0.$$

Strang showed that these conditions were sufficient and Thomée (1965a) has since shown that they are necessary for stability in the maximum norm. Since condition (i) is inapplicable in our present problem, it would seem that (5.19) is not unduly restrictive even in the constant coefficient case, if we are to aim at stability in the maximum norm rather than merely in the L_2 norm. Moreover, Widlund (1966) has given an example of an unstable difference scheme for a coupled pair of parabolic equations with variable coefficients, where the eigenvalues of the principal part of the amplification matrix satisfy condition (ii) above but not the equivalent of (5.19). The scheme is unstable even in the L_2 norm, so (5.19) seems to be the best generally adequate condition that we can expect.

Returning to the variable coefficient problem, we shall in the interest of simplicity, while introducing the main ideas of the argument, consider only the equation

$$\frac{\partial u}{\partial t} = a(x) \frac{\partial^2 u}{\partial x^2}, \quad a(x) \geq \alpha > 0,$$

with an approximating difference scheme (5.11) in which the coefficients c^β are independent of Δt. Although the c^β now depend on x, we can still introduce a local amplification factor, which we shall call the *symbol* of the operator C,

(5.20) $$G(x, \xi) = \sum_{(\beta)} c^\beta(x) e^{i\beta\xi}$$

$$= 1 - a(x)\xi^2 + O(\xi^3).$$

Then we prove: *If the coefficient $a(x)$ is Lipschitz continuous and if*

(5.21) $\quad |G(x, \xi)| < 1 \quad \text{for all } x \text{ and all } \xi \neq 0 \text{ in } (-\pi, \pi),$

with $G(x, 0) = 1$ by consistency, then the difference scheme is stable. From the hypotheses it is clear that, in fact, for some $\delta > 0$,

(5.22) $\quad |G(x, \xi)|^2 \leq 1 - \delta\xi^2, \quad \text{for } |\xi| \leq \pi \quad \text{and all } x.$

We begin by considering stability in the neighborhood of an arbitrary point which, for convenience, we take as the origin. Thus, we restrict our attention to functions u which satisfy

(5.23) $\quad u(x) \equiv 0, \quad \text{for } |x| \geq \zeta > 0.$

Denoting by C_0 the (constant-coefficient) difference operator obtained by replacing $c^\beta(x)$ in (5.11) by $c^\beta(0)$, and by $\hat{u}$ the Fourier transform of u, we have

$$\text{(5.24)} \quad \|C_0 u\|^2 = \int dk |G(0, \xi)|^2 |\hat{u}|^2$$

$$\leq \int dk (1 - \delta \xi^2) |\hat{u}|^2$$

$$\leq \int dk [1 - \delta (2 \sin \tfrac{1}{2}\xi)^2] |\hat{u}|^2$$

$$= \|u\|^2 - \delta \|u\|_1^2;$$

here, we have defined

$$\|u\|_p^2 = \int dk (2 \sin \tfrac{1}{2}\xi)^{2p} |\hat{u}|^2$$

$$= \|\Delta_+^p u\|^2 = \|\Delta_-^p u\|^2,$$

the last line following from the fact that the Fourier transform of the forward (backward) difference $\Delta_\pm$ is $2ie^{\pm \frac{1}{2}i\xi} \sin \tfrac{1}{2}\xi$; it is also clear that

$$\text{(5.25)} \quad \|u\|_p \leq 2^{p-q} \|u\|_q, \quad \text{for } p \geq q.$$

Hence, in order to show that C is a stable operator, we write

$$\|Cu\|^2 = \|C_0 u\|^2 + (\|Cu\|^2 - \|C_0 u\|^2)$$

and attempt to bound the last term by $\delta \|u\|_1^2 + O(\Delta t) \|u\|^2$.

Now, from consistency, corresponding to (5.20) we have

$$\text{(5.26)} \quad C = I + a(x)\Delta_+ \Delta_- + Q,$$

where Q is a difference operator of at least third order, consisting of a finite sum of terms each of which is a product of difference operators Δ_+ or Δ_- and coefficients $a(x)$ in some order; for example, the next term in an expansion would be $\tfrac{1}{2} a \Delta_+ \Delta_- a \Delta_+ \Delta_-$. Thus,

$$\text{(5.27)} \quad \|Cu\|^2 = \|u\|^2 + 2 \operatorname{Re}(u, a\Delta_+\Delta_- u) + \|a\Delta_+\Delta_- u\|^2$$

$$+ 2 \operatorname{Re}(u, Qu) + 2 \operatorname{Re}(a\Delta_+\Delta_- u, Qu + \|Qu\|^2.$$

But, from the elementary identity $y_1 z_1 - y_2 z_2 \equiv y_1(z_1 - z_2) + z_2(y_1 - y_2)$, it is clear that in an expression of the form $(u, \Delta(av))$ the difference operator may be taken past a at the cost of an extra term bounded by $L\Delta x \|u\| \cdot \|v\|$, where L is a Lipschitz constant for $a = a(x)$; such a term we denote by $\varphi(u, v)$. Therefore, for the second term in the above expression for $\|Cu\|^2$, we have

$$(u, a\Delta_+ \Delta_- u) = (u, \Delta_+(a\Delta_- u)) + \varphi(u, \Delta_- u)$$
$$= -(\Delta_- u, a\Delta_- u) + \varphi(u, \Delta_- u),$$

using the summation by parts formula in the last line (see Section 6.2). If we compared this with what would have been obtained from $\|C_0 u\|^2$, we would have a difference of

$$-(\Delta_- u, [a - a(0)]\Delta_- u) + \varphi(u, \Delta_- u).$$

Moreover, since $u(x)$ is zero for $|x| > \zeta$, $\Delta_- u(x)$ is zero for $|x| > 2\zeta$ when $\Delta x < \zeta$. The expression above is then bounded by $2L\zeta \|u\|_1^2 + L\Delta x \|u\| \|u\|_1$.

Other more general terms can be treated similarly. Thus, let us consider the general difference operator $R = a_1 \Delta_1 a_2 \Delta_2 \ldots a_r \Delta_r a_{r+1}$, where a_i are uniformly bounded, uniformly Lipschitz continuous coefficients and the Δ_i are elementary difference operators. Then, by successive interchanges of an a_i with a Δ_j, we obtain operators

$\Delta_1 a_1 R_1$, with $R_1 = a_2 \Delta_2 \ldots \Delta_r a_{r+1}$,

or $S_2 \Delta_2 a_3 R_2$, with $S_2 = \Delta_1 a_1 a_2$, $R_2 = \Delta_3 \ldots a_{r+1}$,

$S_3 \Delta_3 a_4 R_3$, with $S_3 = \Delta_1 a_1 a_2 a_3 \Delta_2$, $R_3 = \Delta_4 \ldots a_{r+1}$,

$\vdots$

$S_{n-1} \Delta_2 a_{r+1} R_{n-1}$, with $S_{n-1} = \Delta_1 a_1 \ldots a_r$, $R_{n-1} = \Delta_3 \ldots \Delta_r$,

$R_n = \Delta_1 a_1 a_2 \ldots a_{r+1} \Delta_2 \Delta_3 \ldots \Delta_r$.

Hence,

$|(u, Ru) - (u, R_n u)|$
$\leq |(u, Ru) - (u, \Delta_1 a_1 R_1 u)| + |(u, S_2 \Delta_2 a_3 R_2 u) - (u, S_3 \Delta_3 a_4 R_3 u)|$
$\quad + \cdots + |(u, S_{n-1} \Delta_2 a_{r+1} R_{n-1} u) - (u, R_n u)|$
$\leq L\Delta x \{\|u\| \|R_1 u\| + \|S_2^* u\| \|R_2 u\| + \cdots + \|S_{n-1}^* u\| \|R_{n-1} u\|\},$

where L is the Lipschitz constant for the a_i and S_i^* denotes the operator adjoint to S_i. Since each of these operators is bounded, the expression in curly brackets could be bounded simply by const. $\|u\|^2$, but we need a more refined estimate here. In fact, each S_i^* ends with Δ_i^* so that all but the first term are bounded by const. $\|u\| \|u\|_1$; and, if $r \geq 2$, an interchange of the last two factors in $R_1 \equiv S_1 \Delta_r a_{r+1}$ yields

$$|\|R_1 u\|^2 - \|S_1 a_{r+1} \Delta_r u\|^2| \leq L \Delta x \|S_1\| \|u\| (\|R_1 u\| + \|S_1 a_{r+1} \Delta_r u\|).$$

Hence,

$$\|R_1 u\| \leq \text{const.} (\|u\|_1 + \Delta x \|u\|).$$

We therefore can say that if R is of at least second order,

(5.28) $\qquad |(u, Ru) - (u, R_n u)| \leq \text{const.} \Delta x \cdot \|u\| (\|u\|_1 + \Delta x \|u\|).$

If in this way we reduce all terms in (5.27) to the form $(\Delta u, (\prod_{(i)} a_i)(\prod_{(j)} \Delta_j) u)$, we clearly obtain, for $\Delta x < \zeta$,

$$|\|Cu\|^2 - \|C_0 u\|^2| \leq M(\zeta) \|u\|_1^2 + K_1 \Delta x \|u\| \|u\|_1 + K_2 (\Delta x)^2 \|u\|^2,$$

where $M(\zeta) \to 0$ as $\zeta \to 0$ and K_1, K_2 are constants. We take ζ so small, $\zeta = \zeta_0$, that $M(\zeta_0) < \delta/4$, and we also have

$$K_1 \Delta x \|u\| \|u\|_1 \leq \frac{K_1^2 (\Delta x)^2}{\delta} \|u\|^2 + \frac{\delta}{4} \|u\|_1^2.$$

Hence, combining this with (5.24), we obtain with $\Delta t = (\Delta x)^2$

(5.29) $\qquad \|Cu\|^2 \leq [1 + (K_2 + K_1^2/\delta) \Delta t] \|u\|^2 - \frac{1}{2} \delta \|u\|_1^2,$

for $\Delta t < \zeta_0^2$, and stability results for a u satisfying (5.23).

To extend the result to a general u, we use a device due to Gårding (1953). Let $\sum_{(j)} d_j^2(x) \equiv 1$ be a smooth partition of unity in which each $d_j(x)$ is a smooth real function of x which vanishes outside some interval I_j and for which, at each fixed point x, only a uniformly bounded number of $d_j(x)$ are non-zero. Each I_j is chosen so small that (5.29) holds for all u whose support is restricted to I_j; the length of I_j need therefore be no smaller than $2\zeta_0$, which was chosen independent of the position of the interval and of Δt. Then we can apply (5.29) to each function $d_j(x) u(x)$, so obtaining an estimate for $\sum_{(j)} \|C d_j u\|^2$, and have to compare this with

SEC. 5.3]   PARABOLIC EQUATIONS   107

$\|Cu\|^2 = \sum_{(j)} \|d_j Cu\|^2$. Thus, it is necessary to estimate the effect of commuting difference operators with the smooth functions d_j. As with the arguments above, one also finds

$$|\|d_j u\|_1^2 - \|d_j \Delta_+ u\|^2| = O(\Delta x)\|u\|(\|d_j \Delta_+ u\| + \|\Delta_+ d_j u\|)$$

$$\leq \tfrac{2}{3}(\|d_j \Delta_+ u\|^2 + \|\Delta_+ d_j u\|^2) + O(\Delta t)\|u\|^2.$$

So,

$$\|d_j u\|_1^2 \geq \tfrac{1}{2}\|d_j \Delta_+ u\|^2 + O(\Delta t)\|u\|^2.$$

Moreover, at any point x, only a finite number of d_j make contributions to the last term, so that an interchange of integration and summation yields

$$\sum_{(j)} \|d_j u\|_1^2 \geq \tfrac{1}{2} \sum_{(j)} \|d_j \Delta_+ u\|^2 + O(\Delta t)\|u\|^2$$

$$= \tfrac{1}{2}\|u\|_1^2 + O(\Delta t)\|u\|^2.$$

Hence, treating the terms arising from the operator C in this manner and using (5.29) we obtain

$$\|Cu\|^2 = \sum_{(j)} \|d_j Cu\|^2$$

$$\leq \sum_{(j)} \|C d_j u\|^2 + O(\Delta x)\|u\|\,\|u\|_1 + O(\Delta t)\|u\|^2$$

$$\leq [1 + O(\Delta t)] \sum_{(j)} \|d_j u\|^2 - \tfrac{1}{2}\delta \sum_{(j)} \|d_j u\|_1^2$$

$$+ \tfrac{1}{4}\delta \|u\|_1^2 + O(\Delta t)\|u\|^2$$

$$\leq [1 + O(\Delta t)]\|u\|^2.$$

This completes the proof of the theorem.

The rather limited result provided by the above theorem may be easily extended in a number of different ways. For example, lower order terms may be included in the differential equation as in (5.10) and, as is suggested by the arguments at the beginning of this section, the hypothesis (5.21) need only be applied then to the amplification matrix with $\Delta t = 0$. Such a result was first proved by Fritz John (1952), using the maximum norm and assuming a_0, $\partial a_0/\partial x$, $\partial^2 a_0/\partial x^2$, a_1, $\partial a_1/\partial x$, and a_2 to be continuous and uniformly bounded: besides proving stability with respect to this norm, he eschewed any assumption that the initial-value problem for the differential equation was properly posed, deriving instead existence, uniqueness, and other properties of the solution from the properties of the difference equation. He went even further by replacing the term $a_2 u$ in (5.10) by a non-linear term $d(x, t, u)$ satisfying a uniform Lipschitz condition in u.

It will also be clear from the next section that extension to more than one space dimension is quite straightforward. This result appears in the paper by Lax and Wendroff (1962), to whom is due the major part of our treatment of the variable coefficient problem. The details of our method are, however, more akin to the approach of Kreiss (1964), as is that of Widlund (1965), who has further extended the results to systems of general parabolic equations approximated by a wide class of implicit multi-step difference methods. Subsequently, Widlund (1966) has carried these general theorems through in the maximum norm so generalizing the work of John, Aronson (1963), and others. In all cases, condition (5.22), applied to the eigenvalues of the symbol of the principal part of the difference operator, plays an essential role. Indeed, Widlund calls difference schemes with this property "parabolic".

5.4. Dissipative Difference Schemes for Symmetric Hyperbolic Equations

We consider the initial-value problem for the hyperbolic system

$$(5.30) \qquad \frac{\partial \mathbf{u}}{\partial t} = \sum_{j=1}^{d} A_j(\mathbf{x}) \frac{\partial \mathbf{u}}{\partial x_j}, \quad -\infty < x_j < +\infty, 0 \leq t \leq 1,$$

where $\mathbf{u} = \mathbf{u}(\mathbf{x}, t)$ is a p-component vector and $A_j(\mathbf{x})$ are $p \times p$ matrices depending smoothly on $\mathbf{x}$ but not on t. Unless specifically stated otherwise, we shall assume that the A_j are hermitian. Then we define a matrix $iP(\mathbf{x}, \mathbf{k})$, where

$$(5.31) \qquad P(\mathbf{x}, \mathbf{k}) = \sum_{j=1}^{d} A_j(\mathbf{x}) k_j, \quad -\infty < k_j < +\infty,$$

which, in the constant coefficient case, represents the Fourier transform of the operator on the right side of (5.30) (compare Section 4.3). The

SEC. 5.4] DISSIPATIVE DIFFERENCE SCHEMES 109

assumption of hyperbolicity is that there exists a non-singular matrix $T = T(\mathbf{x}, \mathbf{k})$ such that

$$(5.32) \qquad TPT^{-1} = \begin{bmatrix} \mu_1 & & 0 \\ & \mu_2 & \\ & & \ddots \\ 0 & & & \mu_p \end{bmatrix},$$

where the $\mu_v = \mu_v(\mathbf{x}, \mathbf{k})$ are real and T, T^{-1} are bounded uniformly in $\mathbf{x}$ and $\mathbf{k}$, i.e.,

$$\|T(\mathbf{x}, \mathbf{k})\|, \|T^{-1}(\mathbf{x}, \mathbf{k})\| \leq K_0.$$

When the A_j are hermitian, T is unitary and these conditions are obviously satisfied.

In the approximating difference schemes, we shall, for convenience, take $\Delta x_j = \Delta t$ for all j and shall consider only explicit schemes

$$(5.33) \qquad \mathbf{u}^{n+1} = C(\Delta t)\mathbf{u}^n,$$

where $C(\Delta t)$ is a finite sum of translation operators with matrix coefficients,

$$(5.34) \qquad (C(\Delta t)\mathbf{u})(\mathbf{x}) \equiv \sum_{(\beta)} c^\beta(\mathbf{x}, \Delta t)\mathbf{u}(\mathbf{x} + \beta \Delta t).$$

We shall assume that when the $A_j(\mathbf{x})$ are hermitian so are the $c^\beta(\mathbf{x}, \Delta t)$. The local amplification matrix, or symbol, G is defined as

$$(5.35) \qquad G(\mathbf{x}, \Delta t, \boldsymbol{\xi}) = \sum_{(\beta)} c^\beta(\mathbf{x}, \Delta t) e^{i\beta \cdot \boldsymbol{\xi}},$$

where $\boldsymbol{\xi} = \Delta t \mathbf{k}$. Denoting the eigenvalues of G by λ_v, we make the following definition:

The difference approximation (5.33) is called dissipative of order $2r$, where r is a positive integer, if there exists a constant $\delta > 0$ such that, for all $\boldsymbol{\xi}$ satisfying $Max_j|\xi_j| \leq \pi$, all $\mathbf{x}$ and all Δt less than some $\tau > 0$,

$$(5.36) \qquad |\lambda_v(\mathbf{x}, \Delta t, \boldsymbol{\xi})| \leq 1 - \delta|\boldsymbol{\xi}|^{2r},$$

where $|\boldsymbol{\xi}|$ denotes the euclidean vector norm of $\boldsymbol{\xi}$.

Then, recalling the definition of order of accuracy (4.12a) or (4.13) we can state the following theorem due to Kreiss (1964):

Suppose the matrices occurring in the differential equation (5.30) and its approximating difference equation (5.33) are hermitian, uniformly bounded, and uniformly *Lipschitz continuous functions of* **x**. *Then, if* (5.33) *is dissipative of order* $2r$ *and accurate of order* $2r - 1$ *for some positive integer* r, *it is stable.*

The proof is quite long but relatively straightforward; several of the improvements over the original proof of Kreiss are due to Parlett (1966). There are three main steps. First we show that the hypotheses are sufficient for stability in the case of constant coefficients, i.e., sufficient for local stability. This is equivalent, as we saw in Section 4.9, to demonstrating the existence, for each fixed **x**, of a positive hermitian matrix $\hat{H} = \hat{H}(\mathbf{x}, \Delta t, \boldsymbol{\xi})$ such that, for some constant $K_2 > 0$,

$$(5.37) \qquad K_2^{-1} I \leq \hat{H} \leq K_2 I,$$

and for which $G^* \hat{H} G \leq \hat{H}$. If $\hat{H}$ were sufficiently smooth in **x**, the subsequent argument would be considerably simpler (see the next section); but this is often not the case and any way is difficult to prove. Hence, we construct $\hat{H}$ so that it will serve for a whole neighborhood of **x**: it will be 2π-periodic in each component of $\boldsymbol{\xi}$ and it is easy to ensure that

$$(5.38) \qquad G^* \hat{H} G \leq (1 - \tfrac{1}{2}\delta |\boldsymbol{\xi}|^{2r}) \hat{H};$$

but in addition we shall need $\hat{H}$ to be of the special form

$$(5.39) \quad \hat{H} = I + \hat{H}_{2r-1}(\mathbf{x}, \Delta t, \boldsymbol{\xi}), \quad \text{where } |\hat{H}_{2r-1}| \leq \text{const.} |\boldsymbol{\xi}|^{2r-1},$$

for $\text{Max}_j |\xi_j| \leq \pi$. The final step of the proof is to use this $\hat{H}$ in the definition of a norm $\| \cdot \|_H$ for which the solution of (5.33) satisfies the strong stability condition

$$(5.40) \qquad \|\mathbf{u}^{n+1}\|_H^2 \leq [1 + O(\Delta t)] \|\mathbf{u}^n\|_H^2.$$

We note first that the dependence of C on Δt, and therefore that of G and $\hat{H}$, comes about mainly through the calculation of $c^{\beta}(\mathbf{x}, \Delta t)$ from values of A_j at points neighboring **x**. Since the A_j are Lipschitz continuous, $\|C(\Delta t) - C(0)\| (\Delta t)^{-1}$ is then bounded and the stability properties of $C(\Delta t)$ and its principal part $C(0)$ are the same. Hence, we shall consider only the latter and drop the explicit dependence on Δt in C, G, $\hat{H}$ and λ in the following.

To carry out the first step of the proof, we drop the assumption that the A_j are hermitian. This goes a little beyond our immediate aim but is valuable in illuminating the role of the various hypotheses. Now, from the alternative definition of order of accuracy, (4.33), we have

(5.41) $$G(\xi) = e^{iP(\xi)} + Q(\xi)$$

where $\|Q\| \leq$ const. $|\xi|^{2r}$ for $\text{Max}_j|\xi_j| \leq \pi$. Moreover, because of the hyperbolicity, this can be transformed by the matrix T of (5.32) into $TGT^{-1} = M + R$, where $M = \exp(iTPT^{-1})$ is diagonal and $\|R\| \leq$ const. $|\xi|^{2r}$. If this is now triangularized by a unitary transformation U, we obtain

(5.42) $$\tilde{G} = UTG\,T^{-1}U^* = UMU^* + URU^* = \Lambda + N,$$

where Λ is the diagonal matrix formed of the eigenvalues λ_j of G and N is the nilpotent matrix of the off-diagonal elements. The important point to notice is that $\|N\| \leq K_1|\xi|^{2r}$ for some real constant $K_1 > 1$. This comes about as follows. The elements of URU^* are bounded by a constant multiple of $|\xi|^{2r}$ and, if $\tilde{G}$ is upper triangular, so then are those elements of UMU^* below the diagonal. But UMU^* is a normal matrix and, therefore, its elements above the diagonal are bounded by a constant multiple of the bound on those below.[3] Hence, the result.

At this point it is easy to see directly that the powers of G are bounded. For, as in Section 5.2,

$$\|G^n\| \leq K_0^2 \|\tilde{G}^n\| \leq K_0^2 \sum_{j=0}^{p-1} \binom{n}{j} \|\Lambda\|^{n-j} \|N\|^j$$

$$\leq K_0^2 (K_1/\delta)^{p-1} \sum_{j=0}^{p-1} \binom{n}{j} (1 - \delta|\xi|^{2r})^{n-j} (\delta|\xi|^{2r})^j$$

$$\leq K_0^2 (K_1/\delta)^{p-1}, \quad \text{for } n \geq p - 1.$$

---

[3] We have

$$A^*A = AA^* \rightarrow \sum_{k=1}^{p} |a_{ik}|^2 = \sum_{k=1}^{p} |a_{ki}|^2.$$

Now suppose that $|a_{ik}| \leq K$ for $i > k$. Then

$$\sum_{k=2}^{p} |a_{1k}|^2 \leq (p-1)K^2, \quad \sum_{k=3}^{p} |a_{2k}|^2 \leq (p-2)K^2 + |a_{12}|^2 \leq (2p-3)K^2,$$

and so on. It clearly follows by induction that

$$|a_{ik}|^2 \leq [1 + (p-2)2^{i-1}]K^2, \quad \text{for } i < k.$$

Less directly, we may construct $\hat{H} = \hat{H}(\xi)$ as in the proof of Kreiss' matrix theorem in the previous chapter. Introducing the diagonal matrix

$$D = \begin{bmatrix} \Delta & & & 0 \\ & \Delta^2 & & \\ & & \ddots & \\ 0 & & & \Delta^p \end{bmatrix}, \text{ with } \Delta > 1,$$

we choose Δ so large that

$$(1 - \tfrac{1}{2}\delta|\xi|^{2r})D - \tilde{G}^*D\tilde{G} \geqq 0,$$

i.e.,

$$(1 - \tfrac{1}{2}\delta|\xi|^{2r})I - \tilde{G}_D^*\tilde{G}_D \geqq 0,$$

where we use the notation R_D for $D^{1/2} R D^{-1/2}$. It is clear that this is possible for, from (5.42), $\tilde{G}_D = (\Lambda + N)_D = \Lambda + N_D$ and, since the $(j,k)^{\text{th}}$ element of N_D is $\Delta^{\frac{1}{2}(j-k)}N_{jk}$, $\|N_D\| \leqq \Delta^{-\frac{1}{2}} \|N\| \leqq K_1\Delta^{-\frac{1}{2}}|\xi|^{2r}$. Hence,

$$\|\tilde{G}_D^*\tilde{G}_D\| \leqq \|\Lambda^*\Lambda\| + 2\|N_D\| + \|N_D\|^2$$

$$\leqq 1 - \delta|\xi|^{2r} + \text{const. } \Delta^{-\frac{1}{2}}|\xi|^{2r}$$

$$\leqq 1 - \tfrac{1}{2}\delta|\xi|^{2r}, \text{ for } \Delta \text{ large enough.}$$

Thus, for a sufficiently large Δ, inequality (5.38) is satisfied with $\hat{H} = T^*U^*DUT$ and $\hat{H}$ satisfies (5.37) with $K_2 \leqq \Delta^p K_0^2$.

To construct an $\hat{H}$ with the additional property (5.39), we use a different approach. In doing so we have to reimpose the restriction that the A_j and the c^β are hermitian. Then we can suppose that P has been diagonalized by a preliminary unitary transformation. We also put $\xi = \sigma\xi_0$, where ξ_0 is some fixed unit vector, and so have

(5.43) $$G = M + \sigma^{2r}R + o(\sigma^{2r}),$$

where $M = \exp(i\sigma P(\xi_0))$ is diagonal and unitary and $R = R(\xi_0)$ is hermitian. Now suppose that two eigenvalues, μ_j and μ_k, of P are equal. There is then some arbitrariness in the unitary transformation diagonalizing P—to the extent, in fact, of any unitary transformation in the

2-dimensional subspace corresponding to these two elements. Since R is hermitian, this can be used to ensure that R_{jk} is made zero in this case.

Supposing this has been done, we define the elements of an hermitian matrix $\hat{H}'$ in terms of R by

(5.44) $$\hat{H}'_{jk} = \begin{cases} 0, & \text{if } \mu_j = \mu_k \\ -2iR_{jk}/(\mu_j - \mu_k), & \text{otherwise,} \end{cases}$$

and thence form the hermitian matrix

(5.45) $$\hat{H} = I + \sigma^{2r-1}\hat{H}'.$$

Checking (5.38), we have, by expanding M in powers of σ,

(5.46) $$G^*\hat{H}G = \hat{H} + \sigma^{2r}[2R + i(\hat{H}'P - P\hat{H}')] + o(\sigma^{2r}).$$

The construction of $\hat{H}$ has ensured that the off-diagonal elements of the expression in square brackets are zero. Moreover, we have the following lemma, whose proof we shall leave for the moment:

Lemma. *For a dissipative difference scheme with G in the above form (5.43),*

(5.47) $$R_{jj} \leq -\delta + o(1), \quad \text{as } \sigma \to 0.$$

Hence, for $\xi = \sigma\xi_0$, we have

$$2R + i(\hat{H}'P - P\hat{H}') \leq -2\delta I + o(1),$$

where $\hat{H}' = \hat{H}'(\xi_0)$ is bounded, although not uniformly so in ξ_0. Thus, there is a finite range $0 \leq \sigma \leq \sigma_0(\xi_0)$ in which (5.38) and (5.39) are satisfied in this direction ξ_0. Further, since P and R are continuous functions of ξ, there is an $\varepsilon > 0$ such that for $|\xi_0 - \xi_1| < \varepsilon, |\xi_1| = 1$

$$2R(\xi_1) + i[\hat{H}'(\xi_0)P(\xi_1) - P(\xi_1)\hat{H}'(\xi_0)] \leq -\delta I;$$

and for $\xi = \sigma\xi_1$, $0 \leq \sigma \leq \sigma_1(\xi_0)$ inequalities (5.38) and (5.39) are satisfied. But by the Heine-Borel theorem, the whole sphere $|\xi| = 1$ can be covered by a finite number of such neighborhoods, in each of which $\hat{H}'$ can be taken as constant. Hence, we can construct an $\hat{H}(\xi)$ in a finite neighborhood $|\xi| \leq \sigma_2$ of the origin satisfying all three requirements (5.37), (5.38), and (5.39). Outside this neighborhood, (5.39) has no force and we may use the $\hat{H}$ constructed in the first step.

To prove the lemma above, we show that the eigenvalues of G equal its diagonal elements up to order σ^{2r}. This results from a stronger form of Gerschgorin's theorem (see footnote 8 in Section 4.9) stating that if s of the Gerschgorin circles of a matrix M are disjoint from the rest, then their union contains precisely s of the eigenvalues of M; this follows easily from a continuity argument based upon the fact that the eigenvalues of a matrix depend continuously on its elements. Suppose then, that μ_k is a simple eigenvalue of P and multiply the kth row of G by $\sigma^{\frac{1}{2}}$ while dividing the kth column by $\sigma^{\frac{1}{2}}$. This is a similarity transformation which leaves the eigenvalues λ_j of G unchanged, but it makes the radius of the kth Gerschgorin circle $O(\sigma^{2r+\frac{1}{2}})$ and those of the remainder $O(\sigma^{2r-\frac{1}{2}})$. Since the centers of these circles are at $G_{jj} = e^{i\sigma\mu_j} + \sigma^{2r}R_{jj} + O(\sigma^{2r+1})$ and $\mu_k \neq \mu_j$ for $j \neq k$, the kth circle is disjoint from the rest for sufficiently small σ. Thus, there is an eigenvalue of G such that

$$\lambda = e^{i\sigma\mu_k} + \sigma^{2r}R_{kk} + o(\sigma^{2r}).$$

But, by (5.36), $|\lambda| \leq 1 - \delta\sigma^{2r}$ so that

$$1 - 2\delta\sigma^{2r} \geq 1 + 2\sigma^{2r}R_{kk} + o(\sigma^{2r})$$

i.e., $R_{kk} \leq -\delta + o(1)$, as $\sigma \to 0$.

When μ_k is not a simple eigenvalue of P, all the rows and columns corresponding to μ_k are treated as above. Because we have ensured that $R_{jk} = 0$ for $\mu_j = \mu_k$, $j \neq k$, the radii of the corresponding Gerschgorin circles are still $O(\sigma^{2r+\frac{1}{2}})$. Therefore, at least one contains an eigenvalue of G so that (5.47) is true for that one. On the other hand, if for some of them $R_{jj} > -\delta + o(1)$, these would become disjoint from those with $R_{jj} \leq -\delta + o(1)$ as $\sigma \to 0$, and a contradiction would follow from the fact that they would then contain an eigenvalue of G.

The final stage in the proof of the main theorem brings us back to the consideration of the variable coefficient problem (5.30), (5.33), and (5.34) with A_j and c^β hermitian. This part of the proof is similar in form to that given for the simpler problem treated in the previous section, except that $\hat{H}$ is now not identically equal to the unit matrix. We again begin by considering stability in the neighborhood of an arbitrary point, which for convenience we take as the origin, and first restrict our attention to functions **u** which satisfy

(5.48) $\qquad \mathbf{u}(\mathbf{x}) \equiv 0, \quad \text{for } |\mathbf{x}| \geq \zeta > 0.$

SEC. 5.4]   DISSIPATIVE DIFFERENCE SCHEMES   115

Then, corresponding to the coefficients $c^\beta(0)$ of the difference operator C_0, there is a local amplification matrix $G = G(0, \xi)$ given by (5.35). Further, we have shown that there exists an hermitian matrix $\hat{H}(0, \xi)$ of the form

(5.49) $\qquad \hat{H} = I + \hat{H}_{2r-1}(0, \xi)$, with $\|\hat{H}_{2r-1}\| \leq K_3 |\xi|^{2r-1}$,

which satisfies (5.37) and (5.38). From this we define an operator H by

(5.50) $\qquad H\mathbf{u}(\mathbf{x}) = (2\pi)^{-d/2} \int d\mathbf{k}\, \hat{H}(0, \Delta t \mathbf{k}) \hat{\mathbf{u}}(\mathbf{k}) e^{i\mathbf{k}\cdot\mathbf{x}}$,

where $\hat{\mathbf{u}}$ is the Fourier transform of $\mathbf{u}$. We define H_{2r-1} similarly and denote $\{(\mathbf{u}, H\mathbf{u})\}^{1/2}$ by $\|\mathbf{u}\|_H$. Because of the properties of $\hat{H}$, this is clearly a proper norm, and we have

(5.51) $\qquad \|C\mathbf{u}\|_H^2 = (C\mathbf{u}, [I + H_{2r-1}]C\mathbf{u}) = E + F$
$\qquad\qquad = (E_0 + F_0) + (E - E_0) + (F - F_0),$

where

$$E = \|C\mathbf{u}\|^2, \qquad F = (C\mathbf{u}, H_{2r-1}C\mathbf{u}),$$

and E_0, F_0 are the same except that C is replaced by C_0. Our main task is to estimate the three terms on the right-hand side of (5.51) in order to arrive at the desired inequality (5.40). We have

(5.52) $\qquad E_0 + F_0 = \|C_0 \mathbf{u}\|_H^2 = \int d\mathbf{k}\, \hat{\mathbf{u}}^* G^* \hat{H} G \hat{\mathbf{u}}$

$\qquad\qquad \leq \int d\mathbf{k}\, \hat{\mathbf{u}}^* (1 - \tfrac{1}{2}\delta |\xi|^{2r}) \hat{H} \hat{\mathbf{u}}$

$\qquad\qquad \leq \int d\mathbf{k}\, \hat{\mathbf{u}}^* (1 - \tfrac{1}{2}\delta \sum_{j=1}^{d} (2 \sin \tfrac{1}{2}\xi_j)^{2r}) \hat{H} \hat{\mathbf{u}}$

$\qquad\qquad = \|\mathbf{u}\|_H^2 - \tfrac{1}{2}\delta K_2^{-1} \|\mathbf{u}\|_r^2,$

where, as in Section 5.3,

(5.53) $\qquad \|\mathbf{u}\|_\rho^2 = \int d\mathbf{k} \sum_{j=1}^{d} (2 \sin \tfrac{1}{2}\xi_j)^{2\rho} |\hat{\mathbf{u}}|^2$

$\qquad\qquad = \sum_{j=1}^{d} \|\Delta^\rho_{+,j}\mathbf{u}\|^2 = \sum_{j=1}^{d} \|\Delta^\rho_{-,j}\mathbf{u}\|^2,$

so that

(5.54) $$\|u\|_p \leq 2^{p-q}\|u\|_q \quad \text{for } p \geq q.$$

To estimate the remaining terms, it is necessary to consider the form of C in more detail. Because it is an accurate representation of the differential operator up to order $2r - 1$, it is uniquely determined to that order. Hence, we may write (compare (5.41) for G)

(5.55) $$C = \sum_{m=0}^{2r-1} \frac{1}{m!} P_\Delta^m + Q_\Delta,$$

where

$$P_\Delta = \sum_{j=1}^{d} A_j(x)\Delta_j^{(2r-1)}$$

and $\Delta_j^{(2r-1)}$ denotes the truncated expansion of $\Delta t(\partial/\partial x_j)$ in powers of the centered difference operator $\Delta_{0j} = \frac{1}{2}(\Delta_{+j} + \Delta_{-j})$:

(5.56) $$\Delta_j^{(2r-1)} = \sum_{m=0}^{r-1} (-1)^m \gamma_{2m+1} \Delta_{0j}^{2m+1},$$

the γ_{2m+1} being certain real constants. The remainder term Q_Δ also consists of a finite sum of products of difference operators, $\prod_{(i)} A \Delta^{p_i}$, where each A represents some one of the $A_j(x)$ and each Δ one of Δ_{+j} or Δ_{-j}; for this term $2r \leq \sum_{(i)} p_i \leq \text{const.}$, since the difference scheme is supposed explicit and of order of accuracy $2r - 1$.

As previously observed, the difference operator in an expression of the form $(u, \Delta A v)$ can be taken past A at the cost of an extra term bounded by $L \Delta t \|u\| \cdot \|v\|$, where L is a Lipschitz constant for A: we shall use the notation $\varphi(u, v)$ to denote such a term in the following. Thus, from the simple summation by parts formula $(u, \Delta_0 A v) = -(\Delta_0 u, A v)$, one obtains, since A is hermitian,

(5.57) $$(u, A\Delta_0 v) = -(A\Delta_0 u, v) + \varphi(u, v).$$

Similar formulas hold for each of the finite number of odd-order differences making up P_Δ, and hence for P_Δ itself. We therefore can write

(5.58) $$\left\|\left(\sum_{m=0}^{2r-1} \frac{1}{m!} P_\Delta^m\right)u\right\|^2 = \left(u, \sum_{l=0}^{2r-1} \frac{1}{l!}(-P_\Delta)^l \left(\sum_{m=0}^{2r-1} \frac{1}{m!} P_\Delta^m\right)u\right) + \varphi(u, u)$$
$$= \|u\|^2 + (u, Ru) + \varphi(u, u).$$

SEC. 5.4] DISSIPATIVE DIFFERENCE SCHEMES

In this expression it is clear that R is a difference operator of the same form as Q_Δ, i.e., of order greater than or equal to $2r$. To estimate terms involving such operators, we make further use of the argument leading to (5.28) in Section 5.3; reordering the factors A and Δ, we can obtain

$$(\mathbf{u}, R\mathbf{v}) = (\Delta^r \mathbf{u}, S\Delta^r \mathbf{v}) + \varphi(\mathbf{u}, \mathbf{v}),$$

where S is a bounded difference operator of order $2r$ less than that of R and Δ^r stands for a product of r factors $\Delta_{\pm j}$. Hence,

$$(5.59) \quad \|C\mathbf{u}\|^2 = \left\|\left(\sum_{m=0}^{2r-1} \frac{1}{m!} P_\Delta^m\right)\mathbf{u}\right\|^2 + 2\operatorname{Re}\left(Q_\Delta \mathbf{u}, \left(\sum_{m=0}^{2r-1} \frac{1}{m!} P_\Delta^m\right)\mathbf{u}\right) + \|Q_\Delta \mathbf{u}\|^2$$

$$= \|\mathbf{u}\|^2 + (\Delta^r \mathbf{u}, S'\Delta^r \mathbf{u}) + \varphi(\mathbf{u}, \mathbf{u})$$

where S', like S, is a bounded difference operator.

But $E - E_0 = \|C\mathbf{u}\|^2 - \|C_0 \mathbf{u}\|^2$ and therefore consists mainly of a term in $S' - S_0'$. However, since the support of $\mathbf{u}$ is limited to a sphere of radius ζ, C may be replaced by $C_0 + g(\mathbf{x})(C - C_0)$, where $g(\mathbf{x})$ is a smoothly varying function satisfying $g(\mathbf{x}) \equiv 1$ for $|\mathbf{x}| \leq \tfrac{3}{2}\zeta$ and $g(\mathbf{x}) \equiv 0$ for $|\mathbf{x}| > 2\zeta$, when Δx is sufficiently small. As a result, the integration in $\|C\mathbf{u}\|^2 - \|C_0\mathbf{u}\|^2$ need be extended only over a sphere of radius 2ζ and, therefore,

$$(5.60) \quad |E - E_0| \leq M_1(\zeta)\|\mathbf{u}\|_r^2 + K_4 \Delta t \|\mathbf{u}\|^2,$$

where $M_1(\zeta) \to 0$ as $\zeta \to 0$ and K_4 is a constant.

For the final term, we make use of the special form of H as derived from (5.49):

$$(5.61) \quad |F - F_0| = |\operatorname{Re}([C + C_0]\mathbf{u}, H_{2r-1}[C - C_0]\mathbf{u})|$$

$$\leq \operatorname{const.} |(\mathbf{u}, H_{2r-1}[C - C_0]\mathbf{u})|,$$

since C is certainly a bounded operator. Now, factorize H_{2r-1} into $H_{2r-1} = H_1 H_2$, where we define the Fourier transform of the factors by

$$\hat{H}_1 = \left(\sum_{j=1}^d (2\sin \tfrac{1}{2}\xi_j)^2\right)^{r/2} I, \quad \hat{H}_2 = \hat{H}_1^{-1} \hat{H}_{2r-1}.$$

It follows that

$$|\xi|^r I \geq \hat{H}_1 \geq \left(\frac{2}{\pi}\right)^r |\xi|^r I \quad \text{and} \quad \|\hat{H}_2\| \leq \left(\frac{\pi}{2}\right)^r K_3 |\xi|^{r-1}.$$

Hence,

$$|(\mathbf{u}, H_{2r-1}[C - C_0]\mathbf{u})| \leq \|H_1\mathbf{u}\| \|H_2[C - C_0]\mathbf{u}\|$$
$$\leq \text{const.} \|\mathbf{u}\|_r \cdot \|[C - C_0]\mathbf{u}\|_{r-1};$$

and, from (5.53), (5.54) and the by now familiar arguments,

$$\|[C - C_0]\mathbf{u}\|_{r-1} \leq \sum_{j=1}^{d} \|\Delta_{+j}^{r-1}[C - C_0]\mathbf{u}\|$$

$$\leq \sum_{j=1}^{d} \|[C - C_0]\Delta_{+j}^{r-1}\mathbf{u}\| + \text{const.} \Delta t \cdot \|\mathbf{u}\|$$

$$\leq M_2(\zeta) \|\mathbf{u}\|_r + K_5 \Delta t \|\mathbf{u}\|,$$

where $M_2(\zeta) \to 0$ as $\zeta \to 0$, since $C - C_0$ is a difference operator of at least first order. Combining this with (5.60) and again noting (5.54), we have, with $M(\zeta) \to 0$ as $\zeta \to 0$ and K_6 a constant,

$$|E - E_0| + |F - F_0| \leq M(\zeta) \|\mathbf{u}\|_r^2 + K_6 \Delta t \|\mathbf{u}\|^2.$$

Hence, by taking $\zeta \leq \zeta_0$ for some $\zeta_0 > 0$, the first term above may be dominated by the negative term in (5.52) and therefore

(5.62) $\quad \|\mathbf{u}^{n+1}\|_H^2 \leq \|\mathbf{u}^n\|_H^2 + \alpha K_2^{-1} \Delta t \|\mathbf{u}^n\|^2 \leq (1 + \alpha \Delta t) \|\mathbf{u}^n\|_H^2,$

where $\alpha = K_2 K_6$.

To obtain such a result for a general $\mathbf{u}^n$ we again use the device of Gårding (1953). Let $\sum_{(i)} d_i^2(\mathbf{x}) \equiv 1$ be a smooth partition of unity in which each $d_i(\mathbf{x})$ is a smooth real function of $\mathbf{x}$ which vanishes outside some sphere S_i and for which, at each fixed point $\mathbf{x}$, only a uniformly bounded number of $d_i(\mathbf{x})$ are non-zero. Each S_i is chosen so small that (5.62) holds for all $\mathbf{u}^n$ whose support is restricted to S_i; thus, the radius of S_i is bounded below by ζ_0. Then, we define H_i at the center of S_i as in the above, and set

$$\|\mathbf{u}\|_H^2 = \sum_{(i)} \|d_i \mathbf{u}\|_{H_i}^2 = \sum_{(i)} (d_i \mathbf{u}, H_i d_i \mathbf{u}),$$

a norm which clearly still satisfies the inequalities common to all H_i. Hence,

$$\|\mathbf{u}^{n+1}\|_H^2 = \sum_{(i)} (d_i C \mathbf{u}^n, H_i d_i C \mathbf{u}^n),$$

SEC. 5.5] FURTHER RESULTS, SYMMETRIC HYPERBOLIC EQUATIONS

and noting that in interchanging d_i and C only a finite number of d_i contribute at each point $\mathbf{x}$, we have

$$\|\mathbf{u}^{n+1}\|_H^2 \leq \sum_{(i)} (Cd_i\mathbf{u}^n, H_i Cd_i\mathbf{u}^n) + \text{const. } \Delta t \|\mathbf{u}^n\|^2$$

$$\leq \sum_{(i)} \|d_i\mathbf{u}^n\|_{H_i}^2 (1 + \alpha \Delta t) + \text{const. } \Delta t \|\mathbf{u}^n\|^2$$

$$\leq \|\mathbf{u}^n\|_H^2 [1 + O(\Delta t)],$$

and the stability of the difference scheme is finally demonstrated.

This completes the proof of Kreiss' theorem. We have given it in some detail, because of both the importance of the result and the general utility of the arguments. It is always valuable to have a result depending only on the eigenvalues of the amplification matrix or symbol G, but also the concept of a dissipative difference scheme appears to be significant for non-linear problems—see, for example, Richtmyer and Morton (1964). In this respect, it is important that a number of similar results can be derived in the same way. In particular, the theorem is not directly applicable to the Lax-Wendroff schemes (see Section 12.7), which are dissipative of order 4 but are accurate only to order 2. This and a number of other cases can be covered by making a change of variables from ξ. Thus, suppose G has the form

$$G(\xi) = \exp\{iP[\mathbf{g}(\xi)]\} + Q(\xi),$$

where $\|Q\| \leq \text{const.} |\xi|^{2r}$ and the vector function $\mathbf{g}(\xi)$ is analytic with $g_j = \xi_j + o(|\xi|)$ as $\xi \to 0$. (In the Lax-Wendroff case, $g_j = \sin \xi_j$.) Then, changing variables to $\mathbf{g}$ enables the arguments to be taken through as before. We may observe that, for this form of G, the difference scheme may be only first-order accurate. In general, however, it appears that the condition of accuracy to order $2r - 1$ cannot be overcome by these methods, although Parlett (1966) has shown that $2r - 2$ is sufficient if P has distinct eigenvalues, i.e., the differential system is *regular* hyperbolic.

5.5. Further Results for Symmetric Hyperbolic Equations

There are several theorems on difference schemes for hyperbolic equations which are similar in kind to that described in the previous section. By making assumptions about the symbol G which are in some sense

120   VARIABLE COEFFICIENTS AND NON-LINEAR PROBLEMS   [CHAP. 5

stronger (although no dissipation is assumed), one need not use as much detailed information about the difference scheme. Then it is more convenient to work with the translation operators $T^\beta \mathbf{u}(\mathbf{x}) = \mathbf{u}(\mathbf{x} + \beta \Delta x)$ rather than the difference operators $\Delta_\pm$.

One of the earliest of these theorems is due to Friedrichs (1954):

Suppose the difference scheme

$$(5.63) \qquad \mathbf{u}^{n+1} = C\mathbf{u}^n \equiv \left[\sum_{(\beta)} c^{(\beta)}(\mathbf{x}) T^\beta\right] \mathbf{u}^n,$$

with $\Delta t = \Delta x$, has the properties that (i) the matrices $c^{(\beta)}(\mathbf{x})$ are hermitian, independent of Δt and Lipschitz continuous in $\mathbf{x}$; and (ii) the $c^{(\beta)}(\mathbf{x})$ are also non-negative; then the scheme is stable.

As pointed out by Lax (1961), condition (ii) effectively limits the difference schemes covered to those of only first order, but it has the effect of allowing a very simple proof of the theorem. We note first that it implies that the symbol of the difference scheme has norm less than or equal to unity:

$$(5.64) \qquad \|G(\mathbf{x}, \boldsymbol{\xi})\| = \left\|\sum_{(\beta)} c^{(\beta)}(\mathbf{x}) e^{i\beta \cdot \boldsymbol{\xi}}\right\| \leq 1, \quad \text{all } \mathbf{x} \text{ and } \boldsymbol{\xi}.$$

For, from the consistency condition, $\sum_{(\beta)} c^{(\beta)}(\mathbf{x})$ equals the identity matrix and, hence, if α_β are any complex numbers of unit modulus, it follows from hypothesis (ii) that

$$\left|\mathbf{v}^*\left(\sum_{(\beta)} \alpha_\beta c^{(\beta)}\right)\mathbf{u}\right| \leq \sum_{(\beta)} |\mathbf{v}^* c^{(\beta)} \mathbf{u}| \leq \tfrac{1}{2} \sum_{(\beta)} (\mathbf{v}^* c^{(\beta)} \mathbf{v} + \mathbf{u}^* c^{(\beta)} \mathbf{u})$$
$$= \tfrac{1}{2} |\mathbf{v}|^2 + \tfrac{1}{2} |\mathbf{u}|^2;$$

putting $\mathbf{v} = \sum \alpha_\beta c^{(\beta)} \mathbf{u}$, then yields the desired result, $|\mathbf{v}|^2 \leq |\mathbf{u}|^2$.

Then we have

$$|(\mathbf{v}, C\mathbf{u})| = \left|\sum_{(\beta)} \int d\mathbf{x}\, \mathbf{v}^*(\mathbf{x}) c^{(\beta)}(\mathbf{x}) \mathbf{u}(\mathbf{x} + \beta \Delta t)\right|$$
$$\leq \tfrac{1}{2} \|\mathbf{v}\|^2 + \tfrac{1}{2} \sum_{(\beta)} \int d\mathbf{x}\, \mathbf{u}^*(\mathbf{x}) c^{(\beta)}(\mathbf{x} - \beta \Delta t) \mathbf{u}(\mathbf{x})$$
$$\leq \tfrac{1}{2} \|\mathbf{v}\|^2 + \tfrac{1}{2} \sum_{(\beta)} \int d\mathbf{x}\, \mathbf{u}^*(\mathbf{x}) c^{(\beta)}(\mathbf{x}) \mathbf{u}(\mathbf{x}) + O(\Delta t)$$
$$= \tfrac{1}{2} \|\mathbf{v}\|^2 + \tfrac{1}{2} \|\mathbf{u}\|^2 + O(\Delta t).$$

SEC. 5.5]   FURTHER RESULTS, SYMMETRIC HYPERBOLIC EQUATIONS   121

Again, putting $\mathbf{v} = C\mathbf{u}$ yields the final result,

(5.65) $$\|C\| \leq 1 + O(\Delta t).$$

We have already noted the early work of Lax on this type of theorem. More recently, see Lax and Nirenberg (1966), this has been extended to give a theorem going straight from the local stability condition (5.64) to global stability (5.65):

Suppose that the real symmetric coefficient matrices $c^{(\beta)}(\mathbf{x})$ are independent of Δt and have bounded second derivatives in $\mathbf{x}$; and suppose (5.64) is satisfied everywhere; then the difference scheme (5.63) for real vector functions $\mathbf{u}^n$ is stable.

The original proof was for only the scalar case and involved a partition of unity depending on Δt; the proof for the non-scalar case is even more delicate. We begin by defining the hermitian matrix

(5.66) $$R(\mathbf{x}, \xi) = I - G^*(\mathbf{x}, \xi)G(\mathbf{x}, \xi),$$

and forming the corresponding difference operator R_Δ. From the Lipschitz continuity of the coefficients $c^{(\beta)}$ it is clear that

(5.67) $$R_\Delta = I - C^*C + O(\Delta t).$$

Thus, from the hypothesis (5.64), i.e., that

(5.68) $$R(\mathbf{x}, \xi) \geq 0 \quad \text{for all } \mathbf{x} \text{ and } \xi,$$

we have to show that

(5.69) $$(\mathbf{u}, R_\Delta \mathbf{u}) \geq - M \Delta t \|\mathbf{u}\|^2,$$

for some constant M. We need the following two lemmas:

Lemma 1. If the symbol R is hermitian and $\varphi(\mathbf{x})$ is a smooth real scalar function with $|\varphi(\mathbf{x})| \leq 1$, then for real $\mathbf{u}$

(5.70) $$|(\varphi \mathbf{u}, R_\Delta \varphi \mathbf{u}) - (\varphi \mathbf{u}, \varphi R_\Delta \mathbf{u})| \leq K(L + L^2)(\Delta t)^2 \|\mathbf{u}\|^2,$$

where the constant K depends only on R and L is the Lipschitz constant for φ. Since R is hermitian, R_Δ must consist of a sum of terms of the form $r(\mathbf{x})[T^\beta \pm T^{-\beta}]$, where r is hermitian when the positive sign is taken and anti-hermitian otherwise. We consider the first case. Then, using the subscript β to denote a shifted argument $x \to x + \beta \Delta t$, the left-hand side of (5.70) for such a term is

$$\left| \int dx (\varphi_\beta - \varphi) \varphi u^T r u_\beta - \int dx (\varphi - \varphi_{-\beta}) \varphi u^T r u_{-\beta} \right|$$

$$= \left| \int\int dx (\varphi_\beta - \varphi)(\varphi u^T r u_\beta - \varphi_\beta u_\beta^T r_\beta u) \right|$$

$$\leq L|\beta|\Delta t \int dx |(\varphi - \varphi_\beta) u^T r u_\beta + \varphi_\beta (u^T r u_\beta - u_\beta^T r_\beta u)|,$$

the superscript on u^T denoting the transpose of u. The first term in the integrand is dominated by $\|r\| L |\beta| \|u\|^2 \Delta t$, and the second, by $|\beta| \|u\|^2 \Delta t$ times the Lipschitz constant for $r(x)$, because $u^T r u_\beta = u_\beta^T r u$; hence, the result follows.

Lemma 2. *If A and B are hermitian and $A + B \geq 0$ and $A - B \geq 0$, then*

(5.71) $\qquad |(u, Bv)| \leq \frac{1}{2} M^2 (u, Au) + \frac{1}{2} M^{-2} (v, Av).$

for any real number M.

One merely has to observe that

$$((f + g), (A + B)(f + g)) + ((f - g), (A - B)(f - g))$$
$$= 2(f, Af) + 2(g, Ag) + 2(f, Bg) + 2(g, Bf) \geq 0,$$

put $f = Mu$, $g = M^{-1} e^{i\theta} v$ and choose θ appropriately.

To prove the main theorem, one proceeds via a basic smooth partition of unity, $\sum_{(i)} d_i^2(x) \equiv 1$, and then forms a scaled family by putting $e_i(x) = d_i((\Delta t)^{-1/2} x)$. As a result, the $e_i(x)$ have support of the order of $(\Delta t)^{1/2}$ and satisfy a Lipschitz condition, $|e_i(x) - e_i(y)| \leq \text{const.} (\Delta t)^{-1/2} |x - y|$. One can then show that

$$\left| (u, R_\Delta u) - \sum_{(i)} (e_i u, R_\Delta e_i u) \right| \leq \text{const.} \Delta t \|u\|^2.$$

For any term $r(x) T^\beta$ of R_Δ leads to a difference of

$$\int dx\, u^T r u_\beta (1 - \sum_{(i)} e_i(x) e_i(x + \beta \Delta t))$$

$$= \int dx\, u^T r u_\beta \sum_{(i)} \frac{1}{2} [e_i(x) - e_i(x + \beta \Delta t)]^2.$$

FURTHER RESULTS, SYMMETRIC HYPERBOLIC EQUATIONS

The bracket in the last expression is non-zero for only a bounded finite number of values of i for any fixed value of $\mathbf{x}$. It is then $O[(\Delta t)^{1/2}]$ and hence the truth of the assertion made above follows. Thus, it remains only to obtain the inequality (5.69) for $\mathbf{v} = e_i\mathbf{u}$, having support of the order of $(\Delta t)^{1/2}$.

Taking the origin as a point in this support, we expand R there in a Taylor series,

$$(5.72) \qquad R(\mathbf{x}, \boldsymbol{\xi}) = R(0, \boldsymbol{\xi}) + \sum_{j=1}^{d} x_j R^{(j)}(\boldsymbol{\xi}) + O(\Delta t),$$

where $R^{(j)}(\boldsymbol{\xi}) = \partial_{x_j} R(\mathbf{x}, \boldsymbol{\xi})|_{\mathbf{x}=0}$; then, writing $R^{(0)}(\boldsymbol{\xi})$ for $R(0, \boldsymbol{\xi})$ and

$$\tilde{R}(\mathbf{x}, \boldsymbol{\xi}) = R^{(0)}(\boldsymbol{\xi}) + \sum_{(j)} x_j R^{(j)}(\boldsymbol{\xi}),$$

it is clear that we can reduce the problem still further, proving that

$$(\mathbf{v}, \tilde{R}_\Delta \mathbf{v}) \geq - M \Delta t \|\mathbf{v}\|^2,$$

i.e.,

$$(\mathbf{v}, R_\Delta^{(0)} \mathbf{v}) + \sum_{(j)} (\mathbf{w}_j, R_\Delta^{(j)} \mathbf{v}) \geq - M \Delta t \|\mathbf{v}\|^2,$$

where $\mathbf{w}_j = x_j \mathbf{v}$ and the suffix Δ is used to denote the difference operator derived from the corresponding symbol. Moreover, from the hypothesis $R(\mathbf{x}, \boldsymbol{\xi}) \geq 0$, there must exist some constant M_1 such that

$$R^{(0)}(\boldsymbol{\xi}) + x_j R^{(j)}(\boldsymbol{\xi}) + M_1 \Delta t I \geq 0, \quad \text{for } |x_j| \leq (\Delta t)^{1/2}.$$

Thus, we can use Lemma 2 with

$$A = R^{(0)}(\boldsymbol{\xi}) + M_1 \Delta t I, \quad B = (\Delta t)^{1/2} R^{(j)}(\boldsymbol{\xi}),$$

to obtain an inequality for the awkward terms in $R^{(j)}$; transferring this straight away into real space, because these are now all constant coefficient operators, gives

$$(\Delta t)^{1/2} |(\mathbf{w}_j, R_\Delta^{(j)} \mathbf{v})| \leq \frac{M_2}{2} [(\mathbf{v}, R_\Delta^{(0)} \mathbf{v}) + M_1 \Delta t \|\mathbf{v}\|^2]$$

$$+ \frac{1}{2M_2} [(\mathbf{w}_j, R_\Delta^{(0)} \mathbf{w}_j) + M_1 \Delta t \|\mathbf{w}_j\|^2].$$

If we take $M_2 = (\Delta t)^{1/2}/d$ and observe that $\|w_j\|^2 \leq O(\Delta t)\|v\|^2$, this becomes

$$|(w_j, R_\Delta^{(j)}v)| \leq \frac{1}{2d}(v, R_\Delta^{(0)}v) + \frac{d}{2\Delta t}(w_j, R_\Delta^{(0)}w_j) + O(\Delta t)\|v\|^2,$$

so that it remains to prove that

$$\tfrac{1}{2}(v, R_\Delta^{(0)}v) - \frac{d}{2\Delta t}\sum_{(j)}(w_j, R_\Delta^{(0)}w_j) \geq -O(\Delta t)\|v\|^2.$$

Now, from Lemma 1 with $\varphi = x_j$, we have

$$|(w_j, R_\Delta^{(0)}w_j) - (x_j^2 v, R_\Delta^{(0)}v)| \leq K(\Delta t)^2\|v\|^2,$$

with K depending only on $R_\Delta^{(0)}$; and therefore the desired result will follow from the fact that

$$([1 - (d/\Delta t)|x|^2]v, R_\Delta^{(0)}v) \geq -O(\Delta t)\|v\|^2,$$

if the support of v is contained in the circle $|x| \leq (\Delta t/2d)^{1/2}$. To see that this is so, we put $\psi(x) = (1 - (d/\Delta t)|x|^2)^{1/2}$ inside this circle and continue it outside as a smooth positive function. Its Lipschitz constant then equals $(d/\Delta t)^{1/2}$ and so by Lemma 1, again,

$$|(\psi^2 v, R_\Delta^{(0)}v) - (\psi v, R_\Delta^{(0)}\psi v)| \leq 2K(\Delta t)^2(d/\Delta t)\|v\|^2$$

while

$$(\psi v, R_\Delta^{(0)}\psi v) \geq 0.$$

5.6. Non-Linear Equations with Smooth Solutions

With non-linear equations we reach the limits of what can be stated for very general classes of difference schemes. Thus, except in special cases, very little has yet been proved about difference schemes for approximating the discontinuous solutions that frequently arise for such equations. It is only if the solution is fairly smooth that an approach is possible via the equations of first variation. In fluid dynamics, for example, this means that the approximation of shocks cannot yet be rigorously treated.

However, even though it excludes many of the physically interesting phenomena, the assumption of a smooth solution to the differential

SEC. 5.6] NON-LINEAR EQUATIONS WITH SMOOTH SOLUTIONS

equation problem does allow us to round off the theory developed in the last three chapters. Although the concept of stability as the boundedness of the solutions of the difference equations is still valid, the Lax equivalence theorem cannot be used of course, and we have to aim directly at a proof of convergence. There are several ways in which this might be done, but the simple treatment given here is due to Strang (1964a).

We consider the hyperbolic initial-value problem for the quasi-linear system

$$(5.73) \qquad \frac{\partial \mathbf{u}}{\partial t} = \sum_{j=1}^{d} A_j(\mathbf{u}, \mathbf{x}) \frac{\partial \mathbf{u}}{\partial x_j}, \quad -\infty < x_j < +\infty, 0 \leq t \leq 1,$$

where $\mathbf{u} = \mathbf{u}(\mathbf{x}, t)$ is a p-component vector whose initial value is $\mathbf{u}(\mathbf{x}, 0) = \mathbf{u}_0(\mathbf{x})$. The approximating difference scheme is written in the form

$$(5.74) \qquad \mathbf{u}^{n+1} = \varphi(\mathbf{u}^n, \mathbf{x}, \Delta t), \quad \mathbf{u}^0(\mathbf{x}, \Delta t) = \mathbf{u}_0(\mathbf{x}),$$

where the right-hand side is a non-linear function of the values $\mathbf{u}^n(\mathbf{x} + \boldsymbol{\beta}\Delta t)$ of $\mathbf{u}^n$ at a finite number of points neighboring $\mathbf{x}$; as in the previous section we take all space mesh widths as equal to Δt. Then, for consistency, we have for any smooth $\mathbf{v}$,

$$(5.75) \qquad \varphi(\mathbf{v}, \mathbf{x}, \Delta t) = \mathbf{v} + \Delta t \sum_{j=1}^{d} A_j(\mathbf{v}, \mathbf{x}) \frac{\partial \mathbf{v}}{\partial x_j} + o(\Delta t);$$

and the order of accuracy is q if, for the solution of (5.73),

$$(5.76) \qquad \|\mathbf{u}(\mathbf{x}, t + \Delta t) - \varphi(\mathbf{u}, \mathbf{x}, \Delta t)\| = O(\Delta t)^{q+1}.$$

Suppose now that we substitute the expansion

$$(5.77) \qquad \mathbf{U}(\mathbf{x}, t, \Delta t) = \mathbf{u}(\mathbf{x}, t) + \sum_{(m)} (\Delta t)^m \mathbf{U}_m(\mathbf{x}, t)$$

into $\mathbf{U}(\mathbf{x}, t + \Delta t, \Delta t) = \varphi(\mathbf{U}(\mathbf{x}, t, \Delta t), \mathbf{x}, \Delta t)$, expand the left-hand side in a Taylor series, and equate to zero the coefficient of each power of Δt, making use of (5.75). The first equation is just the differential equation (5.73) for $\mathbf{u}$; the remainder yield linear equations for the $\mathbf{U}_m$:

$$(5.78) \qquad \begin{aligned} \frac{\partial \mathbf{U}_m}{\partial t} &= \sum_{j=1}^{d} A_j(\mathbf{u}, \mathbf{x}) \frac{\partial \mathbf{U}_m}{\partial x_j} + \alpha(\mathbf{x}, t) \mathbf{U}_m + \boldsymbol{\beta}_m(\mathbf{x}, t) \\ \mathbf{U}_m(\mathbf{x}, 0) &= 0, \quad m = 1, 2, \ldots. \end{aligned}$$

Here, the coefficient α depends on the first derivatives of **u** and A_j (evaluated at **u**); and β_m depends on derivatives up to order $m + 1$ of **u**, φ, and A_j and up to order $m + 1 - j$ of $\mathbf{U}_j$, for $j < m$. Now, as a consequence of the hyperbolicity of (5.73), the solutions of equations (5.78) exist and depend continuously on the inhomogeneous terms β_m and their derivatives up to some order r_0. Thus, if we assume that **u**, φ, and A_j have $r_0 + r + 1$ derivatives, all the $\mathbf{U}_m$ for $m = 1, 2, \ldots, r$ exist: they are called the *principal error terms*[4] and

(5.79) $$\mathbf{U}(\mathbf{x}, t, \Delta t) = \mathbf{u}(\mathbf{x}, t) + \sum_{m=1}^{r} (\Delta t)^m \mathbf{U}_m(\mathbf{x}, t)$$

satisfies

(5.80) $$\mathbf{U}(\mathbf{x}, t + \Delta t, \Delta t) = \varphi(\mathbf{U}(\mathbf{x}, t\,\Delta t), \mathbf{x}, \Delta t) + o[(\Delta t)^{r+1}].$$

Then Strang's central idea is to compare this function **U** instead of **u** with the difference approximation $\mathbf{u}^n$. The advantage is that we can make the perturbation to the difference equation (5.74) arbitrarily small by taking r sufficiently large. The need for this arises because we have to switch from the L_2 to the maximum norm in estimating the difference between φ and its first variation.

Hence, let us denote by $\mathbf{E}^n$ the difference $\mathbf{U}(n\Delta t) - \mathbf{u}^n$ and subtract equation (5.74) from (5.80) at $t = n\Delta t$: we denote by $c^\beta(\mathbf{v})$ the Jacobian of $\varphi(\mathbf{v}, \mathbf{x}, \Delta t)$ with respect to the argument corresponding to the β-neighbor $\mathbf{v}(\mathbf{x} + \beta \Delta t)$ and, using the mean-value theorem, obtain

(5.81)
$$\mathbf{E}^{n+1}(\mathbf{x}) = \sum_{(\beta)} c^\beta(\mathbf{U}(n\Delta t) - \theta \mathbf{E}^n)\mathbf{E}^n(\mathbf{x} + \beta \Delta t) + o[(\Delta t)^{r+1}]$$
$$\mathbf{E}^0(\mathbf{x}) = 0,$$

where θ lies between 0 and 1 and depends on n, β, **x**. Now, the linear difference operator $C^n = C(n\Delta t, \Delta t)$ corresponding to the coefficients c^β evaluated at $\mathbf{u}(\mathbf{x}, n\Delta t)$ is called the *first variation* of the operator φ. We assume it is stable, i.e.,

$$\left\| \prod_{i=1}^{n} C^i \right\| \leq K$$

for $n\Delta t \leq 1$. Then, we show by induction that if the order of accuracy

---

[4] Notice that these error terms are independent of Δt, hence this is true for the bounds on their derivatives required below.

SEC. 5.6] NON-LINEAR EQUATIONS WITH SMOOTH SOLUTIONS

of φ is q and r is chosen to be $q + [(d + 1)/2]$, where [ ] denotes "integral part of",

(5.82) $$\operatorname*{Max}_{\mathbf{x}} |\mathbf{E}^n(\mathbf{x}, \Delta t)| \leq n(\Delta t)^{q+1}, \quad n\Delta t \leq 1.$$

It is certainly true for $n = 0$; assume it is true for $n < N$. Then, for these n, if we write (5.81) in the form

$$\mathbf{E}^{n+1} = (C^n + B^n)\mathbf{E}^n + o[(\Delta t)^{r+1}],$$

it follows from (5.79) and the fact that $q \geq 1$ that

(5.83) $$\|B^n\| \leq \mu\Delta t$$

for some constant μ. Hence, recalling the argument of Section 3.9,

$$\mathbf{E}^N = \sum_{i=1}^{N-1} \prod_{j=i}^{N-1} (C^j + B^j) o[(\Delta t)^{r+1}]$$

and

(5.84) $$\left\| \prod_{j=i}^{N-1} (C^j + B^j) \right\| \leq Ke^{\mu K}.$$

This product, for $i = 1$, represents an infinite matrix with $O(N^d)$ non-zero elements in each row, connecting the value of $\mathbf{E}^N$ at some point at time level N to the truncation errors at $O(N^d)$ points at time level 1. The above bound is on the L_2-norm; the bound on the corresponding maximum norm will therefore be increased by a factor of $O(N^{d/2})$. Thus,

$$\operatorname*{Max}_{\mathbf{x}} |\mathbf{E}^N| \leq NKe^{\mu K}O(N^{d/2})o[(\Delta t)^{r+1}]$$

$$\leq N(\Delta t)^{q+1},$$

for $\Delta t < \delta$, where δ is independent of N. This completes the induction, and the proof that $\mathbf{u}^n(\mathbf{x})$ converges to $\mathbf{u}(\mathbf{x}, n\Delta t)$ as $\Delta t \to 0$. We summarize the result in the theorem: *Suppose that in the equation* (5.74), φ *is a consistent operator and that its first variation is L_2-stable. Then if $\mathbf{u}$, A_j and φ have continuous derivatives up to order $[(d + 1)/2] + r_0 + 2$, the difference approximation $\mathbf{u}^n$ converges to the solution of (5.73) as $\Delta t \to 0$. If φ has order of accuracy q and the number of assumed derivatives is $[(d + 1)/2] + r_0 + q + 1$, then by considering the order of magnitude of the β_k one can show that $\mathbf{u}^n(x) - \mathbf{u}(\mathbf{x}, n\Delta t) = O[(\Delta t)^q]$.*

Despite the generality of this theorem, we must not delude the reader into believing that all is therefore plain sailing with non-linear equations if their solutions are sufficiently smooth. And the fact that this is not so is a measure of the limitation of the ideas of stability and convergence when applied to practical computations. For one thing the effect of boundary conditions has not so far been considered. The following example, given by Richtmyer (1963), will serve to illustrate the sort of situation that can arise even in quite simple cases.

The differential equation

(5.85) $$\frac{\partial u}{\partial t} + \frac{\partial}{\partial x}(\tfrac{1}{2}u^2) = 0, \quad 0 \leq x \leq 1, t \geq 0,$$

with initial and boundary conditions

(5.86) $$u(x, 0) = x \quad \text{for } 0 \leq x \leq 1$$
$$u(0, t) = 0 \quad \text{for } t \geq 0,$$

has the exact solution

(5.87) $$u(x, t) = x/(1 + t).$$

Now, suppose we attempt to solve this problem using the "leap-frog" difference equation

(5.88) $$u_j^{n+1} - u_j^{n-1} + \frac{\lambda}{2}[(u_{j+1}^n)^2 - (u_{j-1}^n)^2] = 0$$

with $\Delta t = \lambda \Delta x$. Then, we can show that this equation has solutions which "explode" like $\exp(\text{const.}/\Delta t)$ as $\Delta t \to 0$ for fixed t, even if the von Neumann condition for the linearized equation,

(5.89) $$\lambda \operatorname*{Max}_{x, t} \{|u|\} < 1,$$

is satisfied. For consider a solution of the form

(5.90) $$u_j^n = C^n \cos \tfrac{1}{2}\pi j + S^n \sin \tfrac{1}{2}\pi j + U^n \cos \pi j + V.$$

Substitution leads to the recurrence relations

(5.91) $$C^{n+1} - C^{n-1} = 2\lambda S^n(U^n - V)$$
$$S^{n+1} - S^{n-1} = 2\lambda C^n(U^n + V)$$
$$U^{n+1} - U^{n-1} = 0.$$

According to the third of these, U^n takes on two constant values, say A and B, on alternate cycles. Then the first two relations give the equation

(5.92) $$C^{n+2} - 2C^n + C^{n-2} = 4\lambda^2(A + V)(B - V)C^n$$

and a similar equation for S^n. If the coefficient on the right lies between -4 and 0, all solutions are bounded; otherwise, there is always a solution C^n such that $|C^n|$ grows exponentially with n, that is, with $t/\Delta t$.

To see when the exponential solution can occur, we note that the von Neumann condition (5.89) implies that

$$\lambda\{|V| + \text{Max}\,(|A|, |B|)\} < 1,$$

which shows that the coefficient of C^n on the right of (5.92) can never be less than -4. If, furthermore,

(5.93) $$|A| < |V| \quad \text{and} \quad |B| < |V|,$$

the coefficient can never be > 0. Therefore, if we regard the constant term V in (5.90) as representing a smooth solution on which the other terms have been superposed as a perturbation, then, if the amplitude of the perturbation is below a limiting value or threshold set by (5.93), it will not grow; but if the perturbation is initially above this threshold, we may expect it to grow exponentially.

Test calculations reported by Richtmyer and Morton (1964) and by Stetter (1965) on the problem given by (5.85) and (5.86) illustrate this and a number of other phenomena. Use of the leap-frog scheme (5.88) in several cases yielded solutions which "exploded" violently while, for the corresponding linearized problem, the scheme either appeared stable or showed only a very slight error growth. The conclusions to be drawn from these examples were not clear-cut due partly to the effect of boundary conditions. But the probable explanation is that the leap-frog scheme provides a convergent approximation in both cases *for sufficiently small* Δt, i.e., when the truncation error in the difference scheme generates sufficiently small perturbations. But in the non-linear problem, the behavior depends markedly on the relative magnitude of the perturbations, as cannot happen in a linear problem. In any case, the practical difference in the two problems is certainly very great. We shall see in the later chapters that in many other problems stability properties as $\Delta t \to 0$ are not always sufficient for practical needs.

It will be noted that the leap-frog scheme, even in the linearized case, is only "marginally stable," i.e., the eigenvalues of its amplification matrix have modulus unity. It therefore is impossible to apply to it the theorems of Kreiss given above. However, extra higher order difference terms may be added to (5.88) to make it dissipative in his sense without affecting the order of accuracy. Richtmyer and Morton then found that the coefficients of these terms could be chosen to ensure stability even in the non-linear case. It should also be noted that although we are dealing with only a scalar variable here, the theorem of Lax and Nirenberg cannot be applied to the linearized case either. The leap-frog scheme is a two-step scheme and, when converted to a one-step scheme (see Chapter 7) the solution vector has two components; the amplification matrix for this does not satisfy the hypothesis of their theorem.

CHAPTER 6

Mixed Initial-Boundary-Value Problems

6.1. Introduction

The numerical analyst, asked to provide guidance in the choice of stable finite difference schemes, has essentially two courses open to him. He can attempt to devise a criterion capable of deciding the stability of any proposed difference scheme, or he can suggest a class of methods whose stability he will guarantee. The former course is the more ambitious and is the one we have adopted so far in our study of stability. But in the last chapter we found it increasingly difficult to maintain. Thus, for linear hyperbolic equations, we found that the theorems of general validity needed to assume either very smooth coefficients plus a stringent local stability condition or a dissipativity condition. In both cases, the proofs of the theorems were of considerable complexity. The dissipativity condition is perhaps acceptable when one is really considering linearized versions of non-linear equations, since numerical experiments, such as those mentioned in Section 5.6, suggest this condition may then be very desirable even if not strictly necessary. But for many linear difference schemes of practical importance it would appear that the hypotheses of these general theorems are too stringent: in fact, one would often be surprised to find that Lipschitz continuous coefficients made any practical difference to stability conditions compared with constant coefficients. Thus, it appears that, as is so often the case, in our aim at generality we are losing some of the precision possible in particular cases.

Hence, in this chapter we turn our attention more to the study of particular classes of difference schemes. In this way we shall in addition be able to start considering problems in bounded domains with non-periodic boundary conditions and to study the effect that the treatment of their boundary conditions has on stability. The introduction of boundaries, besides affecting the analytical tools that can be used, raises two new difficulties in the treatment of stability. First, in showing that no normal mode of the difference equation is magnified from step to step, one has to consider modes other than those which have constant amplitude in the space direction, i.e., the Fourier modes for the constant coefficient problems; for modes which decay away from the boundary may also be acceptable normal modes so long as they satisfy the boundary conditions. Indeed, these latter have to be sufficient to eliminate all such modes that grow with time, if stability (or well-posedness) is to be assured. Second, many difference schemes demand for their solution more boundary conditions than are needed for the corresponding differential equation problem. One is therefore faced with the problem of choosing these extra conditions in such a way that stability is not adversely affected.

6.2. Basic Ideas of the Energy Method

We begin our consideration of these problems by studying in a little more detail than hitherto the so-called energy method. This is a rather confusing name given to a loose collection of techniques based on the concept of strong stability defined and discussed at the end of Section 5.2. The essential idea is to devise a norm for the solution vector which one can then show increases by a factor no greater than $1 + O(\Delta t)$ at each time step. This implies stability in this norm and the argument is then usually rounded off by demonstrating the equivalence of the norm to the L_2 norm. The reason for the name is that in certain fairly simple cases the physical energy of the system provides such a norm.

Such methods have long been used in the theory of differential equations to prove existence and uniqueness of solutions—see, for example, Friedrichs (1954, 1958) and Phillips (1957, 1959), where earlier references are given. Their use for difference equations is also of long standing, beginning with Courant, Friedrichs and Lewy (1928). More recently, they have been used by many authors, including Friedrichs (1954), Lees (1960), Lax (1961) and Kreiss (1963).

The methods can sometimes be used to obtain quite general results, and we used them in this way in the previous chapter. In his early studies of variable coefficient problems, Lax (1960) discussed the range of applicability of the energy method and showed that under certain circumstances

BASIC IDEAS OF THE ENERGY METHOD

a norm with the desired properties always exists, and its form can be written down quite explicitly in terms of the solution vector and a certain number of its differences. The circumstances under which this was possible, however, were rather limited. We know from the Kreiss matrix theorem that such norms exist, at least for constant coefficient problems: the difficulty lies in their explicit representation in terms of the solution and its differences.

We therefore shall regard the energy method rather as a set of techniques to be applied to each difference scheme (or class of difference schemes) individually, so as to obtain a norm with the desired properties for that particular scheme. Thus, for a single equation in one space dimension $B_1 u^{n+1} = B_0 u^n$, the procedure consists basically of two steps. First, by squaring both sides of the equation and summing over mesh points, or by taking appropriate inner products, we look for a relation of the form

$$(6.1) \qquad S_{n+1} = S_n + O(\Delta t)(\|u^n\|^2 + \|u^{n+1}\|^2),$$

where $S_n = S(u^n)$ is some quadratic form in u^n and its differences. Then, we have to show that S_n is positive definite and equivalent to the square of the L_2 norm, $\|u^n\|^2$: it is in doing this that any stability condition on the time step will appear. Note that by working with the difference between u^n and the solution of the differential equation an estimate of the error growth and the rate of convergence can be obtained directly by this method.

It is clear that in this way we shall in general obtain only a sufficient condition for stability. The actual limit to the stability of a method may elude us either because of our lack of ingenuity in devising a norm or because of the inherent limitations of the method. Thus, to some extent the method is complementary to normal mode analysis which tends to yield necessary stability conditions. Compensations for these limitations lie mainly in the ease with which variable coefficients and boundary conditions can be dealt with. In addition, we shall find later examples in which the concept of stability, defined so far as a property holding in the limit as $\Delta t \to 0$, can be misleading in practice. The energy method can answer the more practical question—given a finite Δx, for what values of Δt does the norm increase no faster than exponentially with time t?

In keeping with the spirit of carefully tailoring the norm to the individual difference method, these points are best shown by particular examples, several of which are given in the next section. These will also demonstrate another feature of the use of the energy method, that the difference scheme can (and often should) be chosen so as to facilitate the application of the method. This is particularly so in multi-dimensional problems.

First, we need to introduce some notation and to derive a few basic inequalities and identities. For simplicity, we confine ourselves here to problems in only one space variable x, for which we take as a standard range the interval $0 \leq x \leq 1$; allowing either or both boundaries to be at infinity usually leads only to simplification, and the necessary modifications to the statements below should be clear. We cover the range of x with a mesh of uniform spacing Δx whose position relative to the boundaries now has to be chosen and fixed, with the result that the solution $\mathbf{u}^n(x)$ is obtained only at the mesh points, where it is denoted by $\mathbf{u}_j^n$ with j running from 0 to J. The choice of the mesh at these end points is largely indicated by the boundary conditions that have to be applied and only two arrangements will be considered here:[1] either an end point is a mesh point, then called a *boundary point*, which is useful when the boundary condition has to be imposed on the function value itself; or it lies midway between two mesh points, which is of most use when the boundary condition involves the function derivative. Mesh points in the interior of the range are called *interior points* and, in the second case, it is sometimes useful to introduce fictitious points outside the range of x which are then called *exterior points*. Note that the values that Δx can take on are limited by the choice of boundary arrangement, e.g., when both end points are boundary points, one must have $(\Delta x)^{-1} = J$, a positive integer.

There is a great deal of variation in the notation used in the literature for finite differences. For our purposes it is advantageous to denote differencing by an explicit operator, rather than, for example, adding to the multiplicity of subscripts on a differenced function. Moreover, we prefer to use undivided differences in a notation closely approximating that used in older books on the subject. Although most of the notation

---

[1] In practical problems, the mesh may well differ for different components of the vector $\mathbf{u}_j^n$, depending on the convenience of approximating the differential equations and the boundary conditions, but such complications would be out of place here.

SEC. 6.2]   BASIC IDEAS OF THE ENERGY METHOD

has already been introduced at various points in the book, it is convenient to collect it together here:

(6.2) $$\Delta_+ u_j = u_{j+1} - u_j, \quad \Delta_- u_j = u_j - u_{j-1},$$

(6.3) $$\Delta_0 u_j = \tfrac{1}{2}(\Delta_+ + \Delta_-)u_j = \tfrac{1}{2}(u_{j+1} - u_{j-1}),$$

(6.4) $$\delta^2 u_j = \Delta_+ \Delta_- u_j = \Delta_- \Delta_+ u_j = u_{j+1} - 2u_j + u_{j-1}.$$

When several space variables are involved, they are distinguished by subscripts on the difference operator. In addition, we occasionally make use of the translation operators

(6.5) $$T_+ u_j = u_{j+1}, \quad T_- u_j = u_{j-1},$$

as, for instance, in the identities for differencing a product:

(6.6) $$\Delta_+(au) = a\Delta_+ u + (\Delta_+ a)T_+ u = T_+(a\Delta_- u) + (\Delta_+ a)u,$$

(6.7) $$\Delta_-(au) = a\Delta_- u + (\Delta_- a)T_- u = T_-(a\Delta_+ u) + (\Delta_- a)u,$$

(6.8) $$\Delta_0(au) = a\Delta_0 u + (\Delta_0 a)u + \tfrac{1}{2}\Delta_-[(\Delta_+ a)(\Delta_+ u)].$$

Although these identities are given only in their scalar forms, they are also clearly true when a is a matrix and u is a vector.

With the solution function $\mathbf{u}^n$ no longer defined at every point of the x space, we cannot use the normal integral definition of the L_2 norm. Instead, we use as a norm the following sum over mesh points:

(6.9) $$(\mathbf{u}, \mathbf{v})_\Delta = \Delta x \sum_{(j)} \mathbf{u}_j^* \mathbf{v}_j, \quad \|\mathbf{u}\|_\Delta^2 = (\mathbf{u}, \mathbf{u})_\Delta,$$

where the summation is to be carried out over interior points only: the subscript Δ will be omitted when there is no danger of confusion with the usual L_2 norm. This may seem at variance with the principle stated in Chapter 2 of embedding the mesh function $\mathbf{u}^n$ in the function space for the differential problem. But in reality this is not so. For example, when the mesh is such that there are no boundary points, i.e., the boundaries are either at infinity or lie midway between two mesh points, $\mathbf{u}^n$ may be embedded in this way by considering it as a step function with the value $\mathbf{u}_j^n$ extending half a mesh spacing either side of the jth mesh point: (6.9) is then identical to the L_2 definition. On the other hand, if both $x = 0$ and $x = 1$ are boundary points at each of which the homogeneous condition $\mathbf{u}^n = 0$ is imposed, the sums in (6.9) are equivalent to approximating the corresponding integrals by the trapezoidal rule. But if $\mathbf{u}^n$ had been embedded in this case by linear interpolation, we would have obtained contributions to the integral for the L_2 norm of the form

$$\int_0^1 |\mathbf{u}_j^n + (\mathbf{u}_{j+1}^n - \mathbf{u}_j^n)t|^2 dt = \tfrac{1}{3}(|\mathbf{u}_j^n|^2 + |\mathbf{u}_{j+1}^n|^2 + \operatorname{Re} \mathbf{u}_j^{n*}\mathbf{u}_{j+1}^n),$$

which lies between a sixth and a half of $|\mathbf{u}_j^n|^2 + |\mathbf{u}_{j+1}^n|^2$. Hence, here too the two norms are equivalent, and the same can be shown to be true for intermediate cases. All we have done is to vary the embedding technique with the problem in hand in such a way that (6.9) is equivalent to the L_2 norm.

When differences of $\mathbf{u}$ occur in inner products, the range of summation may have to be modified: it is therefore convenient to extend the notation to

(6.10) $\qquad (\mathbf{u},\mathbf{v})_{r,s} = \Delta x \sum_{j=r}^{s} \mathbf{u}_j^* \mathbf{v}_j, \qquad \|\mathbf{u}\|^2_{r,s} = (\mathbf{u},\mathbf{u})_{r,s}.$

Thus, for example, we can easily show that

(6.11) $\qquad \|\Delta_+^n \mathbf{u}\|^2_{r,s} \leq 4^n \|\mathbf{u}\|^2_{r,s+n}, \qquad \|\Delta_-^n \mathbf{u}\|^2_{r,s} \leq 4^n \|\mathbf{u}\|^2_{r-n,s},$

while

(6.12) $\qquad \|\Delta_0^n \mathbf{u}\|^2_{r,s} \leq \|\mathbf{u}\|^2_{r-n,s+n}.$

Also, if A is a matrix, there follows from (6.6) and (6.7) the three inequalities

(6.13) $\quad |(\mathbf{u}, \Delta_+(A\mathbf{v}))_{r,s} - (\mathbf{u}, A\Delta_+\mathbf{v})_{r,s}| \leq M_+\Delta x \|\mathbf{u}\|_{r,s}\|\mathbf{v}\|_{r+1,s+1},$

(6.14) $\quad |(\mathbf{u}, \Delta_-(A\mathbf{v}))_{r,s} - (\mathbf{u}, A\Delta_-\mathbf{v})_{r,s}| \leq M_-\Delta x \|\mathbf{u}\|_{r,s}\|\mathbf{v}\|_{r-1,s-1},$

(6.15) $\quad |(\mathbf{u}, \Delta_0(A\mathbf{v}))_{r,s} - (\mathbf{u}, A\Delta_0\mathbf{v})_{r,s}| \leq M_0\Delta x \|\mathbf{u}\|_{r,s}\|\mathbf{v}\|_{r-1,s+1},$

where M with an appropriate subscript is a bound for $|\partial A/\partial x|$ over an interval containing all the mesh points involved in the corresponding formula.

The most important formulas in the manipulation involved in the energy method are, however, the summation-by-parts formulas, corresponding to the formula for integration by parts in the differential problem.

SEC. 6.3] SIMPLE EXAMPLES OF THE ENERGY METHOD 137

We have already used them extensively in cases where we could neglect the boundary terms. Here, we have to be more careful, and it is convenient to introduce the further notation

(6.16) $\quad \mathbf{u}_i^* \mathbf{v}_j \big|_r^s = \mathbf{u}_{s+i}^* \mathbf{v}_{s+j} - \mathbf{u}_{r-i}^* \mathbf{v}_{r-j}.$

The formulas themselves may take several forms, each of which can be simply derived by summing one of the identities (6.6) to (6.8): the basic pair are

(6.17) $\quad (\mathbf{u}, \Delta_+ \mathbf{v})_{r,s} = -(\Delta_- \mathbf{u}, \mathbf{v})_{r+1,s+1} + \Delta x \cdot \mathbf{u}_0^* \mathbf{v}_0 \big|_r^{s+1},$

(6.18) $\quad (\mathbf{u}, \Delta_- \mathbf{v})_{r,s} = -(\Delta_+ \mathbf{u}, \mathbf{v})_{r-1,s-1} + \Delta x \cdot \mathbf{u}_0^* \mathbf{v}_0 \big|_{r-1}^s;$

adding these together gives

(6.19) $\quad (\mathbf{u}, \Delta_0 \mathbf{v})_{r,s} = -(\Delta_0 \mathbf{u}, \mathbf{v})_{r,s} + \tfrac{1}{2}\Delta x (\mathbf{u}_1^* \mathbf{v}_0 + \mathbf{u}_0^* \mathbf{v}_1)\big|_r^s.$

A little further manipulation leads to

(6.20) $\quad (\mathbf{u}, \Delta_+ \mathbf{v})_{r,s} = -(\Delta_+ \mathbf{u}, \mathbf{v})_{r,s} - (\Delta_+ \mathbf{u}, \Delta_+ \mathbf{v})_{r,s} + \Delta x \cdot \mathbf{u}_0^* \mathbf{v}_0 \big|_r^{s+1},$

(6.21) $\quad (\mathbf{u}, \Delta_- \mathbf{v})_{r,s} = -(\Delta_- \mathbf{u}, \mathbf{v})_{r,s} + (\Delta_- \mathbf{u}, \Delta_- \mathbf{v})_{r,s} + \Delta x \cdot \mathbf{u}_0^* \mathbf{v}_0 \big|_{r-1}^s.$

6.3. Simple Examples of the Energy Method: Stable Choice of Approximations to Boundary Conditions and to Non-Linear Terms

As a first example, consider the mixed initial-boundary-value problem for the real variable $u(x, t)$:

(6.22) $\quad \dfrac{\partial u}{\partial t} + a(x) \dfrac{\partial u}{\partial x} = 0, \quad 0 \leq x < \infty, \, 0 \leq t \leq 1$

with

$$u(x, 0) = f(x), \quad 0 \leq x < \infty,$$
$$u(0, t) = 0, \quad 0 \leq t \leq 1,$$

where $A \geq a(x) \geq \alpha > 0$. We approximate this with the four-point difference scheme—compare Thomée (1962):

(6.23) $\quad u_j^{n+1} + u_{j-1}^{n+1} + a_{j-\frac{1}{2}} \Delta_- u_j^{n+1} = u_j^n + u_{j-1}^n - a_{j-\frac{1}{2}} \Delta_- u_j^n,$
$\qquad n = 0, 1, 2, \ldots, j = 1, 2, \ldots,$

where, with no loss of generality, we have taken $\Delta x = \Delta t$. To apply the energy method, we square each side and sum over all j. Thus, putting

$$(6.24) \qquad S_n = \sum_{j=1}^{\infty} [(u_j^n + u_{j-1}^n)^2 + a_{j-\frac{1}{2}}^2 (u_j^n - u_{j-1}^n)^2]$$

$$(6.25) \qquad = \|(1 + T_-)u^n\|^2 + \|a_{-\frac{1}{2}}\Delta_- u^n\|^2,$$

where $a_{-\frac{1}{2}}$ is used to denote $(T_-)^{\frac{1}{2}} a$, we obtain

$$S_{n+1} - S_n = -2([1 + T_-]u^n, a_{-\frac{1}{2}}\Delta_- u^n) - 2([1 + T_-]u^{n+1}, a_{-\frac{1}{2}}\Delta_- u^{n+1}).$$

Now it is clear that

$$(\Delta_- w_j)[1 + T_-]w_j = w_j^2 - w_{j-1}^2 = \Delta_- w_j^2,$$

so that

$$S_{n+1} - S_n = -2(\Delta_-[(u^n)^2 + (u^{n+1})^2], a_{-\frac{1}{2}}).$$

After summing by parts, using (6.18) and the boundary condition, this becomes

$$(6.26) \qquad S_{n+1} - S_n = 2((u^n)^2 + (u^{n+1})^2, \Delta_+ a_{-\frac{1}{2}})$$

$$\leq 2L\Delta x(\|u^n\|^2 + \|u^{n+1}\|^2),$$

where L is a Lipschitz constant for the coefficient $a(x)$. Hence, if a is a constant, S_n remains constant and, otherwise, it increases at most exponentially—at least this will be the case if we can show that S_n is equivalent to the L_2 norm. But this is straightforward, for-from (6.24)

$$(6.27) \qquad 4 \operatorname{Min}(1, \alpha^2)\|u^n\|^2 \leq S_n \leq 4 \operatorname{Max}(1, A^2)\|u^n\|^2.$$

It follows then that

$$4 \operatorname{Min}(1, \alpha^2)\|u^n\|^2 \leq S_n \leq S_{n-1} + 2L\Delta x(\|u^n\|^2 + \|u^{n-1}\|^2)$$

$$\leq \cdots \leq S_0 + 4L\Delta x \sum_{m=0}^{n} \|u^m\|^2,$$

i.e.,

$$\|u^n\|^2 \leq K_1 \|u^0\|^2 + K_2 \Delta x \sum_{m=0}^{n} \|u^m\|^2,$$

SEC. 6.3] SIMPLE EXAMPLES OF THE ENERGY METHOD 139

where $K_1 = \text{Max}(1, A^2)/\text{Min}(1, \alpha^2)$ and $K_2 = L/\text{Min}(1, \alpha^2)$. This implies that

$$\|u^n\|^2 \leq (K_1 + K_2 \Delta x)(1 - K_2 \Delta x)^{-n} \|u^0\|^2$$
$$\leq \text{const.} \|u^0\|^2$$

for $n\Delta x = n\Delta t \leq 1$ and sufficiently small Δt.

Thus, we have been able to prove the unconditional stability of a difference scheme which is non-dissipative in the sense of the last chapter, yet involves a variable coefficient. Moreover, this was done for a domain which was bounded on the left; and the proof can easily be extended to the case of the finite interval, $0 \leq x \leq 1$, for which no boundary condition is needed on the right either for the differential or the difference equation. In that case, if $J\Delta x = 1$, the sum for S_n in (6.24) is extended over $j = 1, 2, \ldots, J$, and we take the same range for the norm. The inequalities (6.27) are unchanged except that the lower bound is halved,

$$S_n \geq 2 \text{ Min}(1, \alpha^2) \left[\sum_{j=1}^{J} |u_j^n|^2 + \sum_{j=1}^{J-1} |u_j^n|^2 \right]$$
$$\geq 2 \text{ Min}(1, \alpha^2) \|u^n\|^2;$$

moreover, the inequality (6.26) is preserved because of the sign of a:[2]

$$S_{n+1} - S_n = -2(\Delta_-[(u^n)^2 + (u^{n+1})^2], a_{-\frac{1}{2}})_{1,J}$$
$$= 2((u^n)^2 + (u^{n+1})^2, \Delta_+ a_{-\frac{1}{2}})_{0,J-1} - 2\Delta x[(u_j^n)^2 +$$
$$(u_j^{n+1})^2] a_{J-\frac{1}{2}}$$
$$\leq 2L\Delta x(\|u^n\|^2 + \|u^{n+1}\|^2).$$

The remainder of the argument can then be carried through unchanged.

For future reference, it is worth stating this final argument as a general lemma:

If there exists a sequence of real numbers S_n and a pair of positive constants K_1 and K_2 such that

(6.28) $\qquad K_1^{-1} \|u^n\|^2 \leq S_n \leq K_1 \|u^n\|^2,$

---

[2] Note the correspondence with our physical intuition here. If we had had $a(1) < 0$, the boundary condition $u_j^n = 0$ would have been clearly called for.

and

(6.29) $\quad S_{n+1} - S_n \leq K_2 \Delta t (\|u^n\|^2 + \|u^{n+1}\|^2), \quad n = 0, 1, 2, \ldots,$

then

$$\|u^n\|^2 \leq K_1(K_1 + 2K_2\Delta t)(1 - 2K_1 K_2 \Delta t)^{-n} \|u^0\|^2$$
$$\leq \text{const.} \|u^0\|^2$$

for $n\Delta t \leq T$ and sufficiently small Δt.

As a second example, let us apply the leap-frog scheme to the above problem:

(6.30) $\quad u_j^{n+1} - u_j^{n-1} = -2a_j \Delta_0 u_j^n, \quad j = 1, 2, \ldots.$

This is a three-level or two-step method whose formal theory we do not take up until the next chapter, but we can deal with this simple example in an *ad hoc* manner. Initial values for both u^0 and u^1 have to be provided, then for $n \geq 1$ we multiply (6.30) by $u_j^{n+1} + u_j^{n-1}$ and sum over $j = 1, 2, \ldots$ to get

$$\|u^{n+1}\|^2 - \|u^{n-1}\|^2 = -2(u^{n+1} + u^{n-1}, a\Delta_0 u^n).$$

We take

(6.31) $\quad S_n = \|u^n\|^2 + \|u^{n+1}\|^2 + 2(u^{n+1}, a\Delta_0 u^n)$

and obtain

$$S_n - S_{n-1} = -2(u^{n-1}, a\Delta_0 u^n) - 2(u^n, a\Delta_0 u^{n-1}).$$

Now, from equations (6.6), (6.7) and (6.19),

$$2(u^{n-1}, a\Delta_0 u^n) = 2(u^{n-1}, \Delta_0(au^n)) - (u^{n-1}, (T_+ u^n)\Delta_+ a + (T_- u^n)\Delta_- a)$$
$$= -2(\Delta_0 u^{n-1}, au^n) - (u^{n-1}, (T_+ u^n)\Delta_+ a + (T_- u^n)\Delta_- a).$$

Hence,

(6.32) $\quad S_n - S_{n-1} = (u^{n-1}, (T_+ u^n)\Delta_+ a + (T_- u^n)\Delta_- a)$
$$\leq 2L\Delta x \|u^{n-1}\| \|u^n\|$$
$$\leq L\Delta x (\|u^{n-1}\|^2 + \|u^n\|^2).$$

Further,

(6.33) $\quad 2|(u^{n+1}, a\Delta_0 u^n)| \leq A(\|u^{n+1}\|^2 + \|u^n\|^2);$

SEC. 6.3] SIMPLE EXAMPLES OF THE ENERGY METHOD

so if $A < 1$, which is really the Courant-Friedrichs-Lewy condition $A\Delta t/\Delta x < 1$, we have

(6.34) $\qquad K^{-1}(\|u^n\|^2 + \|u^{n+1}\|^2) \leq S_n \leq K(\|u^n\|^2 + \|u^{n+1}\|^2)$

for some constant $K > 0$. Stability, in the form of an inequality $\|u^n\|^2 \leq$ const. $(\|u^0\|^2 + \|u^1\|^2)$, then follows from arguments like those above. Notice that this time the stability depends on a condition which arises from the need for S_n to be positive definite.

When the finite interval $0 \leq x \leq 1$ is considered with this difference scheme, an extra boundary condition is needed at $x = 1$ to enable the difference equations (6.30) to be solved. To see how this should be chosen, we retrace the steps above with $J\Delta x = 1$ and (6.30) holding for $j = 1, 2, \ldots, J - 1$: norms and inner products are also to be summed over this range of j. Then (6.33) no longer holds, because of the term in u_J^n on the left-hand side, and to preserve (6.34) we define S_n as

$$S_n = \|u^n\|^2 + \|u^{n+1}\|^2 + 2(u^{n+1}, a\Delta_0 u^n) - a_{J-1} u_{J-1}^{n+1} u_J^n \Delta x.$$

In addition, we find that the terms involving u_J in the expression

$$2(u^{n-1}, a\Delta_0 u^n) + 2(u^n, a\Delta_0 u^{n-1})$$

amount to $a_{J-1}(u_{J-1}^{n-1} u_J^n + u_{J-1}^n u_J^{n-1})\Delta x$. Hence, (6.32) becomes

(6.35) $\qquad S_n - S_{n-1} \leq L\Delta x(\|u^{n-1}\|^2 + \|u^n\|^2) - a_{J-1} u_J^n(u_{J-1}^{n-1} + u_{J-1}^{n+1})\Delta x.$

It is clear that putting

(6.36) $\qquad u_J^n = \tfrac{1}{2}(u_{J-1}^{n-1} + u_{J-1}^{n+1})$

will not only allow the difference equations to be solved but will also ensure stability: the equation for the end mesh point reduces to

(6.37) $\qquad u_{J-1}^{n+1} = (1 + \tfrac{1}{2}a_{J-1})^{-1}[(1 - \tfrac{1}{2}a_{J-1})u_{J-1}^{n-1} + a_{J-1} u_{J-2}^n],$

which has the desirable property of having positive coefficients adding up to unity.

It is worth noting that the rather obvious procedure of extrapolating with respect to j to obtain u_J^n could lead to troublesome instability. This would give, instead of (6.37),

(6.38) $\qquad u_{J-1}^{n+1} = u_{J-1}^{n-1} - 2a_{J-1}(u_{J-1}^n - u_{J-2}^n).$

Hence, if $a(x) \equiv 1$ and the initial data satisfies

$$u_j^n = (-1)^{j+n}, \quad n = 0, 1, \quad j = 0, 1, \ldots, J - 1,$$

near the right-hand boundary, the solution would continue as

$$u_j^n = (-1)^{j+n} + (-1)^j f(j + n),$$

where $f(k + J) = 0$ for $k \leq 0$ and substitution into (6.38) gives

$$f(k + J) = (-1)^{k-1} 2k(k + 1) \quad \text{for } k > 0.$$

The energy method can be of great help in avoiding pitfalls of this kind.

Before leaving the simple problem of this section, it should be noted that replacement of $a(x)$ in (6.22) by $u(x, t)$ yields a very commonly occurring non-linear operator. In approximating it by some difference schemes one is led to the choice of whether to replace $u\partial u/\partial x$ by $\tfrac{1}{2}(\Delta x)^{-1} \Delta_0 u^2$ or $(\Delta x)^{-1} u\Delta_0 u$, which are equally attractive from the point of view of accuracy. A preferred linear combination is sometimes indicated by the energy method, however. For example, consider the Crank-Nicholson scheme

$$(6.39) \qquad u_j^{n+1} + \tfrac{1}{2} Q(u_j^{n+1}) = u_j^n - \tfrac{1}{2} Q(u_j^n),$$

where $Q(u)$ is a difference approximation to $(\Delta x) u \partial u/\partial x$. Squaring and summing both sides gives

$$\|u^{n+1}\|^2 + \tfrac{1}{4}\|Q(u^{n+1})\|^2 - [\|u^n\|^2 + \tfrac{1}{4}\|Q(u^n)\|^2]$$
$$= -(u^{n+1}, Q(u^{n+1})) - (u^n, Q(u^n)).$$

If the operator $Q(u)$ is $\tfrac{1}{2}\Delta_0 u^2$ and we neglect boundary terms, we have

$$(u, \tfrac{1}{2}\Delta_0 u^2) = -\tfrac{1}{2}(u^2, \Delta_0 u);$$

while for the other choice $u\Delta_0 u$,

$$(u, u\Delta_0 u) = (u^2, \Delta_0 u).$$

Hence, by taking

$$(6.40) \qquad Q(u) = \tfrac{1}{3}(\Delta_0 u^2 + u\Delta_0 u) = \tfrac{1}{3}(u + T_+ u + T_- u)\Delta_0 u,$$

we have $(u, Q(u)) = 0$ and $\|u^n\|^2 + \tfrac{1}{4}\|Q(u^n)\|^2$ remains constant, thus ensuring stability.

6.4. Coupled Sound and Heat Flow

Lest we give too over-simplified a picture of the use of the energy method, we consider now a more complicated system of equations. In doing so, we shall neglect for the moment the added problems due to boundaries and take for our interval the whole real line $-\infty < x < +\infty$. We shall also try to impart some of the flavor of the actual procedure in a practical case by using the most obvious arguments initially and only refining them later as the need becomes apparent.

The differential equations are those for the small amplitude or acoustic motion of a gas with thermal conductivity σ, isothermal sound speed c and adiabatic gas constant $\gamma > 1$:

(6.41)
$$\frac{\partial u}{\partial t} = c\frac{\partial}{\partial x}(w - \overline{\gamma - 1}e),$$
$$\frac{\partial w}{\partial t} = c\frac{\partial u}{\partial x},$$
$$\frac{\partial e}{\partial t} = \sigma\frac{\partial^2 e}{\partial x^2} - c\frac{\partial u}{\partial x}.$$

We consider the difference approximation

(6.42)
$$u^{n+1} - u^n = \nu(\Delta_+ w^n - \overline{\gamma - 1}\Delta_+ e^n),$$
$$w^{n+1} - w^n = \nu\Delta_- u^{n+1},$$
$$e^{n+1} - e^n = \mu\Delta_-\Delta_+ e^{n+1} - \nu\Delta_- u^{n+1},$$

where $\nu = c\Delta t/\Delta x$, $\mu = \sigma\Delta t/(\Delta x)^2$. In Chapter 10 the stability of this scheme will be discussed in some detail, mainly with the help of Fourier methods. It will be shown there that the stability condition when $\sigma \neq 0$ is $\nu < 1$, while if $\sigma = 0$ it is readily seen that the condition is strengthened to $\nu < \gamma^{-1/2}$. The sudden change indicates a practical limitation of the definition of stability that we have been using, having to do with its dependence only on the behavior at the limit $\Delta t \to 0$. This will appear in the discussion of the energy method given below, which is intended to be complementary to that in Chapter 10.

As the first step in the application of the energy method, we take the inner product of each equation with the sum of the pair of terms on the corresponding left-hand sides, obtaining

$$\|u^{n+1}\|^2 - \|u^n\|^2$$
$$= \nu(u^n + u^{n+1}, \Delta_+ w^n) - \nu(\gamma - 1)(u^n + u^{n+1}, \Delta_+ e^n),$$

$$\|w^{n+1}\|^2 - \|w^n\|^2$$
$$= \nu(\Delta_- u^{n+1}, w^n + w^{n+1}) = -\nu(u^{n+1}, \Delta_+ (w^n + w^{n+1})),$$

$$\|e^{n+1}\|^2 - \|e^n\|^2$$
$$= \mu(\Delta_- \Delta_+ e^{n+1}, e^n + e^{n+1}) - \nu(\Delta_- u^{n+1}, e^n + e^{n+1})$$
$$= -\mu(\Delta_+ e^{n+1}, \Delta_+ (e^n + e^{n+1})) + \nu(u^{n+1}, \Delta_+ (e^n + e^{n+1})).$$

Multiplying the last equation by $(\gamma - 1)$ and adding all three together results in considerable cancellation, which suggests defining S_n as

(6.43) $\quad S_n = \|u^n\|^2 + \|w^n\|^2 + (\gamma - 1)\|e^n\|^2$
$$+ \tfrac{1}{2}(\gamma - 1)\mu\|\Delta_+ e^n\|^2 + \nu(u^n, \Delta_+(w^n - \overline{\gamma - 1}e^n)).$$

One then has

(6.44) $\quad S_{n+1} - S_n$
$$= -\tfrac{1}{2}(\gamma - 1)\mu[\|\Delta_+ e^n\|^2 + 2(\Delta_+ e^n, \Delta_+ e^{n+1}) + \|\Delta_+ e^{n+1}\|^2] \leq 0.$$

The real difficulties start with the attempt to show that S_n is a proper norm when $\nu < 1$; that is, when Δx is regarded as given by $\nu =$ constant so that $\mu = \text{const.}\,(\Delta x)^{-1}$. Clearly,

(6.45) $\quad |(u^n, \Delta_+ w^n)| \leq 2\|u^n\|\,\|w^n\| \leq \|u^n\|^2 + \|w^n\|^2,$

which enables one of the inner products to be dominated; and, for any $\alpha > 0$,

(6.46) $\quad |(u^n, \Delta_+ e^n)| \leq \|u^n\|\,\|\Delta_+ e^n\|$
$$\leq \tfrac{1}{2}(\alpha \Delta x \|u^n\|^2 + (1/\alpha \Delta x)\|\Delta_+ e^n\|^2).$$

Hence,

(6.47) $\quad S_n \geq (1 - \nu)\|w^n\|^2 + (\gamma - 1)\|e^n\|^2$
$$+ [1 - \nu(1 + \tfrac{1}{2}\alpha\overline{\gamma - 1}\Delta x)]\|u^n\|^2 + \tfrac{1}{2}(\gamma - 1)(\mu - (\nu/\alpha\Delta x))\|\Delta_+ e^n\|^2.$$

The last term suggests taking $\alpha = \nu/\mu\Delta x = c/\sigma$, in which case it drops

SEC. 6.4]　　　COUPLED SOUND AND HEAT FLOW　　　145

out. In order to keep the coefficient of $\|u^n\|^2$ positive, however, it is necessary to take

(6.48) $$\Delta x < [2\sigma/cv(\gamma - 1)](1 - \nu).$$

Thus, we shall be able to prove stability for arbitrarily small, but non-zero σ, and values of ν arbitrarily near 1, but only for an increasingly narrow range of Δx.

To complete the argument, we have $S_n \leq S_{n-1} \leq \cdots \leq S_1 \leq S_0$, and we would normally attempt to bound S_0 by a constant multiple of the L_2 norm ($\|u^0\|^2 + \|w^0\|^2 + \|e^0\|^2$). We are frustrated in this, however, by the term $(\Delta x)^{-1}\|\Delta_+ e^0\|^2$. So instead we consider S_1 and make use of the last of the difference equations (6.42) to give

$$\|e^1 - \mu\Delta_-\Delta_+ e^1\|^2 = \|e^0 - \nu\Delta_- u^1\|^2,$$

i.e., $\|e^1\|^2 + 2\mu\|\Delta_+ e^1\|^2 + \mu^2\|\Delta_-\Delta_+ e^1\|^2$

$$= \|e^0 - \nu\Delta_- u^1\|^2 \leq 2\|e^0\|^2 + 8\nu^2\|u^1\|^2.$$

As $\mu = (\sigma\nu/c)(\Delta x)^{-1}$, this provides the needed bound and S_1 can be bounded by a combination of the L_2 norms at $n = 0$ and $n = 1$. Stability then follows in the usual way from a bound for the L_2 norm in terms of S_n.

From the practical point of view, these arguments are of little value when σ is small or it is required to take ν near 1, for then (6.48) is much too severe a restriction on Δx. What is required is an accurate estimate of how large Δt may be taken for any given Δx while the growth rate of some suitable norm remains bounded. In the present case, S_n is always non-increasing, and it is in the estimation of the lower bound for it that more refinement is necessary. Thus, we replace (6.45) by

$$|(u^n, \Delta_+ w^n)| \leq \beta\|u^n\|^2 + \beta^{-1}\|w^n\|^2,$$

where $\beta > 0$ is to be chosen later, and combine (6.46) with a similar estimate,

$$|(u^n, \Delta_+ e^n)| \leq \delta\|u^n\|^2 + \delta^{-1}\|e^n\|^2,$$

in the ratio $(1 - \xi):\xi$, where again ξ and δ are to be chosen later. Then we search for an inequality of the form

(6.49) $$S_n \geq a_1\|u^n\|^2 + a_2\|w^n\|^2 + a_3\|e^n\|^2 + a_4\|\Delta_+ e^n\|^2,$$

in which a_1, a_2 and a_3 are each strictly positive and a_4 is non-negative. We have, from the above,

$$a_1 = 1 - \nu[\beta + \xi(\gamma - 1)\delta + \tfrac{1}{2}(1 - \xi)(\gamma - 1)\alpha\Delta x],$$
$$a_2 = 1 - \nu\beta^{-1},$$
$$a_3 = (\gamma - 1)(1 - \nu\xi\delta^{-1}),$$
$$a_4 = \tfrac{1}{2}(\gamma - 1)[\mu - (1 - \xi)\nu\alpha^{-1}(\Delta x)^{-1}].$$

For the last three to be positive it is necessary and sufficient that

$$\beta > \nu, \quad \delta > \nu\xi, \quad \text{and} \quad \alpha > c(1 - \xi)/\sigma.$$

Combining these with the expression for a_1, it is clear that the desired inequality can be found on condition that

$$\nu^2[1 + (\gamma - 1)(\xi^2 + \tfrac{1}{2}(1 - \xi)^2/\mu)] < 1,$$

which is optimized by the choice of $\xi = (1 + 2\mu)^{-1}$. Hence, as a practical stability criterion one obtains

(6.50) $$\nu < \sqrt{(1 + 2\mu)/(\gamma + 2\mu)},$$

which is supported by numerical evidence in Chapter 10.

6.5. Mixed Problems for Symmetric Hyperbolic Systems

To illustrate the energy method in a more general setting, we consider the mixed problem for a symmetric hyperbolic system satisfying an energy inequality. Suppose the complex p-component vector $\mathbf{u} = \mathbf{u}(x, t)$ is a solution of the system

(6.51) $$\frac{\partial \mathbf{u}}{\partial t} = P(x, \partial/\partial x)\mathbf{u}, \quad 0 \leq x \leq 1, 0 \leq t \leq 1,$$

where we write P in the form

(6.52) $$P(x, \partial/\partial x)\mathbf{u} \equiv \frac{1}{2}\left[\frac{\partial}{\partial x}(A(x)\mathbf{u}) + A(x)\frac{\partial}{\partial x}\mathbf{u}\right] + B(x)\mathbf{u}:$$

here $A(x)$ and $B(x)$ are $p \times p$ matrices depending smoothly on x, but not on t, and $A(x)$ is hermitian. As initial condition we have

(6.53) $$\mathbf{u}(x, 0) = \mathbf{f}(x), \quad 0 \leq x \leq 1,$$

SEC. 6.5] MIXED PROBLEMS FOR SYMMETRIC HYPERBOLIC SYSTEMS

and the boundary conditions are assumed to be of the form

(6.54) $\qquad L(x)\mathbf{u}(x, t) = 0, \quad x = 0, 1, \quad 0 \leq t \leq 1,$

consisting of $\mu(x)$ linearly independent homogeneous equations in the components of $\mathbf{u}$.

Then it is assumed that there exists a constant K independent of $\mathbf{u}$ for which the following *energy inequality* holds

(6.55) $\qquad \dfrac{\partial}{\partial t} \|\mathbf{u}(x, t)\|^2 \leq K \|\mathbf{u}(x, t)\|^2, \quad 0 \leq t \leq 1.$

Such a relation, including the special case of energy conservation in which equality holds with $K = 0$, is very commonly valid in physical problems. When the solution $\mathbf{u}$ is differentiable, it follows that

(6.56) $\qquad \dfrac{\partial}{\partial t} \|\mathbf{u}\|^2 = \left(\mathbf{u}, \dfrac{\partial \mathbf{u}}{\partial t}\right) + \left(\dfrac{\partial \mathbf{u}}{\partial t}, \mathbf{u}\right)$

$\qquad\qquad\qquad = 2 \operatorname{Re} (\mathbf{u}, P\mathbf{u}) \leq K \|\mathbf{u}\|^2.$

Moreover, since

$$2 \operatorname{Re} (\mathbf{u}, P\mathbf{u}) = \mathbf{u}^* A \mathbf{u} |_0^1 + 2 \operatorname{Re} (\mathbf{u}, B\mathbf{u}),$$

the requirement reduces to a condition at the boundary points,

(6.57) $\qquad \mathbf{u}^* A \mathbf{u} |_0^1 \leq \text{const. } \|\mathbf{u}\|^2.$

Kreiss (1960, 1963) has called an operator defined by

$$\mathbf{w} = P(x, \partial/\partial x)\mathbf{v}$$

for all infinitely differentiable functions $\mathbf{v} = \mathbf{v}(x)$ satisfying the boundary conditions (6.54), *semi-bounded* if an inequality

(6.58) $\qquad 2 \operatorname{Re} (\mathbf{v}, P\mathbf{v}) \leq K \|\mathbf{v}\|^2$

holds. Suppose several such operators, based on P but with different sets of boundary conditions, are defined. Then those for which the total number $\mu(0) + \mu(1)$ of boundary conditions is a minimum are called *minimally semi-bounded*. Kreiss showed that the problem defined by (6.51) to (6.54) is well-posed when P and the boundary conditions (6.54) give rise to such an operator. Moreover, the boundary conditions must then satisfy:

1. $\mu(0)$, the number of rows of $L(0)$, must equal the number of negative eigenvalues of $A(0)$;
2. $\mu(1)$, the number of rows of $L(1)$, must equal the number of positive eigenvalues of $A(1)$;
3. for $x = 0$ and 1, we must have

(6.59) $\quad (-1)^{x+1}\mathbf{v}^*A\mathbf{v} \leq 0 \quad$ for every vector $\mathbf{v}$ for which $L(x)\mathbf{v} = 0$.

If $A(x)$ is in diagonal form, to which it can be reduced by a unitary transformation $\hat{\mathbf{u}} = U\mathbf{u}$, these conditions can be given more explicitly. Suppose, for $x = 0$ and 1, $A(x)$ and $\mathbf{u}(x)$ are partitioned into the forms

(6.60) $\quad A(x) = \begin{bmatrix} \Lambda_1(x) & \Phi & \Phi \\ \Phi & -\Lambda_2(x) & \Phi \\ \Phi & \Phi & \Phi \end{bmatrix}, \quad \mathbf{u}(x) = \begin{bmatrix} \mathbf{u}_1(x) \\ \mathbf{u}_2(x) \\ \mathbf{u}_3(x) \end{bmatrix},$

where $\Lambda_1(x)$ and $\Lambda_2(x)$ are diagonal with positive elements and the Φ are null matrices, all dimensions depending on x. Then the allowable boundary conditions are of the form

(6.61) $\quad \mathbf{u}_2(0) = S(0)\mathbf{u}_1(0) \quad \text{and} \quad \mathbf{u}_1(1) = S(1)\mathbf{u}_2(1),$

where the matrices $S(0)$ and $S(1)$ satisfy

(6.62) $\quad -\Lambda_1(0) + S^*(0)\Lambda_2(0)S(0) \leq 0,$

(6.63) $\quad -\Lambda_2(1) + S^*(1)\Lambda_1(1)S(1) \leq 0.$

Now, let us consider an implicit difference approximation to (6.51),

(6.64) $\quad (I - \alpha\Delta t P_\Delta)\mathbf{u}_j^{n+1} = (I + \overline{1 - \alpha}\Delta t P_\Delta)\mathbf{u}_j^n, \quad j = 1, 2, \ldots, J-1,$

where $\frac{1}{2} \leq \alpha \leq 1$ and P_Δ is a difference operator approximating $P(x, \partial/\partial x)$. Boundary conditions are to take the form

(6.65) $\quad \begin{cases} M(0)\mathbf{u}^n \equiv \sum_{j=0}^{J-1} m_j(0, \Delta x)\mathbf{u}_j^n = 0, \\ M(1)\mathbf{u}^n \equiv \sum_{j=0}^{J-1} m_{J-j}(1, \Delta x)\mathbf{u}_{J-j}^n = 0, \end{cases}$

SEC. 6.5] MIXED PROBLEMS FOR SYMMETRIC HYPERBOLIC SYSTEMS 149

where the coefficient matrices $m_j(x, \Delta x)$ are assumed suitably normalized. We recall that, in setting up the general theory of Chapter 3, we absorbed the boundary conditions of the differential equation problem in the specification of the Banach space being used. Thus, for the above approximation to be consistent, we require that, for any solution $\mathbf{u}$ of the differential equation,

(6.66) $\quad L(x)\mathbf{u} = 0 \text{ implies } |M(x)\mathbf{u}| \to 0 \quad \text{as } \Delta x \to 0, \text{ for } x = 0, 1.$

We therefore assume that the number of rows of $M(x)$ is less than or equal to the corresponding number for $L(x)$. This point is simplified if the embedding of the mesh function $\mathbf{u}^n$ into a function $\mathbf{u}$ of the Banach space is so arranged that when the mesh function satisfies $M(x)\mathbf{u}^n = 0$, then its image satisfies $L(x)\mathbf{u} = 0$. For example, when $x = 0$ occurs midway between the mesh points $j = 0, 1$, the boundary condition $\mathbf{u}(0) = 0$ will usually be approximated by $\mathbf{u}_0^n + \mathbf{u}_1^n = 0$: then embedding $\mathbf{u}^n$ into the $\mathbf{u}$ space by linear interpolation has the required property.

In addition, we assume that for any set of values ascribed to the mesh function $\mathbf{u}_j^n$ at the interior points, $\mathbf{u}_0^n$ and $\mathbf{u}_J^n$ can be so defined that the boundary conditions (6.65) are satisfied. Thus, the latter cannot include any equation which involves only the interior points. We can also supplement these equations by further similar equations in such a way that $\mathbf{u}_0^n$ and $\mathbf{u}_J^n$ can be uniquely calculated from $\mathbf{u}_j^n, j = 1, 2, \ldots, J - 1$. Similar supplementary equations can be used for all other exterior mesh points which are involved in $P_\Delta \mathbf{u}_j^n$, when j corresponds to an interior mesh point.

We denote by $\mathcal{N}$ the class of mesh functions $\mathbf{w}$ which satisfy the boundary conditions (6.65) and these supplementary relations. Then we can obtain a result for the stability of the difference equations which is closely analogous to that for the well-posedness of the differential equation problem. Namely, *if there is a constant K such that for every mesh function $\mathbf{w} \in \mathcal{N}$, the inequality*

(6.67) $\quad\quad\quad \text{Re } (\mathbf{w}, P_\Delta \mathbf{w})_\Delta \leq K \|\mathbf{w}\|_\Delta^2, \quad \mathbf{w} \in \mathcal{N}$

holds for all sufficiently small Δx, then for these Δx and for $4K\Delta t < 1$, the functions $\mathbf{u}^n, n = 1, 2, \ldots$, given by (6.64) are uniquely determined within the class $\mathcal{N}$ from the initial conditions, and the difference scheme is stable.

It should be noted that the norm $\|\ \|_\Delta$ involves only the interior mesh points.

The proof is a mere gathering together of the assumptions. For arbitrary right-hand side $\mathbf{f}$ we consider the equations

(6.68) $\qquad (I - \alpha\Delta t P_\Delta)\mathbf{w}_j = \mathbf{f}_j, \quad j = 1, 2, \ldots, J - 1.$

When $\mathbf{w} \in \mathcal{N}$, its values at all but the interior mesh points can be eliminated from the above, and the equations then represent $p(J - 1)$ linear equations in the $p(J - 1)$ unknown components of $\mathbf{w}$ at the interior points. Moreover, for $\mathbf{w} \in \mathcal{N}$ and with $\frac{1}{2} \leq \alpha \leq 1$ and $4K\Delta t < 1$,

(6.69) $\qquad \|\mathbf{f}\|_\Delta^2 = \|(I - \alpha\Delta t P_\Delta)\mathbf{w}\|_\Delta^2 \geq \|\mathbf{w}\|_\Delta^2 - 2\alpha\Delta t \,\text{Re}\,(\mathbf{w}, P_\Delta \mathbf{w})_\Delta$

$\qquad \geq (1 - 2\alpha K\Delta t)\|\mathbf{w}\|_\Delta^2 \geq \frac{1}{2}\|\mathbf{w}\|_\Delta^2.$

It follows that (6.68) determines a unique $\mathbf{w} \in \mathcal{N}$. Hence, for any initial mesh function $\mathbf{u}^0 \in \mathcal{N}$ the difference scheme (6.64) yields a unique sequence $\mathbf{u}^n \in \mathcal{N}$, $n = 1, 2, \ldots$.

To prove stability we take the inner product of both sides of (6.64) with $\mathbf{w} = \alpha \mathbf{u}^{n+1} + \overline{1 - \alpha}\,\mathbf{u}^n$ to get

$$\text{Re}\,(\mathbf{w}, \mathbf{u}^{n+1} - \mathbf{u}^n)_\Delta = \Delta t\,\text{Re}\,(\mathbf{w}, P_\Delta \mathbf{w})_\Delta \leq K\Delta t \|\mathbf{w}\|_\Delta^2.$$

The left-hand side equals

$\alpha\|\mathbf{u}^{n+1}\|_\Delta^2 - (2\alpha - 1)\,\text{Re}\,(\mathbf{u}^{n+1}, \mathbf{u}^n)_\Delta - (1 - \alpha)\|\mathbf{u}^n\|_\Delta^2$

$\qquad \geq [\alpha - \frac{1}{2}(2\alpha - 1)]\|\mathbf{u}^{n+1}\|_\Delta^2 - [1 - \alpha + \frac{1}{2}(2\alpha - 1)]\|\mathbf{u}^n\|_\Delta^2$

$\qquad = \frac{1}{2}\|\mathbf{u}^{n+1}\|_\Delta^2 - \frac{1}{2}\|\mathbf{u}^n\|_\Delta^2.$

In addition, $\|\mathbf{w}\|_\Delta^2 \leq \|\mathbf{u}^{n+1}\|_\Delta^2 + \|\mathbf{u}^n\|_\Delta^2$, so that

$$\|\mathbf{u}^{n+1}\|_\Delta^2 - \|\mathbf{u}^n\|_\Delta^2 \leq 2K\Delta t(\|\mathbf{u}^{n+1}\|_\Delta^2 + \|\mathbf{u}^n\|_\Delta^2).$$

It follows that

(6.70) $\qquad \|\mathbf{u}^n\|_\Delta^2 \leq (1 + 2K\Delta t)^n \cdot (1 - 2K\Delta t)^{-n} \|\mathbf{u}^0\|_\Delta^2$

and, therefore, that the difference scheme is stable.

As a simple example of these results, let us consider the application of the Crank-Nicholson scheme to the example of Section 6.3:

(6.71)
$$\frac{\partial u}{\partial t} + a(x)\frac{\partial u}{\partial x} = 0, \quad 0 \leq x \leq 1, 0 \leq t \leq 1$$
$$u(0, t) = 0, \quad u(x, 0) = F(x),$$

where u and a are real and $a(x) > 0$. The mesh is chosen such that $x = 0, 1$ corresponds to $j = 0, J$, respectively, and the Crank-Nicholson scheme results from putting $\alpha = \frac{1}{2}$ and

$$P_\Delta = -(\Delta x)^{-1} a(x) \Delta_0.$$

There is no boundary condition at $x = 1$, and at $x = 0$,

$$M(0)u^n \equiv u_0^n = 0$$

The class $\mathcal{N}$ will be described by this condition plus those supplementary relations necessary to obtain the inequality (6.67). Now, from (6.15), for any mesh function w

$$|(w, a\Delta_0 w) - (w, \Delta_0(aw))| \leq L\Delta x \|w\| \|w\|_{0,J},$$

where L is a Lipschitz constant for $a(x)$. Also, from (6.19),

$$(w, \Delta_0(aw)) = -(\Delta_0 w, aw) + \tfrac{1}{2}\Delta x[(a_J + a_{J-1})w_J w_{J-1} - (a_0 + a_1)w_0 w_1].$$

Thus, we define $\mathcal{N}$ by

(6.72) $\quad \mathcal{N} = \{w | w_0 = 0, w_J = w_{J-1}, \text{ and } w_j = 0 \text{ for } j < 0 \text{ or } j > J\},$

so that

$$\operatorname{Re}(w, P_\Delta w) \leq L\|w\|^2 - \tfrac{1}{4}(a_J + a_{J-1})(w_{J-1})^2$$
$$\leq L\|w\|^2, \quad \text{for } w \subset \mathcal{N}.$$

Solvability and stability of the difference scheme for $4L\Delta t < 1$ then follows from the theorem above.

6.6. Normal Mode Analysis and the Godunov-Ryabenkii Stability Criterion

We return in this section to the analysis of local normal modes as a means of determining stability. With variable coefficients and non-periodic boundary conditions, we cannot, of course, simply take Fourier transforms to get the well-defined algebraic problem studied in Chapter 4. But even in real space, the calculation of the finite difference approximation at each time step consists, in all the cases we have discussed above, of the solution of a set of linear algebraic equations: the properties of these systems of equations must include all that is needed to ascertain the stability of the difference schemes. The chief problem is that their order grows as $\Delta t \to 0$. As a consequence, for example, the eigenvalues of the associated matrices can be quite misleading guides to stability.

Consider, for instance, our standard problem given in (6.71), and let the approximation be

(6.73) $$u_j^{n+1} = (1 - r_j)u_j^n + r_j u_{j-1}^n, \quad r_j = (\Delta t/\Delta x)a_j.$$

Then, regarding $\{u_j^n, j = 0, 1, \ldots, J\}$ as a vector, the matrix describing the transformation to the next time level is

(6.74) $$C = \begin{bmatrix} 0 & & & & & \\ r_1 & 1 - r_1 & & & 0 & \\ 0 & r_2 & 1 - r_2 & & & \\ & \ddots & \ddots & \ddots & & \\ 0 & & & 0 & r_J & 1 - r_J \end{bmatrix}.$$

Its eigenvalues are 0 and $1 - r_j$, $j = 1, 2, \ldots, J$, and these all lie inside the unit circle if $0 < r_j < 2$; thus, for each fixed Δt, all errors are eventually damped out—a statement that has been used as a definition of stability by some authors. Yet the proper stability condition is the well-known Courant-Friedrichs-Lewy condition $0 < r_j < 1$; outside that range there is no convergence as $\Delta t \to 0$. The distinction lies between what happens as $n \to \infty$ with Δt fixed and the behavior required as $n \to \infty$ with $n\Delta t$ fixed. For instance, with each $r_j = 3/2$ and the u_j^0 of alternating sign, the value at each mesh point grows like $u_j^n = (-2)^n u_j^0$ until the influence of the boundary points is felt. It is this alone that causes eventual damping; but as $\Delta t \to 0$, the growth before this happens becomes unbounded—see Parter (1962) for further examples of this phenomenon.

A much better guide to stability is given by the L_2 norm of the matrices

SEC. 6.6] THE GODUNOV-RYABENKII STABILITY CRITERION

C. By considering vectors of alternating sign it is easy to show that, for $r_j = r$, a constant,

$$\|C\| = r + |1 - r| + O(1/J);$$

indeed, the infinite matrix of the form (6.74) has a spectral radius of $r + |1 - r|$.

Considerations of this kind have led Godunov and Ryabenkii (1963a) to introduce the concept of a spectrum of a family of operators:

Let $\mathscr{F} = \{C(\Delta t)|\Delta t > 0\}$ be a one-parameter family of linear operators defined on a Banach space $\mathscr{B}$. Then the complex number λ is said to be a point of the spectrum of the family $\mathscr{F}$ if there is a sequence $\Delta t^{(J)} > 0$ and a sequence $u^{(J)} \in \mathscr{B}$ such that $\|u^{(J)}\| = 1$, $\Delta t^{(J)} \to 0$ as $J \to \infty$ and

(6.75) $$\|[C(\Delta t^{(J)}) - \lambda I]u^{(J)}\| \to 0 \quad \text{as } J \to \infty.$$

If the family consisted of a single member, the definition would reduce to the normal one. And the family of (6.74) shows that it is not an empty extension, for in the case when all $r_j = r$, Godunov and Ryabenkii showed that the family spectrum consists of the closed disk with center $1 - r$ and radius r. One merely has to consider, for $\lambda = 1 - r + s$ with $|s| \leq r$, the vectors whose components are formed by truncating and normalizing the sequence $(s/r)^{-j}$, $j = 0, 1, \ldots$.

It is clear that a necessary condition for the stability of a family of difference operators is that its spectrum lie in the closed unit disk. This requirement we shall call the *Godunov-Ryabenkii stability criterion*, to be abbreviated to the *G-R criterion*. It has obvious similarities to the von Neumann condition. In applications, one considers all the local normal modes and uses them as the $u^{(J)}$ in (6.75) to see whether they satisfy the criterion. Thus, from considering the local Fourier modes, one obtains a condition on the eigenvalues λ of the amplification matrix; in the limit as $\Delta t \to 0$, they will clearly lie in the family spectrum and, hence, one must have $|\lambda| \leq 1 + O(1)$. This is very much weaker than the von Neumann condition $|\lambda| \leq 1 + O(\Delta t)$. But, on the other hand, the G-R criterion is stronger in the sense that a larger class of normal modes is considered. For instance, near a boundary there may well be modes which decrease away from the boundary and affect the stability of the difference scheme.

A simple example is provided by the wave equation on the interval $0 \le x < \infty$, which may be written as the system

$$(6.76) \qquad \frac{\partial u}{\partial t} = c\frac{\partial v}{\partial x}, \qquad \frac{\partial v}{\partial t} = c\frac{\partial u}{\partial x},$$

and approximated by the difference scheme

$$(6.77) \qquad \begin{aligned} u_j^{n+\frac{1}{2}} - u_j^{n-\frac{1}{2}} &= r\Delta_+ v_{j-\frac{1}{2}}^n, \\ v_{j+\frac{1}{2}}^{n+1} - v_{j+\frac{1}{2}}^n &= r\Delta_+ u_j^{n+\frac{1}{2}}, \quad j, n = 0, 1, 2, \dots \end{aligned}$$

where $r = c\Delta t/\Delta x$; consideration of the amplification matrices leads to the stability criterion $r < 1$ (see Chapter 10). The initial conditions are given by prescribing $u_j^{-\frac{1}{2}}$ and v_j^0, and a single boundary condition is necessary at $x = 0$, its general form (bearing in mind the interchangeability of u and v) being

$$(6.78) \qquad v_{-\frac{1}{2}}^{n+1} + \alpha u_0^{n+\frac{1}{2}} = 0,$$

where α is a constant.

In the search for normal modes, we look for solutions of the form

$$(6.79) \qquad \mathbf{w}_j^n = (\lambda)^n (\mu)^j \tilde{\mathbf{w}},$$

where we have written $\mathbf{w}_j^n$ for the vector $(u_j^{n-\frac{1}{2}}, v_{j-\frac{1}{2}}^n)^T$ and $\tilde{\mathbf{w}} = (\tilde{u}, \tilde{v})^T$ is a constant vector. Substitution into (6.77) gives

$$(6.80) \qquad \begin{aligned} (\lambda - 1)\tilde{u} - r(\mu - 1)\tilde{v} &= 0, \\ (\lambda - 1)\mu\tilde{v} - r\lambda(\mu - 1)\tilde{u} &= 0; \end{aligned}$$

and equating to zero the determinant of this pair of equations yields

$$(6.81) \qquad \mu^2 - \left[2 + \frac{(\lambda - 1)^2}{r^2\lambda}\right]\mu + 1 = 0.$$

For (6.79) to be an acceptable solution, we need $|\mu| \le 1$, and it can only contribute to an instability if $|\lambda| \ge 1$. But if $r < 1$, which we shall assume, the Fourier modes $|\mu| = 1$ give $|\lambda| = 1$ and are stable. Thus, in the region $|\lambda| > 1$, the solution of (6.81), namely

$$(6.82) \qquad \mu = 1 + \frac{(\lambda - 1)^2}{2r^2\lambda} \pm \sqrt{\frac{(\lambda - 1)^2}{r^2\lambda} + \frac{(\lambda - 1)^4}{4r^4\lambda^2}},$$

SEC. 6.6] THE GODUNOV-RYABENKII STABILITY CRITERION

separate into two branches on one of which $|\mu| > 1$ and on the other $|\mu| < 1$. We are therefore interested in the latter and so take the minus sign in (6.82) and agree to remain on the branch for which the radical is > 0 for, say, $\lambda > 1$.

Substitution into (6.80) shows that we can take, for example,

(6.83)
$$\tilde{u} = (\lambda - 1) - \sqrt{(\lambda - 1)^2 + 4\lambda r^2},$$
$$\tilde{v} = 2\lambda r;$$

and putting this into the boundary condition $\tilde{v} + \alpha\tilde{u} = 0$ gives for λ

(6.84)
$$\lambda = (\alpha + r\alpha^2)/(\alpha + r).$$

Now it is not difficult to see that the differential problem is well posed except when $\alpha = 1$.[3] Yet the behavior of (6.84) is as sketched in Figure 6.1. From this it is clear that (6.79) represents an unstable mode except

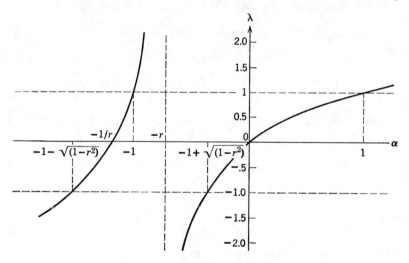

Fig. 6.1. Graph of equation (6.84).

---

[3] It is worth noting that the sufficient conditions for well-posedness given in Section 6.5 yield through (6.59), the inequality $\tilde{u}\tilde{v} \geq 0$ at $x = 0$; that is, that the range $\alpha \leq 0$ leads to well-posed problems. To obtain the full range of possible boundary conditions, one has to choose a different energy or, equivalently, change variables. Thus, putting $\hat{u} = \tilde{u} + \tilde{v}$ and $\hat{v} = \beta(\tilde{u} - \tilde{v})$ with $0 < \beta < \infty$, the acceptable boundary conditions become $\hat{v} = \gamma\hat{u}$ at $x = 0$, where $|\gamma| \leq 1$. This includes all cases of $\tilde{v} + \alpha\tilde{u} = 0$ except $\alpha = 1$.

when

$$1 > \alpha \geqq -1 + \sqrt{1-r^2} \quad \text{or} \quad -1 \geqq \alpha \geqq -1 - \sqrt{1-r^2}:$$

values of α outside these ranges will lead to instability. Note that the usual boundary condition $\tilde{u} - \tilde{v} = 0$ is only on the limit of an acceptable range.

In this example, $(C - \lambda I)u$ actually equals zero for the modes discovered, but if a finite interval $0 \leqq x \leqq 1$ were being considered, u might not satisfy the boundary condition at $x = 1$ except in the limit as $\Delta t \to 0$ and so $J \to \infty$; this was the case with the simple example treated at the beginning of the section. Also, if multiple roots μ of (6.81) occurred, more complicated modes than (6.79), involving polynomials in j, would need to be considered.

6.7. Application of the G-R Criterion to Mixed Problems

Consider a general two-level system of difference equations, on the interval $0 \leqq x \leqq 1$, in the form

$$(6.85) \quad \sum_{k=0}^{K} B_k \mathbf{u}_{j+k}^{n+1} = \sum_{k=0}^{K} A_k \mathbf{u}_{j+k}^{n}, \quad j = 0, 1, \ldots, J - K, \quad n = 0, 1, 2 \ldots;$$

here $\mathbf{u}_j^n$ is, as usual, a p-component vector and A_k, B_k square matrices. At the two boundaries there are boundary conditions

$$(6.86) \quad \sum_{k=0}^{k_0} \beta_{0k} \mathbf{u}_k^{n+1} = \sum_{k=0}^{k_0} \alpha_{0k} \mathbf{u}_k^{n},$$

$$(6.87) \quad \sum_{k=0}^{k_1} \beta_{1k} \mathbf{u}_{J-k}^{n+1} = \sum_{k=0}^{k_1} \alpha_{1k} \mathbf{u}_{J-k}^{n},$$

where the α_{ik} and β_{ik} are rectangular matrices.

Although in a given practical case this is unnecessary, it is convenient in demonstrating the application of the G-R criterion to the above system of equations to reduce them to the special case in which $K = 1$ and $k_0 = k_1 = 0$. This may be accomplished in the following way. We take an interval I_0 in x space which starts at the left-hand mesh point and extends to the right over a total number of mesh points l, where $l \geqq$ Max $(k_0 +$

SEC. 6.7] APPLICATION OF G-R CRITERION TO MIXED PROBLEMS

$1, k_1 + 1, K$). Then, $I_1, I_2, \ldots$ are formed by successively translating I_0 one mesh spacing to the right until all mesh points are covered. All components of vectors $\mathbf{u}_k^n$ whose mesh points lie in I_m are then combined into a new vector $\mathbf{U}_m^n$ which therefore has pl components. As a result, (6.86) becomes a set of homogeneous equations in $\mathbf{U}_0^n$ and $\mathbf{U}_0^{n+1}$, and similarly with (6.87). Also, the components of $\mathbf{U}_m^n$, $\mathbf{U}_{m+1}^n$, $\mathbf{U}_m^{n+1}$ and $\mathbf{U}_{m+1}^{n+1}$ will be connected by a set of pl scalar equations made up of p obtained from (6.85), with some particular value of j, and $p(l-1)$ identification relations due to the overlap of the intervals I_m and I_{m+1}. However, since in general the number of values of m covered in this way will be smaller than the number of values of j for which (6.85) holds, there will be some equations remaining; these will involve the end intervals and must be added to the boundary conditions. The reader will notice in the next chapter that the reduction of multi-level schemes to two-level schemes is accomplished in a manner very similar to the above. In treating the present problem, the two reductions can be carried out simultaneously by using rectangles in the x, t-plane instead of the intervals I_m.

We shall henceforth assume that this reduction has been carried out, so we consider equations (6.85) to (6.87) with $K = 1$ and $k_0 = k_1 = 0$. The coefficient matrices A_k and B_k were assumed by Godunov and Ryabenkii (1963a) in their original presentation to depend piecewise continuously on x and continuously on Δx for some interval $0 \leq \Delta x \leq \tau$. For simplicity, however, we shall treat only the case of constant matrices: the dependence on Δx can be taken into account very easily, and we shall indicate the effect of x dependence at the end of the section.

We assume that when the $\mathbf{u}_j^n (j = 0, 1, \ldots, J)$ are given the $\mathbf{u}_j^{n+1}$ are uniquely determined by the equations. Thus, the number of non-zero rows in the matrix β_{00} plus the number in the matrix β_{10} equals p and there exists a matric C_J of order $p(J+1)$ such that

$$\{\mathbf{u}_j^{n+1} | j = 0, 1, \ldots, J\} = C_J \{\mathbf{u}_j^n | j = 0, 1, \ldots, J\}.$$

It is to this family $\{C_J\}$ of difference operators that we wish to apply the G-R criterion.

A point λ of the complex plane which is not in the spectrum of the family is called a *regular point*. Now, for such points it is clear that the resolvents $(C_J - \lambda I)^{-1}$ exist for all sufficiently large J and have a bound that is independent of J. In terms of the equation above, this means that if we put

(6.88)
$$D_i = \lambda B_i - A_i, \quad i = 0, 1,$$
$$\delta_i = \lambda \beta_{i0} - \alpha_{i0}, \quad i = 0, 1,$$

then the system of equations

(6.89)
$$D_0 \mathbf{u}_j + D_1 \mathbf{u}_{j+1} = \mathbf{f}_{j+\frac{1}{2}}, \quad j = 0, 1, \ldots, J - 1$$
$$\delta_0 \mathbf{u}_0 = \boldsymbol{\varphi}, \quad \delta_1 \mathbf{u}_J = \boldsymbol{\psi}$$

have a unique solution for any right-hand sides and for sufficiently large J, and that

(6.90) $$\text{Max}_{(j)} |\mathbf{u}_j| \leq M \text{ Max } (\text{Max}_{(j)} |\mathbf{f}_{j+\frac{1}{2}}|, |\boldsymbol{\varphi}|, |\boldsymbol{\psi}|),$$

where M does not depend on $\mathbf{f}_{j+\frac{1}{2}}$, $\boldsymbol{\varphi}$, $\boldsymbol{\psi}$, or J; the system (6.89) is then called *well conditioned*. Thus, we may say that *the G-R criterion, which is necessary for stability, is satisfied by the difference scheme if, for all points λ outside the unit disk, equations (6.89) are well conditioned.*

To obtain conditions that these equations be well conditioned, we need the following lemma due to Ryabenkii (1964): *Suppose D_0 and D_1 are $p \times p$ matrices such that $\det(D_0 + \mu D_1) \neq 0$ for all μ such that $|\mu| = 1$; and suppose that μ_0 is a k-fold zero of $\det(D_0 + \mu D_1)$ and $|\mu_0| < 1$. Then there is a k-dimensional sub-space $\mathscr{R}(\mu_0)$ of the p-dimensional vector space such that, if $\mathbf{V}_0 \in \mathscr{R}(\mu_0)$, then the equations*

(6.91) $$D_0 \mathbf{V}_j + D_1 \mathbf{V}_{j+1} = 0, \quad j = 0, 1, 2, \ldots,$$

have a unique solution $\{\mathbf{V}_j\}$ for which $|\mathbf{V}_j| \to 0$ as $j \to \infty$; furthermore, there is an $\alpha > 0$ such that $|\mathbf{V}_j| \leq \text{const.} \, e^{-\alpha j}$ for $j = 0, 1, 2, \ldots$.

Proof. By hypothesis, $D_0 + D_1$ is non-singular so that we may write

$$D_0 + \mu D_1 = (D_0 + D_1)[I + (\mu - 1)E],$$

where $E = (D_0 + D_1)^{-1} D_1$. If, further, Q is a Jordan canonical form for E with $E = PQP^{-1}$, then

(6.92) $$D_0 + \mu D_1 = (D_0 + D_1) P [I + (\mu - 1)Q] P^{-1}$$

and μ_0 is a k-fold zero of $\det[I + (\mu - 1)Q]$. The matrix here is upper

SEC. 6.7] APPLICATION OF G-R CRITERION TO MIXED PROBLEMS 159

triangular so the determinant equals $\Pi_{(i)}[1 + (\mu - 1)\nu_i]$, where ν_i are the eigenvalues of E. Precisely k of these factors must be proportional to $(\mu - \mu_0)$; that is, $1 + (\mu - 1)\nu_i = \nu_i(\mu - \mu_0)$ and hence $\nu_i = (1 - \mu_0)^{-1}$ for k values of i.

We therefore may assume that the rows and columns of Q have been ordered so that the first k diagonal elements are $(1 - \mu_0)^{-1}$. The first k columns of P we then take as a basis for the sub-space $\mathscr{R}(\mu_0)$ and write the sequence of required vectors in this sub-space in the form

(6.93) $$\mathbf{V}_j = P\mathbf{a}_j,$$

where the $\mathbf{a}_j$ are column vectors, all of whose elements except the first k are zero. It follows from the identity (6.92) in μ that

$$D_0 P = (D_0 + D_1)(P - PQ) \quad \text{and} \quad D_1 P = (D_0 + D_1)PQ,$$

and hence that the equations (6.91) become

(6.94) $$(D_0 + D_1)P[I\mathbf{a}_j - Q(\mathbf{a}_j - \mathbf{a}_{j+1})] = 0, \quad j = 0, 1, 2, \ldots.$$

From any initial $\mathbf{V}_0 \in \mathscr{R}(\mu_0)$, providing an initial $\mathbf{a}_0$, these equations may be solved for the sequence $\mathbf{a}_j$ since only the top left $k \times k$ sub-matrix of Q is involved, and this is invertible. Denoting it by $\hat{Q}$ and using the same notation for I and $\mathbf{a}_j$, one obtains

$$\hat{\mathbf{a}}_j = [\hat{Q}^{-1}(\hat{Q} - \hat{I})]^j \hat{\mathbf{a}}_0.$$

The matrix in square brackets is upper triangular and has diagonal elements equal to μ_0; hence its jth power has elements bounded by μ_0^j times a polynomial of degree k in j, and so bounded by const. $e^{-\alpha j}$, where $|\mu_0| < e^{-\alpha} < 1$. The same growth factor clearly obtains for the $\mathbf{V}_j$ given by (6.93).

We denote by $\mathscr{R}$ the direct sum of all the sub-spaces of the same type as $\mathscr{R}(\mu_0)$:

(6.95) $$\mathscr{R} = \sum_{\oplus} \{\mathscr{R}(\mu_i) | \det(D_0 + \mu_i D_1) = 0, |\mu_i| < 1\}.$$

Similarly, there exist sub-spaces $\mathscr{S}(\eta_i)$ corresponding to zeros η_i of det $(\eta D_0 + D_1) = 0$ for which $|\eta_i| < 1$. These sub-spaces contain possible starting vectors for sequences of solutions (6.91) which tend to zero as $j \to -\infty$ and their direct sum we denote by $\mathscr{S}$:

(6.96) $\qquad \mathscr{S} = \sum_{\oplus} \{\mathscr{S}(\eta_i) \mid \det(\eta_i D_0 + D_1) = 0, |\eta_i| < 1\}.$

Since the equation $\det(D_0 + \mu D_1) = 0$ [and the equivalent equation $\det(\eta D_0 + D_1) = 0$] is of formal degree p and there are no roots of unit modulus, we must have $\dim \mathscr{R} + \dim \mathscr{S} = p$.

We have seen that, with $\mathbf{V}_0 \in \mathscr{R}(\mu_0)$, the solution of (6.91) was unique within $\mathscr{R}(\mu_0)$. It is also clear that it is unique within the whole of $\mathscr{R}$, for the principal sub-matrix of Q corresponding to all the zeros μ_i is also invertible. However, if D_1 is singular, then one could add to any $\mathbf{V}_{j+1}$ a vector $\mathbf{W} \in \mathscr{S}$ for which $D_1 \mathbf{W} = 0$. But the sequence of solutions to (6.91) with this starting vector either cannot be continued beyond a finite number of steps or increase exponentially; so the decreasing solution we have found is unique in the whole space and Ryabenkii's lemma is proved.

It is now possible for us to state and prove the main result of Godunov and Ryabenkii:

For the equations (6.89) to be well conditioned, it is necessary and sufficient that the following conditions are satisfied:

(i) $\det(D_0 + \mu D_1) \neq 0$ *for all μ such that $|\mu| = 1$;*

(iia) $\dim \mathscr{R} =$ *number of rows of δ_0 and, for $\mathbf{v} \in \mathscr{R}$, $\delta_0 \mathbf{v} = 0$ implies $\mathbf{v} = 0$;*

(iib) $\dim \mathscr{S} =$ *number of rows of δ_1 and, for $\mathbf{w} \in \mathscr{S}$, $\delta_1 \mathbf{w} = 0$ implies $\mathbf{w} = 0$.*

Here $\mathscr{R}$ and $\mathscr{S}$ are the sub-spaces defined above.[4]

To show the necessity of these conditions, we show that, if any is not satisfied, a sequence of vectors $\mathbf{u}_j$ can be found, with $\text{Max}_{(j)}|\mathbf{u}_j| = 1$, such that the right-hand sides of (6.89) can be made arbitrarily small for sufficiently large J, thus contradicting (6.90). First, if $\det(D_0 + \mu_0 D_1) = 0$ with $|\mu_0| = 1$, we can choose $\mathbf{v}$ such that $(D_0 + \mu_0 D_1)\mathbf{v} = 0$ and $|\mathbf{v}| = 1$ and take $\mathbf{u}_j = 4j(J - j)J^{-2}\mu_0^j \mathbf{v}$. Then $\boldsymbol{\varphi} = \boldsymbol{\psi} = 0$ and $\mathbf{f}_{j+\frac{1}{2}} = O(1/J)$. Next, if $\dim \mathscr{R}$ is greater than the number of rows of δ_0, then there will

---

[4] It should be remembered that we have assumed that D_0 and D_1 are independent of J so that $\mathscr{R}$, $\mathscr{S}$, and the constant α used below are all also independent of J.

SEC. 6.7] APPLICATION OF G-R CRITERION TO MIXED PROBLEMS 161

exist a $\mathbf{u}_0 \in \mathscr{R}$ with $|\mathbf{u}_0| = 1$ and $\delta_0 \mathbf{u}_0 = 0$. The solution of $D_0 \mathbf{u}_j + D_1 \mathbf{u}_{j+1} = 0, j = 0, 1, \ldots, J - 1$, then gives $\boldsymbol{\varphi} = 0$, $\mathbf{f}_{j+\frac{1}{2}} = 0$, and $\boldsymbol{\psi} = O(e^{-\alpha j})$ for some $\alpha > 0$. Finally, if dim $\mathscr{R}$ is less than the number of rows of δ_0, then dim $\mathscr{S}$ must be greater than the number of rows of δ_1 and condition (iib) fails.

The demonstration of the sufficiency of the above conditions requires that we find a suitably bounded solution of the equations (6.89) under each of the following circumstances:

(a) $\boldsymbol{\varphi} = 0$, $\boldsymbol{\psi} = 0$, and $\mathbf{f}_{j+\frac{1}{2}} = 0$ for $j \neq j_0$;
(b) $\boldsymbol{\varphi} = 0$, all $\mathbf{f}_{j+\frac{1}{2}} = 0$;
(c) $\boldsymbol{\psi} = 0$, all $\mathbf{f}_{j+\frac{1}{2}} = 0$.

In case (a) we first observe that there exist unique vectors $\mathbf{v} \in \mathscr{R}$ and $\mathbf{w} \in \mathscr{S}$ such that $D_0 \mathbf{w} + D_1 \mathbf{v} = \mathbf{f}_{j_0 + \frac{1}{2}}$. For we may write $\mathbf{f}_{j_0 + \frac{1}{2}} = \mathbf{f}_0 + \mathbf{f}_1$, where $\mathbf{f}_0 \in \mathscr{R}$ and $\mathbf{f}_1 \in \mathscr{S}$, and this decomposition is unique; then $\mathbf{v}$ and $\mathbf{w}$ are the unique solutions (in their respective sub-spaces) of $D_1 \mathbf{v} = \mathbf{f}_0$ and $D_0 \mathbf{w} = \mathbf{f}_1$. The sequence $\mathbf{u}_j$ is next constructed by putting $\mathbf{u}_{j_0} = \mathbf{w}$ and $\mathbf{u}_{j_0} = \mathbf{v}$ and then solving $D_0 \mathbf{u}_j + D_1 \mathbf{u}_{j+1} = 0$ for $j \neq j_0$. As a result $\mathbf{u}_j = \in \mathscr{R}$ and $|\mathbf{u}_j|$ deceases as j increases for $j \geq j_0 + 1$, while $\mathbf{u}_j \in \mathscr{S}$ and $|\mathbf{u}_j|$ decreases as j decreases for $j \leq j_0$.

This sequence satisfies the main equations of (6.89) but does not satisfy the boundary conditions. To do so, a convergent series of corrections is added so that the complete solution takes the form

(6.97) $$\mathbf{u}_j + \mathbf{v}_j^{(1)} + \mathbf{w}_j^{(1)} + \mathbf{v}_j^{(2)} + \mathbf{w}_j^{(2)} + \cdots.$$

Here, $\mathbf{v}_0^{(1)}$ is chosen in $\mathscr{R}$ to satisfy $\delta_0(\mathbf{u}_0 + \mathbf{v}_0^{(1)}) = 0$, the choice being unique because of hypothesis (iia); and similarly $\mathbf{w}_j^{(1)} \in \mathscr{S}$ is chosen to satisfy $\delta_1(\mathbf{u}_J + \mathbf{w}_J^{(1)}) = 0$, $\mathbf{v}_0^{(2)} \in \mathscr{R}$ to satisfy $\delta_0(\mathbf{w}_0^{(1)} + \mathbf{v}_0^{(2)}) = 0$, and so on. Otherwise, the $\mathbf{v}_j^{(i)}$ are chosen in $\mathscr{R}$ to satisfy $D_0 \mathbf{v}_j^{(i)} + D_1 \mathbf{v}_{j+1}^{(i)} = 0$ and the $\mathbf{w}_j^{(i)}$ is chosen in $\mathscr{S}$ to satisfy the same relation, $D_0 \mathbf{w}_j^{(i)} + D_1 \mathbf{w}_{j+1}^{(i)} = 0$. Since $|\mathbf{v}_J^{(i)}| < \text{const.} \, e^{-\alpha J} |\mathbf{v}_0^{(i)}|$ and $|\mathbf{w}_0^{(i)}| < \text{const.} \, e^{-\alpha J} |\mathbf{w}_J^{(i)}|$, the series clearly converges for sufficiently large J and the boundary conditions are then satisfied.

Solutions of the equations (6.89) in the other cases (b) and (c) are obtained in a similar manner. It then remains for us to show that the general solution obtained by superposing these fundamental solutions satisfies the boundedness condition expressed in (6.90), for the number of terms in the superposition increases as J increases. However, the required constant M can be found, since greatly separated values of j have little influence on one another. The fundamental solution decays exponentially as j departs from j_0 so that the solution at mesh point j_0 resulting from all the right-hand elements $\mathbf{f}_{j+\frac{1}{2}}$ is bounded by

$$\text{const.} \sum_{j=-\infty}^{\infty} \exp(-\alpha|j - j_0|)$$

and this series converges.

We have defined the system of equations (6.89) as being well conditioned and discussed the condition for this in terms of the maximum norm of (6.90) because this seems the most convenient and appropriate. However, it is worth noting that the arguments go through if the L_2 norm $\|\mathbf{u}\|^2 = (1/J)\sum_{(j)}|\mathbf{u}_j|^2$ is used to replace (6.90) by

(6.98) $$\|\mathbf{u}\| \leq \text{Max}\{\|\mathbf{f}\|, |\boldsymbol{\varphi}|, |\boldsymbol{\psi}|\},$$

where M again must not depend on J. In the proof that the conditions of the theorem are necessary, the set of vectors constructed in the case of a root μ_0 with $|\mu_0| = 1$ again causes (6.98) to fail; while, if dim $\mathscr{R}$ is greater than the number of rows of δ_0, taking an initial vector $\mathbf{u}_0 \in \mathscr{R}$ with $|\mathbf{u}_0| = 1$ and $\delta_0 \mathbf{u}_0 = 0$ gives $\|\mathbf{u}\| \geq J^{-\frac{1}{2}}$ and yields $\|\mathbf{f}\| = 0$, $|\boldsymbol{\varphi}| = 0$, and $|\boldsymbol{\psi}| = O(e^{-\alpha J})$, so again (6.98) is violated. In the proof of sufficiency the modifications are also almost trivial.

This completes the proof of the theorem. In practical applications of the G-R criterion, we can work directly with the equations in their general form (6.85) considering, for example, roots of det $(D_0 + \mu D_1 + \ldots + \mu^K D_K) = 0$ where $D_k = \lambda B_k - A_k$. The sub-spaces $\mathscr{R}$ and $\mathscr{S}$ will usually depend not only on λ but also on Δt or Δx. However, if A_k and B_k are assumed to depend continuously on Δt, the conditions of the theorem need only be verified for $\Delta t = 0$. The procedure for doing this will be briefly as follows: We substitute $\mu^j \lambda^n \hat{\mathbf{u}}$ for $\mathbf{u}_j^n$ in (6.85) with $\Delta t = 0$, where $\hat{\mathbf{u}}$ is a constant vector, yielding the condition for a non-trivial solution det $[M(\lambda, \mu)] = 0$, where $M = D_0 + \mu D_1 + \cdots + \mu^K D_K$. If there are any roots $\mu_i(\lambda)$ with $|\mu_i(\lambda)| = 1$ and $|\lambda| > 1$, the equations are unstable because condition (i) of the main theorem fails. Otherwise, for a

SEC. 6.7] APPLICATION OF G-R CRITERION TO MIXED PROBLEMS   163

root $\mu_0(\lambda)$ of multiplicity k with $|\mu_0(\lambda)| < 1$, we look for vectors satisfying $M(\lambda, \mu_0(\lambda))\mathbf{u} = 0$. If we cannot find k linearly independent vectors of this type we look for further vectors satisfying $M^2\mathbf{u} = 0$, and so on until k have been found. These form $\mathcal{R}(\mu_0(\lambda))$.

This process is repeated for all roots $\mu_i(\lambda)$ with $|\mu_i(\lambda)| < 1$, thus building up the whole of $\mathcal{R}$. We check that the dimension of this sub-space equals the number of boundary conditions imposed at the left boundary, and if this is so we apply these conditions to an arbitrary linear combination of the basis of $\mathcal{R}$. The condition that there is a non-trivial solution to this set of equations then yields a determinantal equation in λ. There must be no solution of this for which $|\lambda| > 1$. If this is true at both boundaries, the G-R criterion is satisfied. Note that though $\mathcal{R}$ depends on λ, its dimension is independent of λ for $|\lambda| > 1$ in a stable system. For the roots $\mu_i(\lambda)$ depend continuously on λ and the dimension of $\mathcal{R}$ can only change by having some μ_i change from $|\mu_i| < 1$ to $|\mu_i| > 1$; this cannot happen since $|\mu_i|$ never equals unity when $|\lambda| > 1$.

From the above discussion of well conditioning, one can now see what is the structure of the family spectrum for the difference operators $\{C_j\}$. It consists of three parts, deriving respectively from the main difference equations (6.85) and the boundary conditions (6.86) and (6.87) on the left and right, and resulting in the violation of respectively conditions (i), (iia) and (iib) of the above theorem. The first part therefore consists of the roots λ of $\det [M(\lambda, \mu)] = 0$ with $|\mu| = 1$, and coincides with the spectrum of (6.85) with j running from $-\infty$ to $+\infty$ and no boundary conditions. Thus, these λ are the same as the eigenvalues of the corresponding amplification matrix (with $\Delta t = 0$) and form p closed curves traced out in the complex plane as μ moves (at most p times) around the unit circle. The second part of the spectrum arises from all those points λ for which the difference equations (6.85) applied at $j = 0, 1, 2 \ldots$ and with boundary conditions (6.86) at $j = 0$ have a solution of the form $\mathbf{u}^n = \lambda^n \tilde{\mathbf{u}}$ with a non-trivial bounded $\tilde{\mathbf{u}}$. Thus, this will occur when the dimension of $\mathcal{R}$ is not equal to the number of rows of (6.86) and, considering the distinct connected regions into which the complex plane is divided by the curves in the first part of the spectrum, we have seen that if this inequality occurs at one point of such a region it holds everywhere in the region. The whole region is then added to the family spectrum; for example, in the problem at the beginning of Section 6.6, the whole of the interior of the circle $|\lambda - (1 - r)| = r$ is added to the spectrum by the imposition of the left boundary condition. In addition, there may be a number of points added to the spectrum where the boundary operator $(\lambda \beta_{00} + \alpha_{00}) + \cdots + (\lambda \beta_{0k_0} + \alpha_{0k_0})$ of (6.86) transforms the base of the $\mathscr{R}$ space into a system of linearly dependent vectors. This will either happen identically, in which case the difference scheme cannot be stable, or at most at a finite number of points. Finally, the third part of the spectrum arises from the boundary condition on the right in a similar way.

Although the reader is referred to the papers of Godunov and Ryabenkii for details, he can now understand what happens when the coefficient matrices A_k and B_k are piecewise continuous functions of x. The spectrum will be made up of the following parts:

1. At each point x in $0 \leq x \leq 1$, there will be roots of det $[M(x, \lambda, \mu)] = 0$ with $|\mu| = 1$; at the jump discontinuities, both $M(x + 0, \lambda, \mu)$ and $M(x - 0, \lambda, \mu)$ have to be considered.
2. There will be the points arising from the boundary conditions.
3. At the discontinuities, there will be points for which dim $\mathscr{R}(x + 0, \lambda)$ + dim $\mathscr{S}(x - 0, \lambda) \neq p$ or the two spaces are not linearly independent.

When the whole spectrum has been obtained by considering all these possibilities, the G-R criterion will be satisfied if it lies within the closed unit disk.

6.8. Conclusions

The theory of mixed initial-boundary-value problems is not, at present, in the same healthy state of development as that of pure initial-value problems. The core of the difference lies in the easy availability of Fourier methods to deal with constant coefficient equations in the one case but not in the other. Nevertheless, recent progress has allowed several parts of the picture to be filled in.

The energy method is a powerful tool in dealing with particular equations or particular classes of equations. It can become rather complicated

SEC. 6.8] CONCLUSIONS

to apply, but it can deal effectively with boundary conditions and takes variable coefficients in its stride. Moreover, it can be used not only to prove the stability of a given difference method, but it also indicates the correct choice of method. However, it is limited to providing sufficient conditions for stability and, particularly so far as boundary conditions are concerned, these may be far from what is necessary. The simple example of the wave equation gives some indication of the manipulations involved in choosing an energy that will narrow this gap.

To complement this tool we have the local mode analysis of Godunov and Ryabenkii. Although it is fearfully complicated to apply in any practical case, it is quite general and gives good insight into the phenomena of instability. But by this method one obtains only necessary stability conditions which, unfortunately, a comparison with the von Neumann condition in the initial-value problem suggests may be a good deal too lenient. On the other hand, for our simple model problem of (6.22), Kreiss (1965) has shown that these conditions are sufficient for stability: moreover, with a mild strengthening of the conditions, he has recently extended this result to systems of hyperbolic equations.

There is a further method for treating mixed problems which we have so far not considered, namely, that of Wiener-Hopf factorization which has been used by Strang (1964b), following the work of Krein (1958). This has the merit of being able to handle implicit difference schemes. For the scalar equation with constant coefficients,

$$(6.99) \qquad \sum_{-K}^{K} {}_{(k)} b_k u_{j+k}^{n+1} = \sum_{-K}^{K} {}_{(k)} a_k u_{j+k}^{n},$$

let the system over the whole real line be denoted by

$$(6.100) \qquad Bu^{n+1} = Au^n;$$

and let the restriction of these operators A and B to the positive half-line in which $u_j^n = 0$ for $j \leq 0$, be denoted by A_+ and B_+. Then the stability of the mixed problem on this half-line depends on whether $\|(B_+^{-1}A_+)^n\| \leq$ constant for $n\Delta t \leq 1$. Defining $a(\theta)$ by $\sum_{(k)} a_k e^{ik\theta}$, $b(\theta)$ by $\sum_{(k)} b_k e^{ik\theta}$, and $r(\theta)$ by $a(\theta)/b(\theta)$, Fourier analysis tells us that

$\|(B^{-1}A)\| = \sup_\theta |r(\theta)|$, yielding the von Neumann condition $|r(\theta)| \leq 1$.
The theorem that Strang obtains is as follows:[5]

The operator B_+ is invertible and stability holds if and only if

(i) $b(\theta) \neq 0$ for real θ,
(ii) Index $(b) \equiv (2\pi)^{-1} \int_{-\pi}^{\pi} d[\arg b(\theta)] = 0$,
(iii) $|r(\theta)| \leq 1$ for real θ.

Conditions (i) and (ii) are the conditions that the discrete Wiener-Hopf operator B_+ is invertible; they are equivalent to requiring that $z^K \beta(z) = \sum_{(k)} b_k z^{K+k}$ has precisely K roots in the open disk $|z| < 1$ and K in the exterior $|z| > 1$. This polynomial can then be factorized in the form

$$(6.101) \qquad \beta(z) = s(z)t(z) = \left[\sum_{0}^{K}{}_{(k)} s_k z^k\right]\left[\sum_{-K}^{0}{}_{(k)} t_k z^k\right], \quad s_0 = 1,$$

where $s \neq 0$ for $|z| \leq 1$ and $t \neq 0$ for $|z| \geq 1$; s and t also satisfy (i) and (ii) separately. This factorization is the key to the proof of the theorem. For, just as the polynomials a and b correspond to doubly infinite matrix operators A and B with elements $A_{ij} = a_{j-i}$ and $B_{ij} = b_{j-i}$, so s and t correspond to matrices which are respectively upper and lower triangular. Thus, $B = ST$, $B_+ = S_+ T_+$ and

$$(B_+^{-1} A_+)^n = T_+^{-1} (S_+^{-1} A_+ T_+^{-1})^n T_+$$

Moreover, because[6] of the triangular form of S_+^{-1} and T_+^{-1}, $(S_+^{-1} A_+ T_+^{-1}) = (S^{-1} A T^{-1})_+$; and for the operators on the whole line $S^{-1} A T^{-1} = S^{-1} T^{-1} A = B^{-1} A$. Hence,

$$(B_+^{-1} A_+)^n = T_+^{-1} (B^{-1}A)_+^n T_+$$

so that stability depends on the von Neumann condition (iii).

---

[5] Strang also treats the case in which the coefficients a_k and b_k depend continuously on x and Δx.

[6] For suppose

$$U = \begin{bmatrix} a & b \\ c & d \end{bmatrix}, \quad U_+ = \begin{bmatrix} 0 & 0 \\ 0 & d \end{bmatrix} \text{ and } V = \begin{bmatrix} e & f \\ g & h \end{bmatrix}, \quad V_+ = \begin{bmatrix} 0 & 0 \\ 0 & h \end{bmatrix}.$$

Then

$$(UV)_+ = \begin{bmatrix} 0 & 0 \\ 0 & cf + dh \end{bmatrix}$$

and equals $U_+ V_+$ if either $c = 0$ or $f = 0$, that is, either U is upper triangular or V is lower triangular.

SEC. 6.8] CONCLUSIONS

Note that in this case the conditions imposed by Godunov and Ryabenkii's theorem at the boundary $x = 0$ are rather trivially satisfied. For the number of roots of $z^K[\lambda\beta(z) - \alpha(z)]$ in the unit disk is the same for all $|\lambda| > 1$; that is, it is the same as for β itself, namely K. Further, since for the problem on the half-line, in which (6.99) is satisfied for $j = 1, 2, \ldots$, no boundary condition is imposed on the right, one must have $b_k = 0$ for $k \geq 1$ and can assume $b_{-K} \neq 0$. Thus, at each time step one must impose the conditions $u_0 = u_{-1} = \cdots = u_{-K+1} = 0$ which properly inhibit the propagation of any of the K decreasing modes into the interior.

We have concentrated our attention in this chapter on problems in only one space variable. Most of the methods and results can be readily extended to problems involving more variables when the region treated is still a half-space;[7] but to treat the more complex boundaries that can arise with more space dimensions is very much more difficult and little has so far been achieved in this area.

---

[7] The reader is referred to Hersch (1963) for a treatment of this problem for differential equations.

CHAPTER 7

Multi-Level Difference Equations

7.1. Notation

With partial differential equations, as with ordinary differential equations, one often uses more time levels (or time steps) in the construction of difference equations of high accuracy than the minimum number required by the differential equations. For example, an improvement over the simple implicit equation for the heat flow problem,

$$\frac{u_j^{n+1} - u_j^n}{\Delta t} = \sigma \frac{u_{j+1}^{n+1} - 2u_j^{n+1} + u_{j-1}^{n+1}}{(\Delta x)^2} + O(\Delta t) + O[(\Delta x)^2],$$

is provided by the three-level equation

(7.1) $$\frac{{}^{3}/_{2} u_j^{n+1} - 2u_j^n + {}^{1}/_{2} u_j^{n-1}}{\Delta t}$$
$$= \sigma \frac{u_{j+1}^{n+1} - 2u_j^{n+1} + u_{j-1}^{n+1}}{(\Delta x)^2} + O[(\Delta t)^2] + O[(\Delta x)^2].$$

When this equation is used, one needs initial data at two times, say $t = 0$ and $t = \Delta t$ ($n = 0$ and 1) to start the calculation. These data may be obtained by a simpler equation, perhaps using a finer net, or by a power-series expansion or otherwise; in any case, when these data are available one can advance the solution from cycle to cycle by use of equation (7.1) and appropriate boundary conditions.

More generally, we suppose that an initial-value problem

(7.2) $$\frac{\partial u}{\partial t} = Au, \quad u = u(t),$$

is approximated by a $(q + 1)$-level formula

(7.3) $$B_q u^{n+q} + B_{q-1} u^{n+q-1} + \cdots + B_0 u^n = 0,$$

where $B_q, \ldots, B_0$ denote finite-difference operators, and u^n is an approximation to $u(n\Delta t)$. We suppose that (7.3), together with suitable boundary conditions, is uniquely solvable for u^{n+q} and that the solution depends continuously on $u^{n+q-1}, \ldots, u^n$. That is, (7.3) is equivalent to the equation

(7.4) $$u^{n+q} = C_{q-1}u^{n+q-1} + \cdots + C_0 u^n,$$

where

$$C_j = C_j(\Delta t) = -B_q^{-1} B_j, \quad j = 0, \ldots, q-1,$$

are bounded linear operators. As in previous discussions, we suppose that the space increments, $\Delta x, \Delta y$, etc., are specified functions of Δt.

7.2. Auxiliary Banach Space

A q-component column vector, whose components are elements of $\mathscr{B}$, will be regarded as an element of an auxiliary Banach space $\tilde{\mathscr{B}}$. The norm can be chosen in various ways, for example as

$$\left\| \begin{bmatrix} u \\ v \\ \vdots \end{bmatrix} \right\| = \{\|u\|^2 + \|v\|^2 + \ldots\}^{1/2}.$$

If we define

$$\tilde{\varphi}^n = \begin{bmatrix} u^{n+q-1} \\ \vdots \\ u^n \end{bmatrix}$$

and

(7.5) $$\tilde{C} = \begin{bmatrix} C_{q-1} & C_{q-2} & \cdots & C_0 \\ I & 0 & \cdots & 0 \\ 0 & I & \cdots & 0 \\ \cdot & \cdot & \cdot & \cdot \\ 0 & 0 & \cdots & I & 0 \end{bmatrix} = \tilde{C}(\Delta t)$$

where I denotes the identity operator, then equation (7.4) becomes

(7.6a) $$\tilde{\varphi}^{n+1} = \tilde{C}(\Delta t)\tilde{\varphi}^n.$$

This procedure is equivalent to the introduction of new dependent variables, i.e., denoting u^{n-1} by v^n, u^{n-2} by w^n, etc., whereupon equation (7.4) can be replaced by the system

(7.6b)
$$u^{n+1} = C_{q-1}u^n + C_{q-2}v^n + C_{q-3}w^n + \cdots,$$
$$v^{n+1} = u^n,$$
$$w^{n+1} = v^n,$$
etc.

The problem has thus been reduced to a two-level problem (the only superscripts occurring in these equations are n and $n + 1$).

Primary initial data are assumed given at $t = 0$; the values of u at $t = \Delta t, 2\Delta t, \ldots, (q - 1)\Delta t$ are to be obtained from the primary initial data, by some approximation, to enable one to start the calculation with equation (7.3). For the present we assume that this initial calculation is exact (this requirement will be relaxed later); that is $\tilde{\varphi}^0$ is given by

$$\tilde{\varphi} = \begin{bmatrix} E((q-1)\Delta t)u^0 \\ E((q-2)\Delta t)u^0 \\ \vdots \\ u_0 \end{bmatrix},$$

where the primary initial data are specified by the element u^0 (of $\mathscr{B}$), and $E(t)$ is the solution operator for (7.2). We write this as $\tilde{\varphi}^0 = \tilde{S}\tilde{u}^0$, where

$$\tilde{u}^0 = \begin{bmatrix} u^0 \\ u^0 \\ \vdots \\ u^0 \end{bmatrix}$$

and $\tilde{S}$ is the diagonal matrix

(7.7) $$\tilde{S} = \begin{bmatrix} E((q-1)\Delta t) & & & \\ & E((q-2)\Delta t) & & 0 \\ 0 & & \ddots & \\ & & & I \end{bmatrix}.$$

Thus the approximate solution of the initial-value problem is given by $\tilde{\varphi}^n = \tilde{C}(\Delta t)^n \tilde{S}\tilde{u}^0$. This is expected to be an approximation of the quantity

$$\begin{bmatrix} u((n+q-1)\Delta t) \\ u((n+q-2)\Delta t) \\ \vdots \\ u(n\Delta t) \end{bmatrix} = E(n\Delta t)\tilde{S}\tilde{u}^0.$$

[SEC. 7.3] THE EQUIVALENCE THEOREM

We therefore are interested in comparing $\tilde{C}(\Delta t)^n \tilde{S} \tilde{u}^0$ with $E(n\Delta t)\tilde{S}\tilde{u}^0$.

In the limit $\Delta t \to 0$ we are concerned with elements of $\tilde{\mathscr{B}}$ whose q components are all equal, namely, with

$$\begin{bmatrix} u(t) \\ \vdots \\ u(t) \end{bmatrix} = E(t) \begin{bmatrix} u^0 \\ \vdots \\ u^0 \end{bmatrix}$$

which we write as $\tilde{u}(t) = E(t)\tilde{u}^0$. This is also the limit approached by $E(t)\tilde{S}\tilde{u}$, because $\tilde{S} \to I$ as $\Delta t \to 0$. It is convenient to denote by $\tilde{\mathscr{B}}^0$ the subspace of $\tilde{\mathscr{B}}$ containing elements having equal components, i.e., the element

$$\tilde{\varphi} = \begin{bmatrix} u \\ v \\ w \\ \vdots \end{bmatrix}$$

of $\tilde{\mathscr{B}}$ is an element of $\tilde{\mathscr{B}}^0$ if $u = v = w = \ldots$. Thus, $\tilde{u}^0$ and $\tilde{u}(t)$ lie in the subspace $\tilde{\mathscr{B}}^0$, but the approximation does not—it merely lies close to $\tilde{\mathscr{B}}^0$.

In practice, the starting values are not obtained from an exact calculation $\tilde{\varphi}^0 = \tilde{S}\tilde{u}^0$ but from an approximation $\tilde{\varphi}^0 = \tilde{\psi}$. We shall require convergence to $\tilde{u}(t)$ as $\Delta t \to 0$ and $\tilde{\psi} \to \tilde{u}^0$, independently. We do not require that $\tilde{\psi}$ lies in the subspace $\tilde{\mathscr{B}}^0$, but only that its limit $\tilde{u}^0$ does.

7.3. The Equivalence Theorem

The (multi-level) difference formula (7.6a) is called *stable* if for some positive τ the operators

(7.8) $\qquad \tilde{C}(\Delta t)^n \quad \text{for } \begin{cases} 0 < \Delta t < \tau \\ 0 \leq n\Delta t \leq T \end{cases}$

are uniformly bounded. In other words, the equivalent two-level system (7.6b) is to be stable in precisely the sense of Chapter 3.

The *consistency condition* is that there exists a set U^0, dense in $\mathscr{B}$, of possible initial elements of genuine solutions such that if u^0 is any element of U^0, if $\varepsilon > 0$ and if we call

$$\tilde{u}(t) = \begin{bmatrix} E(t)u^0 \\ \vdots \\ E(t)u^0 \end{bmatrix},$$

then

(7.9) $\quad \|[\tilde{C}(\Delta t) - E(\Delta t)]\tilde{S}\tilde{u}(t)\| < \varepsilon \Delta t \quad \text{for } \begin{cases} \text{sufficiently small } \Delta t \\ 0 \leq t \leq T. \end{cases}$

This is the same as the alternative form given for two-level equations in (3.6b). The $\tilde{u}(t)$ entering the present condition all lie in the subspace $\tilde{\mathscr{B}}^0$.

The approximation is called *convergent* if for any sequences $\Delta_j t$ and n_j such that $\Delta_j t > 0$, $\Delta_j t \to 0$, $n_j \Delta_j t \to t$ as $j \to \infty$ ($0 \leq t \leq T$) and any $\tilde{u}$ in the subspace $\tilde{\mathscr{B}}^0$,

(7.10) $\quad \lim_{\substack{j \to \infty \\ \tilde{\psi} \to \tilde{u}}} \|\tilde{C}(\Delta_j t)^{n_j} \tilde{\psi} - E(t)\tilde{u}\| = 0.$

The equivalence theorem says, as before, that *for a properly posed initial-value problem and an approximation satisfying the consistency condition, stability is necessary and sufficient for convergence.*

The proof is similar to that of Chapter 3, the only real differences arising because of the distinction between the spaces $\tilde{\mathscr{B}}$ and $\tilde{\mathscr{B}}^0$. To prove that convergence implies stability, assume that the approximation is convergent and let $\tilde{\psi}$ be any element of $\tilde{\mathscr{B}}$. We assert that the quantities

$$\|\tilde{C}(\Delta t)^n \tilde{\psi}\| \quad \text{for } \begin{cases} 0 < \Delta t < \tau \\ 0 \leq n\Delta t \leq T \end{cases}$$

are bounded; for, if they were not, we could pick from them a sequence for which $\|\tilde{C}(\Delta_j t)^{n_j} \tilde{\psi}\| \to \infty$ as $j \to \infty$ and then we could pick from this sequence a subsequence for which $n_j \Delta_j t$ has a limit, say t ($0 \leq t \leq T$) as $j \to \infty$. Then the elements

$$\tilde{\psi}_j = \frac{\tilde{\psi}}{\|\tilde{C}(\Delta_j t)^{n_j} \tilde{\psi}\|^{1/2}}$$

would approach the zero element

$$\tilde{u}^0 = \begin{bmatrix} 0 \\ \vdots \\ 0 \end{bmatrix}$$

which lies in the subspace $\tilde{\mathscr{B}}^0$ and we would have $\|\tilde{C}(\Delta_j t)^{n_j} \tilde{\psi}_j\| \to \infty$

SEC. 7.3] THE EQUIVALENCE THEOREM

as $j \to \infty$, whereas according to convergence we should have $\|\tilde{C}(\Delta_j t)^{n_j}\tilde{\psi}_j\| \to \|E(t)\tilde{u}^0\| = 0$ as $j \to \infty$. Therefore there is a bound, $K_1(\tilde{\psi})$ such that

$$\|\tilde{C}(\Delta t)^n \tilde{\psi}\| < K_1(\tilde{\psi}) \quad \text{for} \quad \begin{cases} 0 < \Delta t < \tau \\ 0 \leq n\Delta t \leq T \\ \tilde{\psi} \text{ in } \tilde{\mathscr{B}}, \end{cases}$$

and stability then follows from the principle of uniform boundedness.

To prove conversely that stability implies convergence, let

$$\Delta_j t > 0, \quad \lim_{j \to \infty} \Delta_j t = 0, \quad \lim_{j \to \infty} n_j \Delta_j t = t, \quad 0 \leq t \leq T,$$

and set $\tilde{\chi}_j = [(\tilde{C}(\Delta_j t)^{n_j} - E(n_j \Delta_j t)]\tilde{S}\tilde{u}^0$, where u^0 is a member of the set U^0 of elements occurring in the definition of consistency. Then $\tilde{u}(t) = E(t)\tilde{u}^0$ is a genuine solution and

$$\tilde{\chi}_j = \sum_{0}^{n_j - 1} \tilde{C}(\Delta_j t)^k [\tilde{C}(\Delta_j t) - E(\Delta_j t)] E((n_j - k - 1)\Delta_j t)\tilde{S}\tilde{u}^0.$$

Now $\tilde{S}$ commutes with $E(t)$ (see defining equation (7.7)) and, therefore, according to the consistency condition (7.9), if $\varepsilon > 0$,

$$\|[\tilde{C}(\Delta_j t) - E(\Delta_j t)]E((n_j - k - 1)\Delta_j t)\tilde{S}\tilde{u}^0\| < \varepsilon \Delta_j t$$

for sufficiently small positive $\Delta_j t$. If K_2 denotes the uniform bound of the operators (7.8), we obtain, using the triangle inequality,

$$\|\tilde{\chi}_j\| \leq \sum_{0}^{n_j - 1} K_2 \varepsilon \Delta_j t = K_2 \varepsilon n_j \Delta_j t \leq K_2 \varepsilon T,$$

for sufficiently small positive $\Delta_j t$. Hence, since ε was arbitrary, we have that $\tilde{\chi}_j \to 0$ as $j \to \infty$. As in Chapter 3, we can now show that

$$[\tilde{C}(\Delta_j t)^{n_j} - E(t)]\tilde{S}\tilde{u}^0 \to 0$$

where t is the limit of $n_j \Delta_j t$. Then, also as in Chapter 3, if u is an *arbitrary* element of $\mathscr{B}$, we can find a sequence $\{u_m\}$ of elements of the set U^0 such that $u_m \to u$ as $m \to \infty$, and if we call

$$\tilde{u}_m = \begin{bmatrix} u_m \\ \vdots \\ u_m \end{bmatrix}, \quad \tilde{u} = \begin{bmatrix} u \\ \vdots \\ u \end{bmatrix},$$

then
$$[\tilde{C}(\Delta_j t)^{n_j} - E(t)]\tilde{S}\tilde{u} = [\tilde{C}(\Delta_j t)^{n_j} - E(t)]\tilde{S}\tilde{u}_m$$
$$+ \tilde{C}(\Delta_j t)^{n_j}\tilde{S}(\tilde{u} - \tilde{u}_m) - E(t)\tilde{S}(\tilde{u} - \tilde{u}_m).$$

The last two terms of this equation can be made as small as we please by choosing m sufficiently large, by uniform boundedness[1] of the operators $\tilde{C}(\Delta_j t)^{n_j}$, $E(t)$, and $\tilde{S}$; the first term on the right can then be made as small as we please by choosing j sufficiently large.

To complete the convergence proof we now need a step that was not necessary in Chapter 3, namely,
$$\tilde{C}(\Delta_j t)^{n_j}\tilde{\psi} - E(t)\tilde{u} = [\tilde{C}(\Delta_j t)^{n_j} - E(t)]\tilde{S}\tilde{u}$$
$$+ \tilde{C}(\Delta_j t)^{n_j}(\tilde{\psi} - \tilde{S}\tilde{u}) - E(t)(\tilde{u} - \tilde{S}\tilde{u}).$$

As $\Delta t \to 0$, $\tilde{S} \to I$, hence, if $\tilde{\psi} \to \tilde{u}$, $\|\tilde{\psi} - \tilde{S}\tilde{u}\| \to 0$, $\|\tilde{u} - \tilde{S}\tilde{u}\| \to 0$, and we see that
$$\lim_{\substack{j \to \infty \\ \tilde{\psi} \to \tilde{u}}} \|\tilde{C}(\Delta_j t)^{n_j}\tilde{\psi} - E(t)\tilde{u}\| = 0$$

for arbitrary $\tilde{u}$ in the subspace $\tilde{\mathscr{B}}^0$, which was to be proved.

7.4. Consistency and Order of Accuracy

For the assessment of consistency and accuracy it is convenient to work with the difference equations in their simplest form (7.3), before they have been solved for u^{n+q}. Ambiguity is avoided by assuming that the expression on the left there is an approximation to the operator $\partial/\partial t - A$ when applied to any sufficiently smooth function $u(t)$, that is,

(7.11) $B_q u(t + q\Delta t) + \cdots + B_0 u(t) - \left(\dfrac{\partial}{\partial t} - A\right)u(t) = o(1)$ as $\Delta t \to 0$.

Then we define the *order of accuracy*, as in Chapter 4, as the largest real number ρ for which

(7.12) $\|B_q u(t + q\Delta t) + \cdots + B_0 u(t)\| = O[(\Delta t)^\rho]$,   as $\Delta t \to 0$,

---

[1] Note that $\tilde{S}$ is a function of Δt but is bounded for $0 \leq \Delta t < \tau$ and, in fact, $\tilde{S} \to I$ as $\Delta t \to 0$.

SEC. 7.4]   CONSISTENCY AND ORDER OF ACCURACY   175

for all sufficiently smooth genuine solutions of the differential equation $\partial u/\partial t = Au$. The expression in (7.12) is called the truncation error and can be estimated by expanding u in a Taylor series.

To relate this to the consistency condition (7.9), which is expressed in terms of the operators $C_j = -B_q^{-1}B_j$, we specialize the discussion to the pure initial-value problems with constant coefficients treated in Chapter 4. Then we shall show below that $B_q^{-1} = O(\Delta t)$. This is more difficult to see than in the two-level case but follows from (7.11) together with the boundedness of $C_1, C_2, \ldots, C_{q-1}$. Hence, *if a difference scheme is stable and the truncation error goes to zero as $\Delta t \to 0$, the consistency condition (7.9) is satisfied.* The only nonvanishing component of $[\tilde{C}(\Delta t) - E(\Delta t)]\tilde{S}\tilde{u}(t)$ is the first, so that we have merely to show that, for any $\varepsilon > 0$,

$$\|[C_{q-1}E((q-1)\Delta t) + \cdots + C_0 - E(q\Delta t)]u(t)\| < \varepsilon \Delta t,$$

for any $u(0)$ in some class U^0 dense in $\mathscr{B}$ and for sufficiently small Δt.

As in Chapter 4 the Banach space $\mathscr{B}$ is the class L^2 of functions $\mathbf{u}(\mathbf{x})$ satisfying a periodicity condition, and the norm is the Hilbert norm. Let $\mathscr{B}_m$ be the subspace of trigonometrical polynomials of degree m. If u_0 represents a function $\mathbf{u}_0(\mathbf{x})$ in $\mathscr{B}_m$ and if we take $u(t) = tu_0$ in (7.11), we have

(7.13) $$Lu_0 = o(1) \quad \text{as } \Delta t \to 0,$$

where

(7.14) $$L = \Delta t(qB_q + (q-1)B_{q-1} + \cdots + B_1) - I.$$

Since $\mathscr{B}_m$ has a finite basis, each element of which satisfies (7.13), we can write $\|L\|_m < \varepsilon_1$ for sufficiently small Δt, where ε_1 is any positive number and $\|L\|_m$ denotes the bound of L relative to the subspace $\mathscr{B}_m$. Therefore, $I + L$ has an inverse in this subspace: $(I + L)^{-1} = I + M$, where

$$\|M\|_m < \varepsilon_2 = \frac{\varepsilon_1}{1 - \varepsilon_1};$$

that is, $[qB_q + \cdots + B_1]^{-1} = \Delta t(I + M)$. From the definition $C_j = -B_q^{-1}B_j$, we therefore have

(7.15) $$B_q^{-1} = [qI - (q-1)C_{q-1} - \cdots - C_1]\Delta t(I + M).$$

For the problems under consideration, the solution $u(t)$ lies in $\mathscr{B}_m$ if $u(0)$ does, because A in equation (7.2) is a differential operator with constant coefficients. Furthermore, $\partial u/\partial t - Au = 0$ for this solution, and hence (7.11) gives

(7.16) $\quad \|[B_q E(q\Delta t) + B_{q-1} E((q-1)\Delta t) + \cdots + B_0] u(t)\| < \varepsilon_0,$

for any $\varepsilon_0 > 0$, for sufficiently small Δt.

Multiplying (7.15) by (7.16) we have

(7.17) $\quad \|[E(q\Delta t) - C_{q-1} E((q-1)\Delta t) - \cdots - C_0] u(t)\| < \varepsilon_3 \Delta t,$

where

$$\varepsilon_3 = [\underset{0 < \Delta t < \tau}{\text{Max}} \|qI - (q-1)C_{q-1} - \cdots - C_1\|] \varepsilon_0 \|I + M\|_m.$$

The indicated maximum exists because we have assumed that the difference equations are stable, which implies that the operators $C_{q-1}(\Delta t), \ldots, C_1(\Delta t)$ are uniformly bounded for $0 < \Delta t < \tau$. An inequality of the type (7.17) holds for any solution $u(t)$ in $\mathscr{B}_m$ (this is, of course, a genuine solution). The possible initial elements $u(0)$ of these solutions, taking all the subspaces $\mathscr{B}_m$ into account, are dense in $\mathscr{B}$, and consistency is established.

7.5. Example of Du Fort and Frankel

To illustrate the foregoing, we investigate the consistency of the difference equation

(7.18) $\quad \dfrac{u_j^{n+1} - u_j^{n-1}}{2\Delta t} = \sigma \dfrac{u_{j+1}^n - u_j^{n+1} - u_j^{n-1} + u_{j-1}^n}{(\Delta x)^2}$

proposed by Du Fort and Frankel (1953) for the numerical solution of the diffusion equation:

(7.19) $\quad \dfrac{\partial u}{\partial t} = \sigma \dfrac{\partial^2 u}{\partial x^2}, \quad \sigma = \text{constant} > 0.$

As will be shown below, equation (7.18) is always stable. If we substitute Taylor's series for the various terms in (7.18), e.g.,

$$u_j^{n+1} = \left(u + \Delta t \dfrac{\partial u}{\partial t} + \ldots\right)_j^n \text{etc.,}$$

SEC. 7.5] EXAMPLE OF DU FORT AND FRANKEL 177

and compare with (7.19), we find a truncation error of order $(\Delta t/\Delta x)^2$. Specifically, for any sufficiently often differentiable function $u(x, t)$,

$$\frac{u_j^{n+1} - u_j^{n-1}}{2\Delta t} - \sigma \frac{u_{j+1}^n - u_j^{n+1} - u_j^{n-1} + u_{j-1}^n}{(\Delta x)^2} - \left(\frac{\partial u}{\partial t} - \sigma \frac{\partial^2 u}{\partial x^2}\right)_j^n$$

$$= \sigma \left(\frac{\Delta t}{\Delta x}\right)^2 \left(\frac{\partial^2 u}{\partial t^2}\right)_j^n + O[(\Delta t)^2] + O[(\Delta x)^2] + O\left(\frac{(\Delta t)^4}{(\Delta x)^2}\right).$$

Consistency therefore requires that $\Delta t/\Delta x \to 0$ as $\Delta t \to 0$. That is, (7.18) *is consistent with (7.19) if and only if Δt goes to zero faster than Δx; if, on the other hand, $\Delta t/\Delta x$ is kept fixed, say $= \beta$, then (7.18) is consistent not with the diffusion equation (7.19) but with the hyperbolic equation*

$$\frac{\partial u}{\partial t} - \sigma \frac{\partial^2 u}{\partial x^2} + \sigma \beta^2 \frac{\partial^2 u}{\partial t^2} = 0.$$

We can see that the difference equation (7.18) is always stable by writing he equivalent two-level system:

$$\frac{u_j^{n+1} - v_j^n}{2\Delta t} = \sigma \frac{u_{j+1}^n - u_j^{n+1} - v_j^n + u_{j-1}^n}{(\Delta x)^2},$$

$$v_j^{n+1} = u_j^n.$$

The amplification matrix of this sytem is

(7.20) $$G(\Delta t, k) = \begin{bmatrix} \dfrac{2\alpha}{1 + \alpha} \cos \beta & \dfrac{1 - \alpha}{1 + \alpha} \\ 1 & 0 \end{bmatrix},$$

where $\alpha = 2\sigma\Delta t/(\Delta x)^2$, $\beta = k\Delta x$, and the eigenvalues of G are

$$\lambda = \frac{\alpha \cos \beta \pm \sqrt{1 - \alpha^2 \sin^2 \beta}}{1 + \alpha}.$$

The absolute values of the eigenvalues are plotted schematically against $\cos \beta$ in Figures 7.1 and 7.2, for typical cases with $\alpha > 1$ and $\alpha < 1$, respectively. Condition 2 of Section 4.11, which is sufficient for stability, is satisfied here because one eigenvalue satisfies

$$|\lambda_1| \leq 1$$

and the other satisfies

$$|\lambda_2| \leqq \sqrt{|1 - \alpha|/(1 + \alpha)} < 1.$$

Therefore the Du Fort-Frankel system is stable for any value of α.

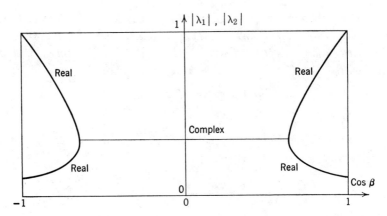

Fig. 7.1. Eigenvalues of the amplification matrix (7.20) of the Du Fort-Frankel difference equation for the heat flow problem in a case in which $\alpha > 1$.

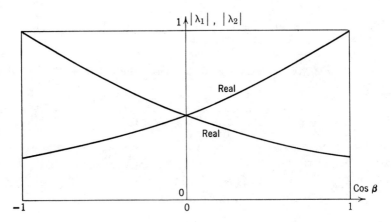

Fig. 7.2. Same as Figure 7.1 but for a case in which $\alpha < 1$.

7.6. Summary

For properly posed problems with constant coefficients, the convergence conditions for multi-level difference equations are practically identical with those for two-level equations. Vanishing of the truncation error in the limit $\Delta t \to 0$ implies consistency of the approximation provided that the difference equations are stable.[2]

To investigate stability one simply writes the equivalent two-level system by introducing further dependent variables and then investigates the stability of this system by any of the methods of Chapter 4. The extra starting values needed can be obtained from any desired approximation which is valid in the limit $\Delta t \to 0$. In particular they can be simply taken equal to the corresponding initial values, $u^1 = u^0$, $u^2 = u^0$, etc., although a better approximation will presumably give greater accuracy for given Δt.

---

[2] This proviso was not necessary for two-level formulas, but presumably consistency is of no interest unless the equations are stable.

PART II
APPLICATIONS

Preface to Part II

This part of the book is devoted to the application of difference methods to certain initial-value problems arising in physical science. The material is organized according to the nature of the physical problems rather than the nature of the mathematical equations, because equations of given mathematical type generally have quite diverse aspects when the detailed considerations of various physical fields are taken into account. For example, the hyperbolic systems that appear in neutron transport theory require different procedures from the ones applicable to the hyperbolic systems that appear in fluid dynamics. Nevertheless, there are many features common to the problems of different areas, and we give here a cross referencing of some of these features.

Methods for *parabolic problems* are discussed mainly in Chapter 8 on diffusion and heat flow, but the equations of elastic vibrations, discussed in Chapter 11, and those of coupled sound and heat flow, discussed in Section 10.4, are also essentially parabolic in the sense that there are infinite signal speeds.

Methods for *linear hyperbolic systems* are extensively discussed in Chapter 9 on the transport equation, especially in Sections 9.6 through 9.16; they appear again in Chapter 10 on sound waves; the equations of the vibration of a bar or wire under tension, discussed in Section 11.6, are very nearly hyperbolic.

Non-linear hyperbolic systems appear in Chapters 12 and 13 on fluid dynamics. Owing to the non-linearity, the methods for these systems are rather complicated and rather specialized to the fluid-dynamic problems. We have not attempted to propose any universal methods applicable to general non-linear hyperbolic systems other than systems of conservation laws.

Methods for the solution of the *systems of linear equations* arising from implicit schemes are discussed in Section 8.5 for the simple tri-diagonal case and in Section 11.5 for the multi-diagonal case.

Implicit schemes for various problems are found in Sections 8.2, 8.5,

8.6 (parabolic in one space variable), 8.8, 8.9 (parabolic in several space variables—alternating-direction and fractional-step methods), 9.9, 10.3 (hyperbolic), 11.3, 11.4, and 11.5 (elastic vibrations). As explained in Section 11.4, the implicit procedure is especially appropriate in problems of dynamic elasticity.

Practical stability criteria for cases in which the usual theory is inadequate for practical purposes are discussed in Sections 9.7, 10.4, 10.5, 11.6, and 12.16. These are cases in which, although the limitation imposed on Δt (for given Δx, etc.) by the usual theory ensures a bounded error growth in the limit $\Delta t \to 0$, for the actual Δt used in practice the error growth may nevertheless be enormous, unless additional limitations are imposed.

CHAPTER 8

Diffusion and Heat Flow

8.1. Examples of Diffusion

An example of the sort of diffusion process we shall consider is the flow of heat in a stationary, isotropic medium (see Carslaw, 1945). An initial temperature distribution is known, heat sources, if any, are known; and the problem is to determine the temperature distribution at later times.

If the temperature distribution at some instant t is denoted by $T = T(\mathbf{x}, t)$, where T denotes temperature and $\mathbf{x}$ the position vector, we may state the law of heat conduction as follows: there exists a positive (scalar) $\varkappa = \varkappa(\mathbf{x}, T)$ called the thermal conductivity of the material, such that for any temperature distribution the heat flux density is given by $\mathbf{F}(\mathbf{x}, t) = -\varkappa(\mathbf{x}, T(\mathbf{x}, t))\nabla T(\mathbf{x}, t)$, which we shall abbreviate as

(8.1) $$\mathbf{F} = -\varkappa \nabla T.$$

The physical significance of the vector field $\mathbf{F}$ is that the integral of the normal component of $\mathbf{F}$ over any surface is equal to the rate of flow of heat through the surface. The law (8.1) says that for a given temperature T at a given point $\mathbf{x}$ the heat flux density is proportional to the temperature gradient there, and has the opposite direction. The proportionality constant $\varkappa$ might also depend on time if chemical reactions occur, but we suppose this is not the case.

Let $E = E(\mathbf{x}, t)$ and $Q = Q(\mathbf{x}, t)$ denote the internal energy per unit volume and the heat source (heat produced per unit time, unit volume), respectively. The divergence of $\mathbf{F}$ is the rate of efflux of energy per unit volume by heat flow; hence, by conservation of energy, we have

$$\frac{\partial E}{\partial t} = \nabla \cdot \varkappa \nabla T + Q.$$

If $a = a(\mathbf{x}, T)$ denotes the specific heat of the material (referred to unit volume), we obtain a partial differential equation for the temperature

(8.2) $$\frac{\partial T}{\partial t} = \frac{1}{a}[\nabla \cdot \varkappa \nabla T + Q].$$

With suitable initial and boundary conditions, this equation provides an initial-value problem of the sort considered in this chapter. The actual examples and stability discussions will be for greatly simplified versions of the problem. The finite-difference equation systems given can, however, be generalized quite easily and have been applied to quite general cases of equation (8.2) in practice.

Other phenomena lead to an equation of the same form. For example, the above equation becomes that of diffusion of a substance through a permeable system if T is interpreted as the concentration of the substance, $\varkappa$ as the diffusion constant, a as unity, and Q as the specific rate of generation of the substance by chemical reactions, if any.

A further example is provided by heat flow in a star, in which energy is transferred by radiation flow as well as by thermal conduction. If the material is at rest, the heat flow throughout the star is governed quite accurately by the above equation, except in a very thin surface layer, but with the following changes of interpretation of the symbols:

α = heat capacity of matter per unit volume + $4a_0 T^3$,

$\varkappa$ = thermal conductivity of matter + $4a_0 c_0 T^3 / 3K\rho$,

where a_0 is the Stefan-Boltzmann constant, c_0 is the speed of light, ρ is the density, and $K = K(\rho, T)$ is the Rosseland mean opacity coefficient for the material of the star. A boundary condition adequate for some purposes is $T = 0$ at the surface of the star.

8.2. The Simplest Heat Flow Problem

Taking $Q = 0$, a and $\varkappa$ as constants, and considering only one space variable (as for the flow of heat along a thin rod) we have

(8.3) $$\frac{\partial u}{\partial t} - \sigma \frac{\partial^2 u}{\partial x^2} = 0, \quad \sigma = \text{const.} > 0.$$
$$u = u(x, t),$$
$$u(x, 0) = u_0(x) \quad \text{(given)},$$

SEC. 8.2] THE SIMPLEST HEAT FLOW PROBLEM 187

and we suppose, as in Chapter 1, that the boundary conditions can be replaced by a periodicity condition.

The most usual four-point and six-point difference equations for (8.3) are included in the scheme

$$(8.4) \qquad \frac{u_j^{n+1} - u_j^n}{\Delta t} - \sigma \frac{\theta(\delta^2 u)_j^{n+1} + (1-\theta)(\delta^2 u)_j^n}{(\Delta x)^2} = 0,$$

where θ is a positive constant; $\theta = 0$ gives the four-point explicit scheme with forward time difference. Other values of θ give implicit schemes; $\theta = \frac{1}{2}$ gives the six-point scheme with centered time difference, and $\theta = 1$ gives the four-point scheme with backward time difference. As in the previous chapters, the symbol δ stands for the centered space difference. A simple algorithm for solving the implicit equations is given in Section 8.5, below.

The stability condition for (8.4) was derived in Chapter 1—equation (1.24)—but it can also be obtained from the formalism of Chapter 4. The amplification matrix $G(\Delta t, k)$ has in this case just one element (because there is only one dependent variable) and is in fact just the growth factor denoted by $\xi(m)$ in Chapter 1 and given by (1.23). The necessary condition and the sufficient condition for stability given in Chapter 4 coincide and are in agreement with the condition found in Chapter 1:[1]

$$2\sigma \Delta t / (\Delta x^2) \leq 1/(1 - 2\theta) \quad \text{if } 0 \leq \theta < \tfrac{1}{2},$$

no restriction $\qquad$ if $\tfrac{1}{2} \leq \theta \leq 1$.

According to Chapter 1, the truncation error is defined by:

$$e[u] = \frac{u_j^{n+1} - u_j^n}{\Delta t} - \sigma \frac{\theta(\delta^2 u)_j^{n+1} + (1-\theta)(\delta^2 u)_j^n}{(\Delta x)^2},$$

where $u = u(x, t)$ is an exact solution. By Taylor's series expansion about the point $(x = j\Delta x, t = n\Delta t)$, we find that $e[u] = O(\Delta t) + O[(\Delta x)^2]$. Since this goes to zero as the net is refined, it follows from the considerations of Chapters 4 and 7 that the consistency condition is satisfied by the difference equations (8.4) as an approximation to the heat flow equation.

---

[1] Conditional stability also occurs for $\theta < 0$ and unconditional stability for $\theta > 1$, but generally these values of θ are not of interest.

If $u(x, t)$ is sufficiently differentiable, the truncation error can be written as

$$(8.5) \quad e[u] = \sigma \frac{\partial^4 u}{\partial x^4} [\sigma \Delta t (\tfrac{1}{2} - \theta) - \tfrac{1}{12}(\Delta x)^2] + O[(\Delta t)^2] + O[(\Delta x)^4],$$

by use of the fact that $\partial^2 u/\partial t^2 = \sigma^2 \partial^4 u/\partial x^4$. Therefore, if the constants θ, Δt, Δx are so chosen as to make the square bracket on the right side of (8.5) equal to zero, the approximation achieves a higher order of accuracy than otherwise. This occurs (assuming $0 \leq \theta < \tfrac{1}{2}$) when Δt is taken equal to one third of the maximum value permitted by the stability condition. It is seen also that the highest order of accuracy is not in general obtained by centering the time difference—i.e., by taking $\theta = \tfrac{1}{2}$—but rather by taking $\theta = \tfrac{1}{2} - (\Delta x)^2/12\sigma\Delta t$ (the approximation is stable for this value of θ). It has been pointed out by Saul'ev (1958) that in the special case where $(\Delta x)^2/\sigma\Delta t = \sqrt{20}$, there is a further reduction of the truncation error to $O[(\Delta x)^6]$.

This method of achieving improved accuracy can be somewhat generalized for problems with variable coefficients. If $\sigma = \sigma(x, t)$ varies with x and t, the estimate (8.5) of the truncation error is no longer strictly correct, but if $\sigma(x, t)$ is a slowly varying function, considerable improvement of accuracy can still be obtained by taking

$$\theta = \theta(x, t) = \frac{1}{2} - \frac{(\Delta x)^2}{12\sigma(x, t)\Delta t}.$$

In this case, it is more convenient to rewrite the equation in the equivalent form, listed as number 12 in the table, from which θ has been eliminated. (See Section 8.3.)

Table 8.I summarizes various difference systems for the simple diffusion equation. The mnemonic diagram at the left indicates the type of differencing. The points shown are those points of the net in the x, t-plane used in one application of the formula; the t-axis points to the top of the page and the x-axis horizontally to the right. Where three points are connected by a horizontal line, they are used in forming the second space difference, and where two points are connected by a vertical line they are used in forming the time difference. When there are two or more sets of

TABLE 8.I
Finite-difference Approximations to

$$\frac{\partial u}{\partial t} = \sigma \frac{\partial^2 u}{\partial x^2}, \quad \sigma = \text{const.} > 0$$

1.	⊥	1.	$\dfrac{u_j^{n+1} - u_j^n}{\Delta t} = \sigma \dfrac{(\delta^2 u)_j^n}{(\Delta x)^2}$
			$e = O(\Delta t) + O[(\Delta x)^2]$
			explicit, stable if $\sigma \Delta t/(\Delta x)^2 = \text{const.} \leq \tfrac{1}{2}$ as $\Delta t, \Delta x \to 0$.
2.	⊤⊥ ½ ½	2.	$\dfrac{u_j^{n+1} - u_j^n}{\Delta t} = \sigma \dfrac{(\delta^2 u)_j^n + (\delta^2 u)_j^{n+1}}{2(\Delta x)^2}$
			Crank and Nicholson (1947)
			$e = O[(\Delta t)^2] + O[(\Delta x)^2]$
			implicit, always stable.
3.	⊤	3.	$\dfrac{u_j^{n+1} - u_j^n}{\Delta t} = \sigma \dfrac{(\delta^2 u)_j^{n+1}}{(\Delta x)^2}$
			Laasonen (1949)
			$e = O(\Delta t) + O[(\Delta x)^2]$
			implicit, always stable.
4.	(Special)	4.	Same as 1, but with $\sigma \Delta t/(\Delta x)^2 = \tfrac{1}{6}$
			$e = O[(\Delta t)^2] = O[(\Delta x)^4]$
			special case of 1, stable.
5.	⊤⊥ θ 1−θ	5.	$\dfrac{u_j^{n+1} - u_j^n}{\Delta t} = \sigma \dfrac{\theta(\delta^2 u)_j^{n+1} + (1-\theta)(\delta^2 u)_j^n}{(\Delta x)^2}$
			where $\theta = \text{const.}, 0 \leq \theta \leq 1$
			$e = O(\Delta t) + O[(\Delta x)^2]$
			for $0 \leq \theta < \tfrac{1}{2}$, stable if $\sigma \Delta t/(\Delta x)^2 = \text{const.} \leq 1/(2 - 4\theta)$; for $\tfrac{1}{2} \leq \theta \leq 1$, always stable
			includes 1, ..., 4 as special cases.

(continued)

TABLE 8.I *(continued)*

6.   ½ − $(\Delta x)^2/12\sigma\Delta t$
     ½ + $(\Delta x)^2/12\sigma\Delta t$

6. Same as 5, but with $\theta = \frac{1}{2} - (\Delta x)^2/12\sigma\Delta t$
   $e = O[(\Delta t)^2] + O[(\Delta x)^4]$
   stable.

7.

7. $\dfrac{u_j^{n+1} - u_j^{n-1}}{2\Delta t} = \sigma\dfrac{(\delta^2 u)_j^n}{(\Delta x)^2}$
   always unstable.

8.

8. $\dfrac{u_j^{n+1} - u_j^{n-1}}{2\Delta t} = \sigma\dfrac{u_{j+1}^n - u_j^{n+1} - u_j^{n-1} + u_{j-1}^n}{(\Delta x)^2}$
   where $\Delta t/\Delta x \to 0$ as $\Delta t, \Delta x \to 0$
   $e = O[(\Delta t)^2] + O[(\Delta x)^2] + O\left[\left(\dfrac{\Delta t}{\Delta x}\right)^2\right]$
   Du Fort and Frankel (1953)
   explicit, always stable.

9.   ³⁄₂
     −½

9. $\dfrac{3}{2}\dfrac{u_j^{n+1} - u_j^n}{\Delta t} - \dfrac{1}{2}\dfrac{u_j^n - u_j^{n-1}}{\Delta t} = \sigma\dfrac{(\delta^2 u)_j^{n+1}}{(\Delta x)^2}$
   $e = O[(\Delta t)^2] + O[(\Delta x)^2]$
   always stable.

10. $1 + \theta$
      $-\theta$

10. $(1 + \theta)\dfrac{u_j^{n+1} - u_j^n}{\Delta t} - \theta\dfrac{u_j^n - u_j^{n-1}}{\Delta t} =$
    $\sigma\dfrac{(\delta^2 u)_j^{n+1}}{(\Delta x)^2}$
    where $\theta = $ const. $\geqq 0$, $\sigma\Delta t/(\Delta x)^2 = $ const.
    $e = O(\Delta t) + O[(\Delta x)^2]$
    always stable
    contains 3, 9 as special cases.

(continued)

TABLE 8.I (*continued*)

11. $\frac{3}{2} - (\Delta x)^2/12\sigma\Delta t$
    $-\frac{1}{2} + (\Delta x)^2/12\sigma\Delta t$

11. Same as 10 but with
$$\theta = \frac{1}{2} + \frac{(\Delta x)^2}{12\sigma\Delta t}$$
$$e = O[(\Delta t)^2] = O[(\Delta x)^4]$$
always stable.

12. $\frac{1}{12}$ $\frac{5}{6}$ $\frac{1}{12}$; $\frac{1}{2}$, $\frac{1}{2}$

12. $$\frac{1}{12}\frac{u_{j+1}^{n+1} - u_{j+1}^{n}}{\Delta t} + \frac{5}{6}\frac{u_j^{n+1} - u_j^n}{\Delta t}$$
$$+ \frac{1}{12}\frac{u_{j-1}^{n+1} - u_{j-1}^n}{\Delta t} = \sigma\frac{(\delta^2 u)_j^{n+1} + (\delta^2 u)_j^n}{2(\Delta x)^2},$$
$$e = O[(\Delta t)^2] + O[(\Delta x)^4]$$
always stable
identical with 6.

13. $\frac{1}{12}$ $\frac{5}{6}$ $\frac{1}{12}$; $\frac{3}{2}$, $-\frac{1}{2}$

13. $$\frac{1}{12}\frac{\frac{3}{2}u_{j+1}^{n+1} - 2u_{j+1}^n + \frac{1}{2}u_{j+1}^{n-1}}{\Delta t}$$
$$+ \frac{5}{6}\frac{\frac{3}{2}u_j^{n+1} - 2u_j^n + \frac{1}{2}u_j^{n-1}}{\Delta t}$$
$$+ \frac{1}{12}\frac{\frac{3}{2}u_{j-1}^{n+1} - 2u_{j-1}^n + \frac{1}{2}u_{j-1}^{n-1}}{\Delta t}$$
$$= \sigma\frac{(\delta^2 u)_j^{n+1}}{(\Delta x)^2}$$
$$e = O[(\Delta t)^2] + O[(\Delta x)^4]$$
always stable.

14. (b), (a)

14a. $u_j^{n+1} - u_j^n = \alpha(u_{j+1}^n - u_j^n - u_j^{n+1} + u_{j-1}^{n+1})$,

14b. $u_j^{n+2} - u_j^{n+1}$
$$= \alpha(u_{j+1}^{n+2} - u_j^{n+2} - u_j^{n+1} + u_{j-1}^{n+1}),$$
where $\alpha = \sigma\Delta t/(\Delta x)^2$

Saul'ev (1957)—see text.

points of either kind, the weights used are indicated at the side or underneath.

A few comparisons are in order. Schemes 2 and 9 are formally of the same accuracy; but scheme 2 is preferred when one is dealing with smooth functions because the coefficient of $(\Delta t)^2$ in its truncation error is smaller, while scheme 9 is preferred for rapidly varying or discontinuous initial data because it damps the short-wavelength components (components with $k\Delta x \approx \pi$) more rapidly. There is a similar relationship between schemes 6 and 11 and between schemes 12 and 13.

In the method of Saul'ev (1957), different equations are used on alternate time steps, as given in entry 14 of Table 8.I. The scheme is effectively explicit: although each equation of the system

$$u_j^{n+1} - u_j^n = \alpha(u_{j+1}^n - u_j^n - u_j^{n+1} + u_{j-1}^{n+1}) \quad (j = 1, \ldots, J-1)$$

contains two of the unknowns $u_0^{n+1}, \ldots, u_{J-1}^{n+1}$, the left boundary condition gives u_0^{n+1} so that the equations can be solved in the order $j = 1, \ldots, J-1$. Similarly, the second system of equations can be solved for the u_j^{n+2} in the order $j = J-1, J-2, \ldots, 1$. The overall amplification factor for the two time steps is easily found to be

$$g = \frac{[1 - \alpha(1 - \cos \beta)]^2 + \alpha^2 \sin^2\beta}{[1 + \alpha(1 - \cos \beta)]^2 + \alpha^2 \sin^2\beta},$$

where $\alpha = \sigma\Delta t/(\Delta x)^2$, $\beta = k\Delta x$; since $1 - \cos \beta \geqq 0$, g lies in the interval $[0, 1]$, and the scheme is unconditionally stable. An estimate of the accuracy can be obtained by eliminating the u_j^{n+1}; this gives the fully implicit scheme

$$u^{n+2} - (\alpha + \alpha^2)\delta^2 u^{n+2} = u^n + (\alpha - \alpha^2)\delta^2 u^n.$$

If the terms α^2 were missing, this would be the Crank-Nicholson scheme, whose truncation error is $O[(\Delta t)^2] + O[(\Delta x)^2]$; the contribution of the α^2 terms to $(u^{n+2} - u^n)/2\Delta t$ is

$$O\left([\alpha\Delta x]^2 \frac{\partial^3 u}{\partial x^2 \partial t}\right) = O\left(\frac{(\Delta t)^2}{(\Delta x)^2} \frac{\partial^3 u}{\partial x^2 \partial t}\right).$$

For any fixed α, the truncation error is $O(\Delta t)$; whereas if Δt and Δx are regarded as independent, then, just as for the Du Fort-Frankel scheme, the

SEC. 8.3]   VARIABLE COEFFICIENTS

Saul'ev scheme is consistent with the heat flow equation only if $\Delta t/\Delta x \to 0$ as the net is refined.

8.3. Variable Coefficients

Most of the schemes of Table 8.I can be generalized to the case of variable coefficients.[2] When schemes of high accuracy are generalized, one must use care to avoid loss of accuracy, unless the coefficients are nearly constant. We illustrate by showing how to generalize scheme number 12. We first consider the equation

$$(8.6) \qquad \frac{\partial u}{\partial t} = \sigma \frac{\partial^2 u}{\partial x^2}, \quad \sigma = \sigma(x) \geq a > 0,$$

and assume that $\sigma(x)$ is a twice continuously differentiable function. It is convenient to introduce the space differencing first:

$$\frac{\delta^2 u}{(\Delta x)^2} = \frac{\partial^2 u}{\partial x^2} + \frac{(\Delta x)^2}{12} \frac{\partial^4 u}{\partial x^4} + O[(\Delta x)^4];$$

therefore,

$$\frac{(\delta^2 u)_j}{(\Delta x)^2} = \left(\frac{1}{\sigma}\frac{\partial u}{\partial t}\right)_j + \frac{(\Delta x)^2}{12}\left[\frac{\partial^2}{\partial x^2}\left(\frac{1}{\sigma}\frac{\partial u}{\partial t}\right)\right]_j + O[(\Delta x)^4]$$

$$= \left(\frac{1}{\sigma}\frac{\partial u}{\partial t}\right)_j + \frac{(\Delta x)^2}{12}\left[\left\{\frac{\delta^2\left(\frac{1}{\sigma}\frac{\partial u}{\partial t}\right)}{(\Delta x)^2}\right\}_j + O[(\Delta x)^2]\right] + O[(\Delta x)^4]$$

$$= \frac{5}{6}\left(\frac{1}{\sigma}\frac{\partial u}{\partial t}\right)_j + \frac{1}{12}\left(\frac{1}{\sigma}\frac{\partial u}{\partial t}\right)_{j+1} + \frac{1}{12}\left(\frac{1}{\sigma}\frac{\partial u}{\partial t}\right)_{j-1} + O[(\Delta x)^4].$$

The time differencing is now introduced simply, and we arrive finally at the formula

$$(8.7) \quad \frac{1}{12}\frac{u^{n+1}_{j+1} - u^n_{j+1}}{\sigma_{j+1}\Delta t} + \frac{5}{6}\frac{u^{n+1}_j - u^n_j}{\sigma_j \Delta t} + \frac{1}{12}\frac{u^{n+1}_{j-1} - u^n_{j-1}}{\sigma_{j-1}\Delta t}$$

$$= \frac{(\delta^2 u)^{n+1}_j + (\delta^2 u)^n_j}{2(\Delta x)^2} + O[(\Delta t)^2] + O[(\Delta x)^4].$$

---

[2] Scheme number 4 cannot be so generalized, because the special condition $\sigma \Delta t/(\Delta x)^2 = \frac{1}{6}$ requires σ to be constant.

For the more general equation,

$$\frac{\partial u}{\partial t} = \mu \frac{\partial}{\partial x} \nu \frac{\partial u}{\partial x} \quad \text{where} \quad \begin{cases} \mu = \mu(x) \geq a_1 > 0 \\ \nu = \nu(x) \geq a_2 > 0, \end{cases}$$

the formulas are slightly more complicated, and we will not write them down. But we note that such a formula can be readily obtained by making a transformation

$$y = \int \frac{1}{\nu(x)} dx,$$

whereupon the equation becomes

$$\frac{\partial u}{\partial t} = \frac{\mu}{\nu} \frac{\partial^2 u}{\partial y^2},$$

which has the same form as (8.6) and can be similarly treated. In terms of the original variable x, this scheme has unequal spacing of the net points.

Formulas of this type have been used in calculations with the Fermi age-diffusion equations for nuclear reactors and found to give accurate results with less calculation than required by the simpler formulas.

As regards the stability of these difference schemes with variable (but Lipschitz continuous) coefficients, we showed in Section 5.3 that this follows for explicit schemes if the amplification factor satisfies condition (5.22), namely,

$$|G(x, \xi)| \leq 1 - \delta|\xi|^2, \quad \text{for all } x \text{ and } |\xi| \leq \pi,$$

where $\xi = k\Delta x$ and the constant $\delta > 0$. We also quoted some results for implicit schemes. To see what this condition means, consider scheme 5 of Table 8.I, for which

$$G = \frac{1 - 4\alpha(1 - \theta) \sin^2 \frac{1}{2}\xi}{1 + 4\alpha\theta \sin^2 \frac{1}{2}\xi},$$

where $\alpha(x, \Delta t) = \sigma(x)\Delta t/(\Delta x)^2$. Thus,

$$-1 + \frac{2 - 4\alpha(1 - 2\theta) \sin^2 \frac{1}{2}\xi}{1 + 4\alpha \sin^2 \frac{1}{2}\xi} = G = 1 - \frac{4\alpha \sin^2 \frac{1}{2}\xi}{1 + 4\alpha\theta \sin^2 \frac{1}{2}\xi}$$

and the condition is satisfied if (i) $\alpha(x, \Delta t)$ is uniformly bounded away from zero for all x as $\Delta t \to 0$; and (ii) $2\alpha(x, \Delta t)(1 - 2\theta) \leq \gamma < 1$. Since γ can be arbitrarily close to 1, there is no limitation in practice on the range of Δt that can be used (compare the discussion in the next section).

It is also worth noting here that the proof of stability for variable coefficient schemes, as given in Section 5.3, used the explicitness of the scheme in only a very minor way. If in the sum, $u^{n+1}(x) = \sum_{(\beta)} c^\beta u^n(x + \beta \Delta x)$, there is absolute convergence both of the c^β and their Lipschitz constants, then the proof needs very little modification.

8.4. Effect of Lower Order Terms on Stability

We now consider the equation

$$(8.8) \qquad \frac{\partial u}{\partial t} = \sigma \frac{\partial^2 u}{\partial x^2} + a \frac{\partial u}{\partial x} + bu,$$

where σ, a, b are constants, and $\sigma > 0$. In forming finite-difference equations, each term in (8.8) may be treated in various ways, and the possible combinations are very numerous. We shall give two examples. They (and others) support the conclusion that for diffusion problems, at least, stability is practically unaffected by the lower order terms. By "practically" is meant that in some cases a restriction $\Delta t \leq \ldots$ has to be replaced by $\Delta t < \ldots$, but that there are no other changes. (In Section 5.3 this was shown for quite general explicit schemes for equation (8.8) by arguments essentially the same as those used in the examples below.) However, the presence of low-order terms may force one to use a smaller value of Δt; for example, a large value of b in (8.8) would cause a rapidly changing u even if u is independent of x—a small Δt is then required, just as for the corresponding finite-difference approximation to the ordinary differential equation $du/dt = bu$.

As the first example, we take the space-centered scheme

$$(8.9) \qquad \frac{u_j^{n+1} - u_j^n}{\Delta t} = \sigma \frac{\theta (\delta^2 u)_j^{n+1} + (1 - \theta)(\delta^2 u)_j^n}{(\Delta x)^2} + a \frac{u_{j+1}^n - u_{j-1}^n}{2 \Delta x} + b u_j^n,$$

where θ is a constant ($0 \leq \theta \leq 1$). We call $\sigma \Delta t / (\Delta x)^2 = \alpha$ and suppose

α constant as $\Delta t, \Delta x \to 0$, so that $\Delta x = \sqrt{\sigma \Delta t / \alpha}$. The amplification factor $G = G(\Delta t, k)$ is

$$G = \frac{1 - 4\alpha(1 - \theta)\sin^2 \beta + ia\sqrt{\alpha \Delta t / \sigma} \sin 2\beta + b\Delta t}{1 + 4\alpha\theta \sin^2 \beta}$$

where $\beta = k\Delta x/2$. Therefore,

$$|G| = \frac{\sqrt{[1 - 4\alpha(1 - \theta) \sin^2 \beta + b\Delta t]^2 + a^2(\alpha \Delta t/\sigma) \sin^2 2\beta}}{1 + 4\alpha\theta \sin^2 \beta}$$

which we write as

$$\sqrt{f_0(\beta) + \Delta t f_1(\beta) + (\Delta t)^2 f_2(\beta)},$$

where $f_0(\beta)$, $f_1(\beta)$ and $f_2(\beta)$ are bounded functions. Clearly, we must require $|f_0(\beta)| \leq 1$, and this gives just the condition arrived at for $a = b = 0$, namely,

(8.10) $\qquad \alpha \leq 1/(2 - 4\theta) \quad$ if $0 \leq \theta < \frac{1}{2}$,

$\qquad$ no restriction on $\Delta t \quad$ if $\frac{1}{2} \leq \theta \leq 1$.

If this condition is satisfied, and if m_1 and m_2 are the maxima with respect to β of $|f_1(\beta)|$ and $|f_2(\beta)|$, we have

$$|G| \leq \sqrt{1 + m_1 \Delta t + m_2 \Delta t^2} = 1 + O(\Delta t).$$

Therefore, the stability condition is precisely as before.

As the second example, and a possibly less satisfactory one from the point of view of the truncation error, we take the (uncentered) scheme with uncentered first difference:

(8.11) $\quad \dfrac{u_j^{n+1} - u_j^n}{\Delta t} = \sigma \dfrac{\theta(\delta^2 u)_j^{n+1} + (1 - \theta)(\delta^2 u)_j^n}{(\Delta x)^2} + a\dfrac{\Delta_- u_j^n}{\Delta x} + bu_j^n,$

where, as in earlier chapters, $\Delta_- u_j^n$ is to be interpreted as $u_j^n - u_{j-1}^n$ and again we suppose that $\sigma \Delta t/(\Delta x)^2 = \alpha$ has a fixed value as $\Delta t, \Delta x \to 0$. We first consider the case $a \geq 0$.

We now have

(8.12) $\quad G = \dfrac{1 - 4\alpha(1 - \theta) \sin^2 \beta + a\sqrt{\alpha \Delta t/\sigma} \, (2 \sin^2 \beta + i \sin 2\beta) + b\Delta t}{1 + 4\alpha\theta \sin^2 \beta}.$

SEC. 8.4] EFFECT OF LOWER ORDER TERMS ON STABILITY

It is again clear that (8.10) is at least a necessary condition for stability. We shall show that it is also sufficient.

Stability concerns only what happens in the limit $\Delta t, \Delta x \to 0$. Therefore it is only necessary to consider small Δt. Suppose, then, that

$$2a\sqrt{\alpha \Delta t/\sigma} \leq 4\alpha, \quad \text{i.e.,} \quad \Delta x \leq 2\sigma/a.$$

If (8.10) is satisfied, we have

(8.13)
$$-1 \leq \frac{1 - 4\alpha(1 - \theta)\sin^2 \beta}{1 + 4\alpha\theta \sin^2 \beta}$$

$$\leq \frac{1 - 4\alpha(1 - \theta)\sin^2 \beta + 2a\sqrt{\alpha \Delta t/\sigma}\sin^2 \beta}{1 + 4\alpha\theta \sin^2 \beta} \leq 1.$$

Designating the third member of this inequality by A, we have

$$|G|^2 = \left(A + \frac{b\Delta t}{1 + 4\alpha\theta \sin^2 \beta}\right)^2 + \left(\frac{a\sqrt{\alpha \Delta t/\sigma}\sin 2\beta}{1 + 4\alpha\theta \sin^2 \beta}\right)^2;$$

and, therefore,

$$|G|^2 \leq 1 + O(\Delta t) \text{ uniformly in } \beta,$$

which proves stability.

We must now discuss the opposite case: $a \leq 0$. G is still given by (8.12) but, since $a \leq 0$, the inequality between the second and third members of (8.13) is reversed, and there is danger of A being < -1. To avoid this, we must modify the first line of (8.10) and require

(8.14) $\quad \alpha < 1/(2 - 4\theta) \quad$ if $0 \leq \theta < \frac{1}{2}$,

$\qquad$ no restriction $\quad$ if $\frac{1}{2} \leq \theta \leq 1$.

(The only difference is that we no longer permit $\alpha = 1/(2 - 4\theta)$.) In either case there is for given α, θ, a constant B such that

$$-1 < B \leq \frac{1 - 4\alpha(1 - \theta)\sin^2 \beta}{1 + 4\alpha\theta \sin^2 \beta}.$$

Since A differs from this last quantity only by a negative term proportional to $\sqrt{\Delta t}$, it is clear that by taking Δt sufficiently small, we can insure that $-1 \leq A$. We then have $-1 \leq A \leq 1$, and from here on the argument is just as for the case $a \geq 0$. If we take the forward difference, $\Delta_+ u_j^n$ instead of $\Delta_- u_j^n$ in (8.11), the result is the same except that the roles of $a \geq 0$ and $a \leq 0$ are interchanged.

8.5. Solution of the Implicit Equations

If the equation
$$\frac{\partial u}{\partial t} = \frac{\partial}{\partial x} \sigma \frac{\partial u}{\partial x}, \quad \sigma = \sigma(x) > 0,$$

(to take a typical case) is approximated by the finite-difference system
$$\frac{u_j^{n+1} - u_j^n}{\Delta t} = \frac{\theta[\delta(\sigma\delta u)]_j^{n+1} + (1-\theta)[\delta(\sigma\delta u)]_j^n}{(\Delta x)^2}$$

analogous to system 5 of Table 8.I, the equations to be solved at each time step are:
$$-\theta\alpha_{j+\frac{1}{2}}u_{j+1}^{n+1} + (1 + \theta\alpha_{j+\frac{1}{2}} + \theta\alpha_{j-\frac{1}{2}})u_j^{n+1} - \theta\alpha_{j-\frac{1}{2}}u_{j-1}^{n+1} = ---,$$
$$j = 1, 2, \ldots,$$

where the dashes on the right indicate known quantities, and $\alpha = \alpha(x)$ stands for $\sigma(x)\Delta t/(\Delta x)^2$. These equations must be solved subject to boundary conditions of some sort at, say, $j = 0$ and $j = J$. We rewrite the equations as

(8.15) $$-A_j u_{j+1} + B_j u_j - C_j u_{j-1} = D_j,$$

where A_j, B_j, C_j are the coefficients appearing above, and where the superscript $n+1$ has been deleted. If we had taken the equation $\partial u/\partial t = \sigma \partial^2 u/\partial x^2$, the coefficients would be slightly different. As a second example, if this latter equation is approximated by (8.7) (system 12 of Table 8.I), which is equivalent to permitting θ of the first example to be variable and to assume at each net point the special value $\theta = \frac{1}{2} - (\Delta x)^2/12\sigma\Delta t$ of system 6 of Table 8.I, then

$$B_j = 1 + \frac{5}{6\alpha_j},$$
$$A_j = \frac{1}{2} - \frac{1}{12\alpha_{j+1}},$$
$$C_j = \frac{1}{2} - \frac{1}{12\alpha_{j-1}}.$$

SEC. 8.5] SOLUTION OF THE IMPLICIT EQUATIONS

If it is assumed that in the first example $\theta > 0$ and that in the second $\alpha_j > 1/6$ for all j, then, for both examples,

(8.16) $\qquad A_j > 0, \qquad B_j > 0, \qquad C_j > 0,$

and furthermore the inequality

(8.17) $\qquad\qquad B_j > A_j + C_j$

is satisfied by quite a margin.

To illustrate the method of solution of a recurrence relation like (8.15), we shall suppose that the boundary conditions are

(8.18) $\qquad\qquad u_0 = 0,$

(8.19) $\qquad\qquad u_J = 0.$

The reader will notice that by slight changes in the procedure other boundary conditions can be accommodated as well.

The equations to be solved are linear, and any of the standard methods for solving a linear system could be used. But this system of equations is a very special one, because all the elements of the corresponding matrix vanish except those on three diagonals, and because of inequalities (8.16) and (8.17). For this system the method to be described is very efficient and suitable for automatic computation.[3]

First note that the following method, which naturally occurs to one, is *not* a good method: Take any solution, say $u^{(1)}$, of (8.15) together with the left-hand boundary condition (8.18) and also any solution, say $u^{(2)}$, of the corresponding homogeneous equation (i.e. (8.15) with $D_j = 0$) together with the left-hand boundary condition, and then simply add a suitable multiple of $u^{(2)}$ to $u^{(1)}$ so as to satisfy the right-hand boundary condition. The reason why this is bad is that in general $u^{(1)}$ and $u^{(2)}$ grow very rapidly, more or less exponentially, as j increases from 0 to J, and therefore the true solution has to be obtained by subtracting two very large and very nearly equal quantities. This would require not only floating point operation but also multiple precision. In the method described below, all quantities stay nicely in scale.

---

[3] Apparently this method has been used independently by many people; it is, of course, merely a special adaptation of the Gauss elimination procedure.

In describing the method, we shall nonetheless have occasion to discuss the set C of solutions of the difference equation (8.15) and the left-hand boundary condition (8.18): C is a one-parameter family of solutions, because any value of u_j at $j = 1$ determines such a solution. In fact, we seek two sets of quantities, E_j and F_j, such that for any member of the set C,

(8.20) $$u_j = E_j u_{j+1} + F_j.$$

In particular, if this is to be true for any member of the family, the boundary condition (8.18) shows that we must have

(8.21) $$E_0 = 0, \quad F_0 = 0.$$

We shall see later that $E_j > 0$ for $j > 0$. Therefore, (8.20) also has a one-parameter family of solutions, since u_1 may be chosen arbitrarily, and then (8.20) fixes all the u_j ($j > 1$). Our task is to determine the E_j and F_j so that this family is the same as the set C.

To achieve this, substitute $E_{j-1} u_j + F_{j-1}$ for u_{j-1} into (8.15). The result is a relation between u_j and u_{j+1} which can be written as

$$u_j = \frac{A_j}{B_j - C_j E_{j-1}} u_{j+1} + \frac{D_j + C_j F_{j-1}}{B_j - C_j E_{j-1}}.$$

If we equate this to the right member of (8.20) and then recall that the result must hold for a one-parameter set of values of u_{j+1}, it is clear that we can identify:

(8.22) $$E_j = \frac{A_j}{B_j - C_j E_{j-1}}, \quad j \geq 1,$$

(8.23) $$F_j = \frac{D_j + C_j F_{j-1}}{B_j - C_j E_{j-1}}, \quad j \geq 1.$$

From these equations together with (8.21), one can calculate the E_j and F_j inductively in order of increasing j ($j = 0, 1, \ldots, J - 1$). Now u_{j+1} is given for $j = J - 1$ by the right-hand boundary condition (8.19); therefore we can now calculate the u_j inductively from (8.20) in order of decreasing j ($j = J - 1, J - 2, \ldots, 1$). This completes the calculation.

To see that E_j and F_j stay nicely in scale, observe first that by (8.22) and the inequality $B_j > A_j + C_j$, (8.17), we have:

$$\text{if } E_{j-1} \leq 1, \quad 0 < E_j \leq \frac{A_j}{B_j - C_j} < \frac{A_j}{A_j} = 1.$$

Since $E_0 = 0$, this establishes that E_j lies between 0 and 1 for all j. Second, note (this is an observation of D. Mussman) that if E_j and the desired solution u_j are reasonably bounded, then it is clear from (8.20) that F_j is also reasonably bounded.

To see the efficiency of this method, note that it requires only three multiplications and two divisions per space point, per time step, apart from computation of the coefficients. This may be contrasted with a procedure that has been considered, for problems in which the coefficients are independent of time—namely to invert, once and for all, the matrix of the set of simultaneous equations (8.15), (8.18), (8.19), and then use the inverse matrix to obtain the solution for each time step in turn. Quite apart from the labor of inverting the matrix, this requires multiplying a matrix by a vector—therefore J multiplications—for each space point, each time step!

8.6. A Non-Linear Problem

The application of the foregoing ideas to non-linear problems will be illustrated by considering the equation

$$(8.24) \qquad \frac{\partial u}{\partial t} = \frac{\partial^2}{\partial x^2}(u^5)$$

for which some trial calculations were performed on the Univac.

The initial-boundary-value problem chosen for study is one whose solution is known in analytic form so that the accuracy obtained by the numerical method could be determined by comparison. The running-wave solutions of (8.24) can be obtained by looking for solutions in which u depends on x and t only through the combination $x - vt$, where v is a constant. The general solution of this type is given implicitly by the equation

$$(8.25) \quad \tfrac{5}{4}(u - u_0)^4 + \tfrac{20}{3}u_0(u - u_0)^3 + 15u_0^2(u - u_0)^2 + 20u_0^3(u - u_0)$$
$$+ 5u_0^4 \ln(u - u_0) = v(vt - x + x_0),$$

where u_0, x_0, and v are constants. This is a wave running to the right if $v > 0$; its shape is shown in Figure 8.1. At the left it is approximately a

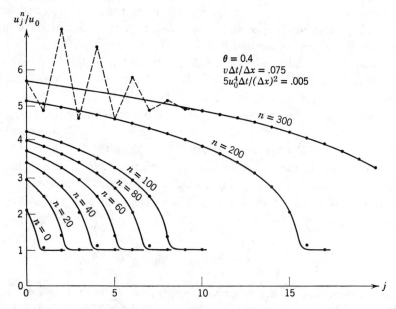

Fig. 8.1. Running-wave solutions of the non-linear equation $\partial u/\partial t = \partial^2(u^5)/\partial x^2$. The curves show the exact solution, given by equation (8.25) and the dots show the solution of the difference equation (8.27) with $\theta = 0.4$ and with Δt and Δx so chosen that $v\Delta t/\Delta x = 0.075$ and $5u_0^4\Delta t/(\Delta x)^2 = 0.005$. The numbers on the curves are cycle numbers.

quartic curve and at the right u approaches u_0 more or less exponentially with increasing x, because for $u - u_0 \ll u_0$, the logarithm predominates on the left side of (8.25). Without loss of generality we can take $u_0 = 1$. Let then the function inverse to the left member of (8.25) be denoted by $\psi(\xi)$, so that the running wave solution is

$$u = \psi(v(vt - x + x_0));$$

A NON-LINEAR PROBLEM

$\psi(\xi)$ can be obtained graphically or by Newton-Raphson solution of (8.25). Initial values and boundary values were taken from the running-wave solution. That is,

(8.26) initial condition: $u(x, 0) = \psi(v(x_0 - x))$,
boundary conditions: $\begin{cases} u(0, t) = \psi(v(vt + x_0)) \\ u(L, t) = \psi(v(vt - L + x_0)). \end{cases}$

Without loss of generality we can take the interval length $L = 1$. By the uniqueness of the solutions of (8.24) the numerical method, if accurate, should reproduce the running-wave solution for $0 \leq x \leq 1, t \geq 0$.

We chose as finite-difference system a generalization of the general implicit system 5 of Table 8.I, so as to be able to study accuracy and stability for various values of θ. The direct generalization of the difference equation,

$$\frac{u_j^{n+1} - u_j^n}{\Delta t} = \frac{\theta[\delta^2(u^5)]_j^{n+1} + (1 - \theta)[\delta^2(u^5)]_j^n}{(\Delta x)^2}$$

would not be suitable, because for it the simultaneous equations to be solved would be non-linear. A linearized version of the equation can be obtained in various ways, and one which is suitable for the problem at hand is obtained by the approximation:

$$(u^5)_j^{n+1} \approx (u^5)_j^n + 5(u^4)_j^n(u_j^{n+1} - u_j^n).$$

If we call $w_j = u_j^{n+1} - u_j^n$, we obtain the equations

(8.27) $\frac{w_j}{\Delta t} - \frac{5\theta}{(\Delta x)^2}[(u^4)_{j+1}^n w_{j+1} - 2(u^4)_j^n w_j + (u^4)_{j-1}^n w_{j-1}]$

$$= \frac{(u^5)_{j+1}^n - 2(u^5)_j^n + (u^5)_{j-1}^n}{(\Delta x)^2},$$

together with the boundary conditions,

$$w_0 = u_0^{n+1} - u_0^n,$$
$$w_J = u_J^{n+1} - u_J^n,$$

whose right members are known. These equations can be solved by the method of the preceding section; then use of the equation $u_j^{n+1} = u_j^n + w_j$ completes the calculation for the time step.

TABLE 8.II

A Portion of the Numerical Solution

| Imp 01 | cycle 0180 | June 1, 1954 | |
j	theory	expt.	stability no.
00	04.9885291	04.9885812	00.0000000
01	04.8892890	04.8892689	01.1428989
02	04.7840560	04.7841171	01.0477001
03	04.6722668	04.6722014	00.9530582
04	04.5525326	04.5524622	00.8590384
05	04.4235425	04.4234839	00.7657584
06	04.2835565	04.2833949	00.6732604
07	04.1300680	04.1297196	00.5817199
08	03.9594097	03.9589363	00.4912986
09	03.7667231	03.7658970	00.4022574
10	03.5437619	03.5423956	00.3149394
11	03.2767332	03.2745918	00.2299584
12	02.9386536	02.9355146	00.1485191
13	02.4614235	02.4580572	00.0730170
14	01.5669351	01.4850563	00.0097235
15	01.0000000	01.0122755	00.0020943
16	01.0000000	01.0000249	00.0020000
17	01.0000000	01.0000000	00.0020000
18	01.0000000	01.0000000	00.0020000
19	01.0000000	01.0000000	00.0020000
20	01.0000000	01.0000000	00.0020000

The problem was so coded that the machine obtained the initial and boundary values by the Newton-Raphson solution of (8.25) and solved the finite-difference equations as indicated above. After each set of twenty time steps the machine printed out the values of the u_j^n so obtained in a column, for $j = 0, 1, \ldots, J$, and beside them, in a parallel column, the exact values, obtained by the Newton-Raphson solution of (8.25). A sample of such a printout is reproduced here as Table 8.II. It is seen that the agreement is good, generally. In Figure 8.1, a number of sets

A NON-LINEAR PROBLEM

of such values are shown graphically, for one of the sample problems; one with $\Delta t/(\Delta x)^2 = .001$, $\theta = 0.4$. The running-wave solution is reproduced quite accurately, except in two portions of the graph:

1. Near the foot of each wave, where the true solution varies most rapidly, there are slight discrepancies.

2. In the upper left-hand portion of the graph the last two curves have oscillations suggestive of instability.

The speed with which the wave moves to the right, estimated from abscissas about half way up the graph, agrees with the correct value to a fraction of a percent. All in all, the result seems quite satisfactory considering the coarseness of the net.

The heuristic approach to the stability question is as follows: The effective diffusion constant σ for this problem is $5u^4$, as can be seen by writing the differential equation as

$$\frac{\partial u}{\partial t} = \frac{\partial}{\partial x} 5u^4 \frac{\partial u}{\partial x},$$

and therefore σ varies with x and t. Noting that instabilities, when they occur, manifest themselves by rapid oscillations of a local character, one would expect that the method would be stable up to time $t = t_0$ if, and only if,

(8.28)
$$\frac{5u^4 \Delta t}{(\Delta x)^2} < \frac{1}{2 - 4\theta} \quad \text{for } 0 \leq \theta < \tfrac{1}{2},$$

$$\text{for all } x, t \text{ in } \begin{cases} 0 \leq x \leq L \\ 0 \leq t \leq t_0 \end{cases}$$

no restriction for $\tfrac{1}{2} \leq \theta \leq 1$.

Note that *for non-linear problems stability depends not only on the structure of the finite-difference system but also generally on the solution being obtained; and for a given solution, the system may be stable for some values of t and not for others.* Compare F. John's result (1952) described in Section 5.3 for explicit systems.

In the present problem, with $\theta = 0.4$, $\Delta t/(\Delta x)^2 = 0.001$, we expect stability until a time when u (which is constantly increasing everywhere, in the running-wave solution) reaches the value $u = (500)^{1/4} \approx 4.7$, and that instability should develop soon after that. This is just what happened in the sample calculation, as can be seen in Figure 8.1. It might be mentioned that for this problem the explicit system would have become unstable already at about the 40th time step, for the mesh used, instead the 250th. A few further sample calculations were made with different θ and different Δt. In all cases the predictions of the heuristic stability argument seemed to be verified.

In the practical solution of non-linear partial differential equations it is necessary, or at least highly desirable, for the machine to keep a constant check on stability (if the difference equations are not unconditionally stable) by making tests, such as of the inequality for Δt in (8.28), and either stop when the condition is not satisfied or automatically alter Δt to restore stability.

8.7. Problems in Several Space Variables

As a first example we consider the parabolic equation for a function $u = u(x, y, t)$:

$$(8.29) \qquad \frac{\partial u}{\partial t} = A \frac{\partial^2 u}{\partial x^2} + 2B \frac{\partial^2 u}{\partial x \partial y} + C \frac{\partial^2 u}{\partial y^2},$$

where

$$(8.30) \qquad A > 0, \quad C > 0, \quad AC - B^2 > 0.$$

The inequalities (8.30) assure the parabolic character of the equation: when they are satisfied, the initial-value problem of (8.29) is properly posed.

For writing the difference equation we introduce the abbreviations

$$(8.31) \qquad u_{jl}^n = u(j\Delta x, l\Delta y, n\Delta t),$$

$$(8.32) \qquad \Phi_{jl}^n = \frac{A}{(\Delta x)^2} (u_{j+1\,l}^n - 2u_{jl}^n + u_{j-1\,l}^n)$$

$$+ \frac{B}{2\Delta x \Delta y} (u_{j+1\,l+1}^n - u_{j-1\,l+1}^n - u_{j+1\,l-1}^n + u_{j-1\,l-1}^n)$$

$$+ \frac{C}{(\Delta y)^2} (u_{jl+1}^n - 2u_{jl}^n + u_{jl-1}^n).$$

Given a constant θ ($0 \leq \theta \leq 1$), we write the finite-difference approximation to (8.29) as

(8.33) $$\frac{u_{jl}^{n+1} - u_{jl}^n}{\Delta t} = \theta \Phi_{jl}^{n+1} + (1 - \theta) \Phi_{jl}^n.$$

If $\theta > 0$, these equations are implicit; for each n one must solve a system of simultaneous equations similar to those encountered in the numerical solution of an elliptic problem for a function $u(x, y)$. There is no simple generalization of the algorithm of Section 8.5 for solving those equations, and one generally has recourse to a relaxation method. This being the case, it might be thought that it would be cheaper to use the explicit equation ($\theta = 0$) and to take Δt sufficiently small that those equations are stable. However, some very effective relaxation methods have been invented,[4] and u_{jl}^n usually provides an excellent first approximation to u_{jl}^{n+1}. In numerical calculations of nuclear reactors[5] it has been found worth-while to use the implicit schemes for solution of the two-dimensional age-diffusion equation which is similar in form to (8.29).

Substitution of $\exp[ik_1 x + ik_2 y]$ for $u^n(x, y)$ into Φ_{jl}^n given by (8.32) and multiplication by Δt give $\Delta t \Phi_{jl}^n = \psi \cdot \exp[ik_1 j \Delta x + ik_2 l \Delta y]$, where

(8.34) $$\psi = -2\left[\frac{A\Delta t}{(\Delta x)^2}(1 - \cos k_1 \Delta x)\right.$$
$$\left. + \frac{B\Delta t}{\Delta x \Delta y} \sin k_1 \Delta x \sin k_2 \Delta y + \frac{C\Delta t}{(\Delta y)^2}(1 - \cos k_2 \Delta y)\right].$$

Therefore, the amplification factor for (8.33) is given by

(8.35) $$G(\Delta t, \mathbf{k}) = \frac{1 + (1 - \theta)\psi}{1 - \theta \psi}.$$

To determine the stability condition one needs to know the range of ψ, i.e., the maximum and minimum of ψ as a function of the real variables $k_1 \Delta x$ and $k_2 \Delta y$. This function is analytic and periodic and hence its extrema occur when

$$\frac{\partial \psi}{\partial (k_1 \Delta x)} = \frac{\partial \psi}{\partial (k_2 \Delta y)} = 0.$$

---

[4] Shortley and Weller (1938), D. M. Young (1950), Jim Douglas, Jr. (1955), Peaceman and Rachford (1955), S. P. Frankel (1950), Liebman (1918), Shortley (1952), Stark (1956).
[5] The basis for such calculations is described in Ehrlich and Hurwitz (1954).

A direct calculation shows that if the inequalities (8.30) hold this extremum can occur only when $\sin k_1\Delta x = \sin k_2\Delta y = 0$. We therefore have to try only the four possibilities given by $\cos k_1\Delta x = \pm 1$ and $\cos k_2\Delta y = \pm 1$. In this way we readily establish that

$$-4\left[\frac{A\Delta t}{(\Delta x)^2} + \frac{C\Delta t}{(\Delta y)^2}\right] \leq \psi \leq 0.$$

Equation (8.35) then shows: (1) that G is real; (2) that G is always ≤ 1; (3) that if $\theta \geq \frac{1}{2}$, G is also always ≥ -1; (4) that if $0 \leq \theta < \frac{1}{2}$, the stability requirement $G \geq -1$ imposes a restriction on Δt. In this way we arrive at the stability condition:

(8.36)
$$\begin{aligned}&\text{no restriction} &&\text{if } \tfrac{1}{2} \leq \theta \leq 1, \\ &\frac{A\Delta t}{(\Delta x)^2} + \frac{C\Delta t}{(\Delta y)^2} \leq \frac{1}{2-4\theta} &&\text{if } 0 \leq \theta < \tfrac{1}{2}.\end{aligned}$$

This is a natural generalization of the stability condition found for the corresponding one-dimensional problem (8.4).

To provide further examples, we now consider generalizations of some of the high-accuracy schemes of Table 8.I to the two-dimensional diffusion equation

$$(8.37) \qquad \frac{\partial u}{\partial t} = \sigma\left(\frac{\partial^2 u}{\partial x^2} + \frac{\partial^2 u}{\partial y^2}\right) = \sigma\nabla^2 u.$$

To get an accurate expression[6] for the Laplacian, we use a triangular net in the x, y-plane, whose meshes are equilateral triangles, as shown in Figure 8.2. The six nearest neighbors of a point (x, y) are the points $(x \pm h, y)$, $(x \pm \frac{1}{2}h, y \pm \frac{1}{2}\sqrt{3}h)$ where h is the distance between nearest neighbors. If the net points are designated by indices j, l, such that $x_j = jh$, $y_l = \frac{1}{2}\sqrt{3}lh$, then points in alternate rows have half-odd integer values of j, and the six neighbors of the point (j, l) are the points $(j \pm 1, l)$ and $(j \pm \frac{1}{2}, l \pm 1)$. We denote by $(\sum u)_{jl}$ the sum of the values of u over the six neighbors, viz.,

$$(\textstyle\sum u)_{jl} = u_{j+1,l} + u_{j-1,l} + u_{j+\frac{1}{2},l+1} + u_{j-\frac{1}{2},l+1} + u_{j+\frac{1}{2},l-1} + u_{j-\frac{1}{2},l-1}.$$

---

[6] Compare L. Collatz (1951), p. 314 ff.

SEC. 8.7]   PROBLEMS IN SEVERAL SPACE VARIABLES

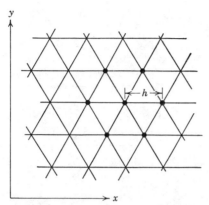

Fig. 8.2. The hexagonal net of points in the (x, y) plane for use in equations (8.38), (8.39) and (8.40).

By Taylor's series expansions it is readily seen that

$$(\sum u)_{jl} = 6u_{jl} + \tfrac{3}{2}h^2(\nabla^2 u)_{jl} + \tfrac{3}{32}h^4(\nabla^4 u)_{jl} + O(h^6).$$

(If the four nearest neighbors in a square net had been used, a similar formula would have been obtained, but the fourth order terms in this formula are $(1/12)h^4(\partial^4 u/\partial x^4 + \partial^4 u/\partial y^4)$ which *cannot* be expressed in terms of the Laplacian of u.) Therefore,

$$\frac{(\sum u)_{jl} - 6u_{jl}}{\tfrac{3}{2}h^2} = (\nabla^2 u)_{jl} + \tfrac{1}{16}h^2(\nabla^4 u)_{jl} + O(h^4).$$

Consequently, if u satisfies $(\partial u/\partial t) = \sigma \nabla^2 u$, we have:

$$\sigma \frac{(\sum u)_{jl} - 6u_{jl}}{\tfrac{3}{2}h^2} = \left(\frac{\partial u}{\partial t}\right)_{jl} + \frac{1}{16}h^2 \left(\nabla^2 \frac{\partial u}{\partial t}\right)_{jl} + O(h^4)$$

$$= \left(\frac{\partial u}{\partial t}\right)_{jl} + \frac{1}{16}h^2 \frac{\left(\sum \frac{\partial u}{\partial t}\right)_{jl} - 6\left(\frac{\partial u}{\partial t}\right)_{jl}}{\tfrac{3}{2}h^2} + O(h^4)$$

$$= \frac{3}{4}\left(\frac{\partial u}{\partial t}\right)_{jl} + \frac{1}{24}\left(\sum \frac{\partial u}{\partial t}\right)_{jl} + O(h^4).$$

The weights $\tfrac{3}{4}$ and $\tfrac{1}{24}$ correspond to the weights $\tfrac{5}{6}$ and $\tfrac{1}{12}$ appearing in examples 12 and 13 of Table 8.I for the one-dimensional diffusion equation.

The analogues of equations 12 and 13 of Table 8.I are:

$$(8.38) \quad \frac{1}{24} \frac{(\sum u)_{jl}^{n+1} - (\sum u)_{jl}^n}{\Delta t} + \frac{3}{4} \frac{u_{jl}^{n+1} - u_{jl}^n}{\Delta t}$$

$$= \frac{\sigma}{2} \frac{(\sum u)_{jl}^{n+1} - 6u_{jl}^{n+1}}{{}^3/_2 h^2} + \frac{\sigma}{2} \frac{(\sum u)_{jl}^n - 6u_{jl}^n}{{}^3/_2 h^2}$$

and

$$(8.39) \quad \frac{1}{24} \frac{[{}^3/_2 (\sum u)_{jl}^{n+1} - 2(\sum u)_{jl}^n + {}^1/_2 (\sum u)_{jl}^{n-1}]}{\Delta t}$$

$$+ \frac{3}{4} \frac{[{}^3/_2 u_{jl}^{n+1} - 2u_{jl}^n + {}^1/_2 u_{jl}^{n-1}]}{\Delta t} = \sigma \frac{[(\sum u)_{jl}^{n+1} - 6u_{jl}^{n+1}]}{{}^3/_2 h^2}$$

We shall refer to these as the 14-point formula and the 21-point formula, respectively (there are seven points in each plane $t = $ constant in each case). The truncation error is $O[(\Delta t)^2] + O(h^4)$ for both.

The analogue of the less accurate equation 9 of Table 8.I is the 9-point formula

$$(8.40) \quad \frac{{}^3/_2 u_{jl}^{n+1} - 2u_{jl}^n + {}^1/_2 u_{jl}^{n-1}}{\Delta t} = \sigma \frac{(\sum u)_{jl}^{n+1} - 6u_{jl}^{n+1}}{{}^3/_2 h^2}.$$

Its truncation error is $O[(\Delta t)^2] + O(h^2)$.

These formulas are consistent with (8.37) and unconditionally stable as $\Delta t \to 0$ and $h \to 0$. There is a relationship between the 14-point formula (8.38) and the 21-point formula (8.39) which is similar to the relationship noted earlier between schemes 12 and 13 of Table 8.I; namely, that the former is slightly preferred for smooth initial data and the latter for rapidly varying or discontinuous initial data.

In extensive numerical tests these formulas were found to give quite accurate results. The implicit equations were solved by a relaxation technique at each cycle. In the next section it will be shown how alternating-direction modifications of these equations can be constructed.

The 21-point formula is somewhat more accurate than the 9-point formula, but perhaps not as much as one might have hoped. There is a need for more accurate formulas, for such applications as integration of the age-diffusion equation in two or three space variables for nuclear reactors. Practical considerations of the speed and capacity of computing machines dictate the use of a quite coarse net for such calculations and reduction of the truncation error is highly desirable. For the simple diffusion equation with constant diffusion coefficient, very accurate difference equations can be obtained from the exact relation

$$u(\mathbf{x}, t + \Delta t) = \left(\frac{1}{4\pi\sigma\Delta t}\right)^{3/2} \int\int\int u(\mathbf{y}, t) \exp\left[-\frac{(\mathbf{x} - \mathbf{y})^2}{4\sigma\Delta t}\right] d^3\mathbf{y}$$

by approximation of the integral by a suitable numerical quadrature; but it seems difficult to generalize this approach to the practical problems having variable coefficients.

8.8. Alternating-Direction Methods

A method for multidimensional heat flow and diffusion problems, developed by Peaceman and Rachford (1955) and by Douglas (1955), combines unconditional stability with calculational simplicity. By now there are many versions of the method, with applications to elliptic and hyperbolic problems as well as to systems of parabolic equations—see Douglas and Gunn (1964) and the papers referred to by them.

For the simple diffusion equation (8.37), two difference approximations on a square net, with $\Delta x = \Delta y$, are used alternately on successive time steps of $\frac{1}{2}\Delta t$ each; they are

(8.41a) $\qquad u^{n+\frac{1}{2}} - u^n = \frac{1}{2}\alpha(\delta_x^2 u^{n+\frac{1}{2}} + \delta_y^2 u^n),$

(8.41b) $\qquad u^{n+1} - u^{n+\frac{1}{2}} = \frac{1}{2}\alpha(\delta_x^2 u^{n+\frac{1}{2}} + \delta_y^2 u^{n+1}),$

where u^n stands for u_{jk}^n, etc., $\delta_x^2 u$ for $u_{j+1\,k} - 2u_{jk} + u_{j-1\,k}$, $\delta_y^2 u$ for $u_{j\,k+1} - 2u_{jk} + u_{j\,k-1}$, and α for $\sigma\Delta t/(\Delta x)^2$. Assuming, for simplicity, that boundary values are given on some rectangle in the x, y-plane, equations (8.41) can be solved by the method of Section 8.5; for any fixed k, the matrix of the unknowns $u_{jk}^{n+\frac{1}{2}}$ appearing in (8.41a) is tridiagonal, as is also for fixed j the matrix of the unknowns u_{jk}^{n+1} in (8.41b).

The two successive time steps constitute a cycle of the calculation, and the amplification factor g is a product $g'g''$ of two factors, one coming from each equation. If β_1 and β_2 denote $k_1\Delta x$ and $k_2\Delta x$, where k_1 and k_2 are the wave numbers of a Fourier component $\exp[i(k_1 x + k_2 y)]$, then

$[1 + \alpha(1 - \cos \beta_1)]g' = 1 - \alpha(1 - \cos \beta_2)$, and $[1 + \alpha(1 - \cos \beta_2)]g'' = 1 - \alpha(1 - \cos \beta_1)$, so that

$$(8.42) \qquad g = \frac{1 - \alpha(1 - \cos \beta_1)}{1 + \alpha(1 - \cos \beta_1)} \frac{1 - \alpha(1 - \cos \beta_2)}{1 + \alpha(1 - \cos \beta_2)};$$

since $\alpha(1 - \cos \beta_{1,2}) \geq 0$, g lies between 0 and 1, and unconditional stability follows.

To find the order of accuracy, (8.41b) is subtracted from (8.41a) to give

$$2u^{n+1/2} = u^{n+1} + u^n - \tfrac{1}{2}\alpha\delta_y^2(u^{n+1} - u^n);$$

this is then substituted into the sum of the two equations to give

$$(8.43) \qquad \frac{u^{n+1} - u^n}{\Delta t} = \sigma \frac{(\delta_x^2 + \delta_y^2)(u^{n+1} + u^n)}{2(\Delta x)^2} - \sigma^2 \Delta t \frac{\delta_x^2 \delta_y^2(u^{n+1} - u^n)}{4(\Delta x)^4};$$

the last term is $O[(\Delta t)^2]$ for a smooth function $u(x, y, t)$, and the other terms are centered as in the Crank-Nicholson equation; hence, $e = O[(\Delta t)^2] + O[(\Delta x)^2]$.

In the most obvious generalization to three dimensions, by three equations containing respectively only u^n and $u^{n+1/3}$, $u^{n+1/3}$ and $u^{n+2/3}$, and $u^{n+2/3}$ and u^{n+1}, the equations being identical except for a cyclic shifting of the implicit term among the x-, y-, and z-derivatives, the unconditional stability is lost, and the accuracy drops to $O(\Delta t) + O[(\Delta x)^2]$. Therefore, a different procedure was invented; it is a special case of a very general method for deriving alternating-direction schemes from fully implicit ones, described by Douglas and Gunn (1964). The procedure is best described in terms of a succession of approximate solutions u^{*n+1}, u^{**n+1}, etc. of the Crank-Nicholson equation. The last of them, u^{***n+1} for three dimensions, is then renamed simply u^{n+1} for use in the next cycle. The equations are

$$u^{*n+1} - u^n = \frac{\alpha}{2}[\delta_x^2(u^{*n+1} + u^n) + \delta_y^2(2u^n) + \delta_z^2(2u^n)],$$

$$(8.44) \quad u^{**n+1} - u^n = \frac{\alpha}{2}[\delta_x^2(u^{*n+1} + u^n) + \delta_y^2(u^{**n+1} + u^n) + \delta_z^2(2u^n)],$$

$$u^{n+1} - u^n = \frac{\alpha}{2}[\delta_x^2(u^{*n+1} + u^n) + \delta_y^2(u^{**n+1} + u^n) + \delta_z^2(u^{n+1} + u^n)].$$

ALTERNATING-DIRECTION METHODS

The obvious generalization works in any number of space variables; in the case of two variables x and y, it gives the scheme (8.41) again, in disguise.

The amplification factor for a complete cycle is readily calculated and found to be

$$(8.45) \qquad g = \frac{1 - a - b - c + ab + ac + bc + abc}{(1 + a)(1 + b)(1 + c)},$$

where a, b, and c stand for $\alpha(1 - \cos k_i \Delta x)$, $i = 1, 2, 3$. The denominator of (8.45), when multiplied out, contains exactly the same terms as the numerator, except that the signs are all $+$, so again $0 \leq g \leq 1$, and unconditional stability follows. It is interesting to note that if u^{*n+1} is replaced by u^{**n+1} in the x-derivative term in the third equation of (8.44) (one might think it wise to use always the most recent approximation), then the unconditional stability is again lost (g can be < -1).

The evaluation of the u_{jkl}^{n+1} by (8.44) consists again in solving a succession of tridiagonal linear systems. The work is somewhat simplified if one subtracts the second equation from the first and the third from the second, and then uses the resulting new equations in place of the last two of (8.44).

The accuracy can be found by eliminating u^{*n+1} and u^{**n+1}, much as $u^{n+\frac{1}{2}}$ was eliminated from (8.41). The result is

$$(8.46) \qquad \frac{u^{n+1} - u^n}{\Delta t} = \sigma \frac{(\delta_x^2 + \delta_y^2 + \delta_z^2)(u^{n+1} + u^n)}{2(\Delta x)^2}$$

$$- \sigma^2 (\Delta t) \frac{(\delta_x^2 \delta_y^2 + \delta_x^2 \delta_z^2 + \delta_y^2 \delta_z^2)(u^{n+1} - u^n)}{4(\Delta x)^4}$$

$$+ \sigma^3 (\Delta t)^2 \frac{\delta_x^2 \delta_y^2 \delta_z^2 (u^{n+1} - u^n)}{8(\Delta x)^6},$$

which shows that here, also, $e = O[(\Delta t)^2] + O[(\Delta x)^2]$.

In the general formulation of the alternating-direction procedure, the difference equation (3.4) is written as

$$(8.47) \qquad (I + A)u^{n+1} + Bu^n = 0,$$

where

$$A = \sum_{i=1}^{q} A_i,$$

each A_i being a difference operator (like δ_x^2, δ_y^2, etc.) such that $I + A_i$ is always easy to invert in the sense that a tridiagonal linear system is easy to solve. The successive approximations to u^{n+1} are called $u^{n+1}(j)$, where j corresponds to the number of asterisks appearing in (8.44). The generalization of (8.44) is then

$$u^{n+1}(1) + A_1 u^{n+1}(1) + \sum_{i=2}^{q} A_i u^n + B u^n = 0,$$

$$\vdots$$

(8.48) $\quad u^{n+1}(j) + \sum_{i=1}^{j} A_i u^{n+1}(i) + \sum_{i=j+1}^{q} A_i u^n + B u^n = 0,$

$$\vdots$$

$$u^{n+1}(q) + \sum_{i=1}^{q} A_i u^{n+1}(i) + B u^n = 0,$$

with $u^{n+1}(q)$ renamed simply u^{n+1} to start the next cycle. In the general equation of (8.48), the unknowns appear only in the first term and in the term with $i = j$. The formulation has been further generalized by Douglas and Gunn (1964) to mildly non-linear equations (a non-linear undifferentiated term is permitted in the original differential equations) and to multi-level schemes of the kind discussed in Chapter 7.

Much effort has been made to find the effect of boundary conditions and of variable coefficients on stability and accuracy. Under quite general circumstances, it can be shown that the scheme (8.48) is consistent with the differential equation if the original scheme (8.47) was. For a rather wide class of parabolic problems, if $q = 2$ (this restricts us to two space variables, but we note that even for two space variables it is sometimes necessary to take $q > 2$, as in the example below), stability of (8.47) implies the stability of (8.48) for any fixed value of $\Delta t[(\Delta x)^{-2} + (\Delta y)^{-2}]$. For q not necessarily equal to 2, under very special circumstances (the difference operators A_i, B are assumed to commute and be hermitian, with the A_i positive definite—this practically restricts us to the simple heat flow problem in a rectangular domain), the stability of (8.47) implies the stability of (8.48). Also, under these same circumstances, it can be shown that the loss in accuracy in going from (8.47) to (8.48) is at most $O[(\Delta t)^2]$. Under certain slightly more general circumstances it can be shown that there exists a function g such that (8.48) is stable for $\Delta t \leq g(\Delta x)$ (we assume $\Delta x = \Delta y = $ etc.) or that (8.48) is stable for a sufficiently small fixed value of $\Delta t/(\Delta x)^4$. For a discussion of these results and applications of the method, see Douglas and Gunn (1964). Intuition suggests that the method is stable and accurate under a wider class of circumstances than those for which rigorous proofs have been given.

As an illustration, we apply the general procedure to the implicit equation (8.38) for the hexagonal point net. We define second-difference operators δ_1^2, δ_2^2, δ_3^2 as follows:

(8.49) $\quad \delta_1^2 u = u(x + h, y) - 2u(x, y) + u(x - h, y),$

$\delta_{2,3}^2 u = u(x + \tfrac{1}{2}h, y \pm \tfrac{1}{2}\sqrt{3}h)$
$\qquad - 2u(x, y) + u(x - \tfrac{1}{2}h, y \mp \tfrac{1}{2}\sqrt{3}h).$

Then, (8.38) can be written as

(8.50) $\qquad u^{n+1} - u^n = \alpha \delta^2 (\theta u^{n+1} + (1 - \theta) u^n),$

where $\delta^2 = \delta_1^2 + \delta_2^2 + \delta_3^2$, $\alpha = 2\sigma \Delta t / 3(\Delta x)^2$, and $\theta = \tfrac{1}{2} - 1/24\alpha$. The equations are then identical with (8.44), except that δ_x, δ_y, and δ_z are replaced by δ_1, δ_2, and δ_3, and the mean $\tfrac{1}{2}u^{n+1} + \tfrac{1}{2}u^n$ is replaced by the weighted mean $\theta u^{n+1} + (1 - \theta) u^n$. In other words, in (8.48) we now have $q = 3$, $A_i = -\alpha \theta \delta_i^2 (i = 1, 2, 3)$, and $B = -I - \alpha(1 - \theta) \delta^2$. Note that $q(=3)$ exceeds the number ($= 2$) of space variables.

To find the amplification factor, we use a method that is applicable generally to the scalar case of the system (8.48) when the operators A_i are positive definite. (We should perhaps remind the reader that we are dealing here, as always in the von Neumann theory, with a pure initial-value problem with constant coefficients. It was pointed out by Birkhoff and Varga (1959) that boundary conditions on other than rectangular regions generally destroy the commutativity of the A_i regarded as operators whose domains consist of functions satisfying the boundary conditions; then, as for problems with variable coefficients, the amplification factor, as we have defined it, is a local property of the system, defined for an interior region of the physical system where the coefficients are slowly varying. As noted above, the general theory is not yet in a very satisfactory state from the point of view of practical applications.) The effect of the operator A_j on the function $\exp[ik_1 x + ik_2 y]$ is to multiply it by the factor $2\alpha\theta(1 - \cos\beta_j)$, where

$$\beta_1 = k_1 h, \quad \beta_{2,3} = \tfrac{1}{2}k_1 h \pm \tfrac{1}{2}\sqrt{3}k_2 h;$$

this factor is non-negative; hence, A_j is positive definite. The intermediate quantities $u^{n+1}(j)(j = 1, \ldots, q - 1)$ can be eliminated from (8.48). For $q = 3$, the result is

(8.51) $(I + A)u^{n+1} + Bu^n$
$+ (A_1 A_2 + A_2 A_3 + A_1 A_3 + A_1 A_2 A_3)(u^{n+1} - u^n) = 0.$

If we denote by $a = a(\mathbf{k})$, $b = b(\mathbf{k})$, and $\Lambda = \Lambda(\mathbf{k})$ the factors by which $\exp(i\mathbf{k}\cdot\mathbf{x})$ becomes multiplied under the action of the operators A, B, and $(A_1 A_2 + \cdots + A_1 A_2 A_3)$ in the above equation, then the amplification factor g is

(8.52) $\qquad g = (-b + \Lambda)/(1 + a + \Lambda).$

According to (8.47), the amplification factor of the original scheme—in the present case, (8.38)—is $g_0 = -b/(1 + a)$. Since $\Lambda \geq 0$, the inequality $-1 \leq g_0 \leq 1$ implies $-1 \leq g \leq 1$; hence, the alternating-direction version of (8.38) is always stable.

8.9. Splitting and Fractional-Step Methods

Methods closely related to (in some cases identical with) the alternating-direction methods were developed at about the same time by Soviet mathematicians,[7] and are known as methods of splitting or methods of fractional steps. The basic idea, as set forth by Bagrinovskii and Godunov (1957) and analyzed more generally by Yanenko (1964), is this: if $\partial u/\partial t = Au$, where $A = A_1 + \cdots + A_q$, one chooses a difference scheme and then replaces A in it successively by $qA_1, qA_2, \ldots, qA_q$, each for a fraction

---

[7] Bagrinovskii and Godunov (1957); Yanenko (1959); D'yakonov (1962); Sofronov (1962); Konovalov (1964); Samarskii (1964); D'yakonov (1964); Marchuk and Yanenko (1965); there is an extensive bibliography in the last of these references and in the recent book by Yanenko (1966).

SEC. 8.9]   SPLITTING AND FRACTIONAL-STEP METHODS

$\Delta t/q$ of the time step Δt. Thus, one can replace a complicated (say multi-dimensional) problem by a succession of simpler (say one-dimensional) ones.

For example, the Crank-Nicholson scheme for multi-dimensional heat flow,

$$\frac{u^{n+1} - u^n}{\Delta t} = \sigma \frac{(\delta_x^2 + \delta_y^2 + \cdots + \delta_w^2)(u^{n+1} + u^n)}{2(\Delta x)^2}$$

$$= -(A_1 + \cdots + A_q)\left(\frac{u^{n+1} + u^n}{2}\right),$$

where $A_i = -(\sigma/(\Delta x)^2)\delta_{x_i}^2$ etc., where we have supposed $\Delta x = \Delta y = \ldots$, and where, as usual, δ_x^2, δ_y^2 etc. stand for second difference operators with respect to the indicated variables, is replaced by the sequence of equations

$$(8.53) \quad \frac{u^{n+i/q} - u^{n+(i-1)/q}}{\Delta t} = -A_i \frac{u^{n+i/q} + u^{n+(i-1)/q}}{2} \quad (i = 1, 2, \ldots, q).$$

In contrast to equation (8.48) of Douglas, the terms A_j of the operator with $j \neq i$ are dropped completely. Consequently, for any scalar equation, the amplification factor is simply the product of the amplification factors of the one-dimensional equations (8.53). In the present case these are of Crank-Nicholson type, and unconditional stability results. To find the accuracy, we write

$$\left(I + \frac{\Delta t}{2} A_i\right) u^{n+i/q} = \left(I - \frac{\Delta t}{2} A_i\right) u^{n+(i-1)/q};$$

since the A_i commute in our example,

$$\left(I + \frac{\Delta t}{2} A_1\right) \cdots \left(I + \frac{\Delta t}{2} A_q\right) u^{n+1} = \left(I - \frac{\Delta t}{2} A_1\right) \cdots \left(I - \frac{\Delta t}{2} A_q\right) u^n.$$

Multiplying out gives

$$u^{n+1} - u^n + \Delta t A \frac{u^{n+1} + u^n}{2}$$

$$= -\frac{(\Delta t)^2}{4} (A_1 A_2 + A_1 A_3 + \cdots + A_{q-1} A_q)(u^{n+1} - u^n) + O[(\Delta t)^3];$$

since $u^{n+1} - u^n$ is $O(\Delta t)$, and since the left member represents the Crank-Nicholson scheme, we see (after dividing by Δt) that the truncation error is $O[(\Delta t)^2] + O[(\Delta x)^2]$.

CHAPTER 9

The Transport Equation

9.1. Physical Basis

Problems of the class discussed in this chapter arise primarily in two fields: neutron diffusion and radiation transfer. The term "neutron diffusion" is often used (unfortunately) to describe neutron migration and multiplication, even when the diffusion approximation is invalid. The basic integro-differential equation for these processes is generally called the "transport equation," although astrophysicists use the slightly more appropriate term "transfer equation." The somewhat misleading term "Boltzmann equation" is sometimes used.

The transport equation describes the motion of particles (neutrons, protons, etc.) which travel in straight lines with constant speed between collisions, but which, in their passage through matter, are constantly subject to a certain probability of colliding with the nuclei or atoms of the matter and being deflected, slowed down, absorbed, or multiplied. The equation treats so large an aggregate or population of such particles that they may be regarded as a continuum, ignoring statistical fluctuations. In the Boltzmann equation of statistical mechanics, the scattering particles are the same as the scattered ones, whereas in the transport equation the distribution and properties of the scattering centers are assumed given; consequently the transport equation is linear while the Boltzmann equation is quadratic.

At any instant the motion of a neutron can be represented by a point in a six-dimensional phase space whose coordinates are the components of the position and velocity vectors of the neutron. The whole population is described by its distribution in this space. The initial-value problem is to find how the distribution changes with time if it is known initially.

A similar situation is found in radiation transfer problems, except that the phase space is based on position and *momentum* vectors (this *could*

SEC. 9.2]   THE GENERAL NEUTRON TRANSPORT EQUATION   219

have been done for the neutrons, too), because the speed of a photon is taken in these problems to be independent of its energy, $h\nu$. In astrophysics the main interest is in steady-state problems, but time dependent problems arise in connection with pulsating or otherwise variable stars.

9.2. The General Neutron Transport Equation

Suppose that the neutron-reacting system is contained in a convex region $\mathscr{R}$ bounded by a surface $\mathscr{S}$. Let $\mathbf{x} = (x, y, z)$ and $\mathbf{v} = (v_x, v_y, v_z)$ denote the position and velocity vectors. For $\mathbf{x}$ in $\mathscr{R}$, let

$$\psi(\mathbf{x}, \mathbf{v}, t)d\mathbf{x}\,d\mathbf{v}$$

be the number of neutrons in $(x, x + dx), (y, y + dy), \ldots, (v_z, v_z + dv_z)$ at time t, where $d\mathbf{x}$ is an abbreviation for $dx\,dy\,dz$ and $d\mathbf{v}$ is an abbreviation for $dv_x dv_y dv_z$.

As in the usual statistical mechanical treatments, we suppose that $d\mathbf{x}$ and $d\mathbf{v}$ can be chosen so as to be small compared to regions in which appreciable variation of conditions occur but still large enough that $\psi d\mathbf{x}\,d\mathbf{v}$ represents a large number of neutrons. We are not interested here in fluctuation phenomena.

For present purposes neutrons are classical point particles whose collisions with nuclei proceed according to laws of chance, so that we must deal with averages and probabilities. Let $v\sigma(\mathbf{v}, \mathbf{x})$ denote the probability per unit time for collision of a neutron whose velocity is $\mathbf{v}$, where v denotes the magnitude of $\mathbf{v}$. $\sigma(\mathbf{v}, \mathbf{x})$, is the total nuclear cross-section per unit volume of material at point $\mathbf{x}$. If the material is isotropic, σ is independent of the direction of $\mathbf{v}$, and if it is homogeneous, σ is independent of $\mathbf{x}$; therefore, $\sigma(v)$ will be written in place of $\sigma(\mathbf{v}, \mathbf{x})$, although in many interesting problems σ depends on $\mathbf{x}$ at least in the sense that the system consists of several regions in which $\sigma(v)$ has different values.

If a neutron with velocity $\mathbf{v}'$ collides, we let $K(\mathbf{v}', \mathbf{v})d\mathbf{v}$ denote the probability that after collision it has velocity components in the ranges $(v_x, v_x + dv_x), (v_y, v_y + dv_y), (v_z, v_z + dv_z)$. For an isotropic medium, K may depend on the speeds v', v, and on the angle between $\mathbf{v}'$ and $\mathbf{v}$, but not otherwise on the directions of $\mathbf{v}'$ and $\mathbf{v}$. For given $\mathbf{v}'$, $\mathbf{x}$, the integral $\int K d\mathbf{v}$ over all velocities of the emergent neutron, denoted by $1 + f$, is the average number of neutrons emergent from a collision, so that $f = 0$ for a purely scattering medium, $f < 0$ for an absorbing medium and $f > 0$ for a multiplying medium. In the last case one considers multiplicative processes such as fission and $(n, 2n)$ processes from which two or more neutrons may emerge. In this case, the word "it" in the above definition of K should be replaced by "a neutron," not necessarily identified with the incident neutron.

The rate of change of $\psi(\mathbf{x}, \mathbf{v}, t)$ is governed by the *transport equation* which is

$$(9.1) \quad (\partial/\partial t + \mathbf{v}\cdot\nabla)\psi(\mathbf{x}, \mathbf{v}, t) = -v\sigma(v)\psi(\mathbf{x}, \mathbf{v}, t) + \int v'\sigma(v')K(\mathbf{v}', \mathbf{v})\psi(\mathbf{x}, \mathbf{v}', t)d\mathbf{v}'$$

where $v\cdot\nabla$ stands for $v_x\partial/\partial x + v_y\partial/\partial y + v_z\partial/\partial z$ and

$$\int \ldots d\mathbf{v}' \text{ stands for } \int_{-\infty}^{\infty}\int_{-\infty}^{\infty}\int_{-\infty}^{\infty} \ldots dv'_x dv'_y dv'_z.$$

The boundary condition is that no neutrons are incident on the system from outside. We assume that the region $\mathscr{R}$ is convex (otherwise one would have to take into account neutrons leaving the system at one place and reentering elsewhere) and that the surface $\mathscr{S}$ has a normal almost everywhere. If this is not the case one can of course always replace $\mathscr{R}$ by a larger region having the required properties, the extra space being assigned zero cross section—and in fact $\mathscr{R}$ may as well be taken as a sphere. If $\mathbf{n}(\mathbf{x})$ is the outward normal to $\mathscr{S}$ at point $\mathbf{x}$, the boundary condition is

$$(9.2) \quad \psi(\mathbf{x}, \mathbf{v}, t) = 0 \quad \text{for } \mathbf{x} \text{ on } \mathscr{S} \text{ and } \mathbf{v}\cdot\mathbf{n}(\mathbf{x}) < 0.$$

The interpretation of the transport equation (or rather the physical picture from which it is usually derived) is as follows: The neutron population is thought of as consisting of the superposition of a very large number (in fact a five-parameter family) of narrow beams of neutrons travelling together. $\psi(\mathbf{x}, \mathbf{v}, t)$ is the intensity at the point $\mathbf{x}$ of one such beam, and the left member of (9.1) is the time rate of change of this intensity as we follow the particles in their motion. The first term of the

SEC. 9.2] THE GENERAL NEUTRON TRANSPORT EQUATION

right member is the rate at which particles are removed from the beam by collisions and the last term (the integral) is the rate at which particles are put back into the beam by other collisions.

If $\psi(\mathbf{x}, \mathbf{v}, 0)$ is given, equations (9.1) and (9.2) provide a linear initial-value problem. We may write $\partial \psi / \partial t = A\psi$, where A is a linear operator given by (9.1), with a suitably defined domain restricted to functions satisfying the boundary condition (9.2).

Because of the large number of independent variables, numerical methods have not been applied directly to the full transport equation. Instead, simplifying assumptions are made leading to various special cases. We have already assumed that $\sigma(v)$ and $K(\mathbf{v}', \mathbf{v})$ are time independent, and our notation implies that the material is isotropic and homogeneous. Additional assumptions that may be made (not all at once) are:

1. Finite number of neutron groups.
2. Slab symmetry.
3. Spherical symmetry.
4. Cylindrical symmetry.
5. Axial symmetry.
6. No space dependence.
7. $1/\sigma$ very small.
8. Emergent neutrons isotropic.

Under 1 is meant that the speed v is restricted to a finite number of values $v_1, v_2, \ldots, v_G$ instead of being allowed to vary continuously and that integration over v is replaced by a summation over the groups. In particular, if one is willing to assume that all fission neutrons have roughly the same energy (and hence the same speed) and that all scatterings are without energy change, there is only one group. Under 2 is meant first all quantities are independent of two cartesian coordinates, which we call x and y, and second that the velocity distribution is symmetrical about the z direction; in other words, the system and the neutron population in it are invariant under translations and rotations in the x,y-plane. 3 means invariance under all rotations about a point. 4 means invariance under all translations parallel to an axis. 5 means invariance under all rotations about an axis. 6 means invariance under all rotations and translations of space. 7 means that the mean free path, $1/\sigma$, is small compared to all other relevant lengths associated with the system or with the neutron distribution in it (this is the diffusion-theory limit). 8 means that $K(\mathbf{v}', \mathbf{v})$ is independent of the direction of $\mathbf{v}$, and therefore of the angle $\angle \mathbf{v}', \mathbf{v}$.

We shall not consider all possible combinations of these assumptions, but only a few typical ones.

9.3. Homogeneous Slab: One Group

All neutrons have speed v, and the slab is assumed to fill the region $-a \leq z \leq a$. The probability that in a scattering the emergent direction will lie in a unit solid angle making an angle α with the incident direction is denoted by $(1/4\pi)p(\cos \alpha)$—this takes the place of the kernel $K(\mathbf{v}', \mathbf{v})$ of the general equation, and the isotropic case is obtained by setting $p(\cos \alpha) = 1$.

If θ is the angle between the velocity vector $\mathbf{v}$ and the z-axis, and $\mu = \cos \theta$, we write the distribution function as $\psi(z, \mu, t)$—it is the number of neutrons per unit volume of space and unit solid angle of direction of motion. The transport equation, after dividing through by v, is

$$(9.3) \quad \left(\frac{1}{v}\frac{\partial}{\partial t} + \mu \frac{\partial}{\partial z} + \sigma\right)\psi(z, \mu, t) = \sigma \frac{1+f}{4\pi} \int_{-1}^{1} d\mu' \int_{0}^{2\pi} d\varphi \, p(\cos \alpha)\psi(z, \mu', t),$$

$$\text{for } -a \leq z \leq a, \; t \geq 0,$$

where

$$(9.4) \qquad \cos \alpha = \mu\mu' + \sqrt{(1-\mu^2)(1-\mu'^2)} \cos \varphi.$$

Here α is the angle between the incident and emergent directions; the polar angles of these directions may be taken as $(\theta, 0)$ and (θ', φ) where $\mu = \cos \theta$ and $\mu' = \cos \theta'$.

The boundary condition is

$$(9.5) \qquad \psi(z, \mu, t) = 0 \quad \text{for } z = a, \mu < 0, \text{ and for } z = -a, \mu > 0$$

which says that no neutrons are incident on the right face of the slab from the right or on the left face of the slab from the left.

In many situations the directional distribution of scattering is isotropic.

That is, there is no correlation between the incident and scattered directions, and $p(\cos \alpha) = 1$. This is true, for example, in the scattering of low energy neutrons by heavy atoms. Then the transport equation becomes

$$(9.6) \quad \left(\frac{1}{v}\frac{\partial}{\partial t} + \mu\frac{\partial}{\partial z} + \sigma\right)\psi(z, \mu, t) = \sigma\frac{(1 + f)}{2}\int_{-1}^{1} d\mu'\, \psi(z, \mu', t).$$

Note that isotropy of the scattering law does not imply isotropy of the solution: ψ still depends on μ even though $p(\cos \alpha)$ is a constant.

9.4. Homogeneous Sphere: One Group

Let a be the radius of the sphere, r the radial coordinate, $\mu = \cos \theta$ where θ is the angle between the radius vector from the origin to the neutron and the velocity vector of the neutron. The transport equation is

$$(9.7) \quad \left(\frac{1}{v}\frac{\partial}{\partial t} + \mu\frac{\partial}{\partial r} + \frac{1-\mu^2}{r}\frac{\partial}{\partial \mu} + \sigma\right)\psi(r, \mu, t)$$

$$= \sigma\frac{1 + f}{4\pi}\int_{-1}^{1} d\mu' \int_{0}^{2\pi} d\varphi\, p(\cos \alpha)\psi(r, \mu', t)$$

$$\text{for } 0 < r \leq a, \quad t \geq 0,$$

where $\cos \alpha$ is as before (equation 9.4), and the boundary condition is

$$(9.8) \quad \psi(r, \mu, t) = 0 \quad \text{for } r = a, \mu < 0.$$

It might be thought that a mathematical boundary condition would be needed at $r = 0$, as so often happens when physical problems are expressed in polar coordinates, because of the idiosyncrasies of the coordinate system at the origin. This is however not the case in the present problem. It can be shown that even without any such boundary condition, the solution of the initial-value problem is unique, if it exists at all in the L_2 sense, except on the line $\mu = 1$. But this line has measure zero in velocity space, and two distribution functions which agree except on a set of measure zero represent the same physical distribution. This argument supposes ψ to be finite-valued and therefore excludes "delta functions" and the like from consideration.

9.5. The "Spherical Harmonic" Method

The method which bears this name (see Chandrasekhar, 1944a) has not been much used for time-dependent problems but, because of its great success in stationary neutron problems, we shall consider a number of simple difference equation systems for the time-dependent spherical-harmonic equations and discuss their stability. We shall confine attention to one-group problems for spherical and slab symmetry, although generalizations to multigroup problems and to problems in axial-cylindrical symmetry are straightforward, just as in stationary cases.

In this method the angular dependence of the distribution at a given point of space is represented by an expansion[1] of $\psi(r, \mu, t)$ in Legendre polynomials in μ with coefficients depending on r and t. Substitution of this expansion into the transport equation leads to simultaneous partial differential equations for the coefficients. Specifically, let

$$(9.9) \qquad \psi(r, \mu, t) = \sum_{0}^{\infty}{}_{(l)} \sqrt{2l+1}\, \psi_l(r, t) P_l(\mu),$$

and

$$(9.10) \qquad p(\cos \alpha) = \sum_{0}^{\infty}{}_{(l)} p_l P_l(\cos \alpha).$$

The extra factor $\sqrt{2l+1}$ appearing in (9.9) is purely arbitrary—its introduction induces a convenient normalization of the $\psi_l(r, t)$; the coefficients p_l in the scattering law (9.10) are assumed known from physical measurements or otherwise—this is equivalent to assuming that $p(\cos \alpha)$ is a known function.

The addition theorem[2] for Legendre polynomials states that if $\cos \alpha$ is given by equation (9.4), then

$$(9.11) \qquad P_l(\cos \alpha) = P_l(\mu)P_l(\mu') + 2\sum_{1}^{l}{}_{(k)} P_l^k(\mu) P_l^k(\mu') \frac{(l-k)!}{(l+k)!} \cos k\varphi$$

---

[1] See Courant-Hilbert, vol. I (1953), p. 513. This expansion was used for transport problems by several investigators during the war. The first published account appears to be that of Chandrasekhar (1944a).

[2] Whittaker and Watson (1927), **p. 395.**

where P_l^k denotes the associated Legendre polynomial. If this expression for $P_l(\cos \alpha)$ is inserted into the expansion of $p(\cos \alpha)$ given in (9.10) we see that, on integration over φ, all terms of the above summation will vanish, leaving only the product $P_l(\mu)P_l(\mu')$. That is,

$$(9.12) \qquad \int_0^{2\pi} d\varphi \, p(\cos \alpha) = 2\pi \sum_{(l)0}^{\infty} p_l P_l(\mu) P_l(\mu').$$

According to equation (9.7), this result must now be multiplied by the expansion of $\psi(r, \mu', t)$, which may be written (compare 9.9) as

$$\psi(r, \mu', t) = \sum_{(j)0}^{\infty} \sqrt{2j+1} \, \psi_j(r, t) P_j(\mu'),$$

and integrated over μ' from -1 to $+1$.

The result can be simplified because of the orthogonality and normalization[3] relations of the Legendre functions, which are

$$\int_{-1}^{1} d\mu' \, P_l(\mu') P_j(\mu') = \begin{cases} 0 & \text{if } l \neq j \\ \dfrac{2}{2j+1} & \text{if } l = j. \end{cases}$$

In consequence, the entire right member of (9.7) reduces to

$$(9.13) \qquad \sigma(1+f) \sum_{(j)0}^{\infty} \frac{p_j \psi_j(r, t) P_j(\mu)}{\sqrt{2j+1}}.$$

We wish to express the left member of (9.7) also in the form of a sum of terms, each consisting of a Legendre polynomial in μ multiplied by a function of r and t only—then we will be able to equate coefficients of $P_j(\mu)$ and obtain a set of simultaneous equations for the $\psi_j(r, t)$. To achieve this we substitute the expansion of $\psi(r, \mu, t)$, but we must also get rid of the explicit occurrence of μ and of $(1 - \mu^2)\partial/\partial\mu$ appearing in (9.7), and this we can do by using certain recurrence relations.[4] One of these is

$$\mu P_l(\mu) = \frac{(l+1)P_{l+1}(\mu) + l P_{l-1}(\mu)}{2l+1},$$

---

[3] Whittaker and Watson (1927), p. 305.

[4] These are similar to, and derivable from, recurrence relations in Whittaker and Watson (1927), pp. 308 and 309.

from which

$$(9.14) \quad \mu\psi(r, \mu, t) = \sum_{(l)}^{\infty} \psi_l(r, t) \frac{(l+1)P_{l+1}(\mu) + lP_{l-1}(\mu)}{\sqrt{2l+1}}$$

$$= \sum_{(j)}^{\infty} \left[\frac{j}{\sqrt{2j-1}} \psi_{j-1}(r, t) + \frac{j+1}{\sqrt{2j+3}} \psi_{j+1}(r, t) \right] P_j(\mu).$$

The last expression results from a rearrangement of the terms of the sum (assumed permissible): in the first term j has been identified with $l+1$ and in the second with $l-1$; we furthermore define $\psi_{-1}(r, t)$ to be identically zero.

A second recurrence relation is

$$(1 - \mu^2) \frac{d}{d\mu} P_l(\mu) = \frac{l(l+1)}{2l+1} [P_{l-1}(\mu) - P_{l+1}(\mu)]$$

which may be used in a similar way to obtain the result

$$(9.15) \quad (1 - \mu^2) \frac{\partial}{\partial \mu} \psi(r, \mu, t)$$

$$= \sum_{(j)}^{\infty} \left[-\frac{j(j-1)}{\sqrt{2j-1}} \psi_{j-1}(r, t) + \frac{(j+1)(j+2)}{\sqrt{2j+3}} \psi_{j+1}(r, t) \right] P_j(\mu).$$

We now substitute (9.13), (9.14), (9.15) into the transport equation (9.7) and equate coefficient of $P_j(\mu)$ in the two sides as promised. The result is a system of partial differential equations for the $\psi_j(r, t)$. The equation containing the time derivative of ψ_j contains not only ψ_j but also ψ_{j+1} and ψ_{j-1} because of the expressions (9.14) and (9.15), so that if the equations are written in sequence, each one is coupled to the one before it and the one after it.

The *spherical harmonic approximations of order L* is the result of solving the first $L+1$ of the equations for the $L+1$ unknown functions $\psi_0, \psi_1,$ ..., after ψ_{L+1} has been set equal to 0 in the last of these equations.

Much of the foregoing is heuristic. Justification is needed for the expansions used, for the rearrangement of series, and for the supposition that the spherical harmonic approximation approaches the solution of the transport equation as $L \to \infty$. Bengt Carlson (1946) has provided the justification in a few special cases by exhibiting the spherical harmonic approximation analytically and examining the limiting process $L \to \infty$. More recently, Keller and Wendroff have proved the convergence of the so-called discrete-ordinate methods. One of these, the Wick-Chandrasekhar method, is equivalent to the spherical harmonic method, and this establishes the convergence of the latter. (See Sections 9.10 and 9.11.) However, it is not our present purpose to give a rigorous derivation of the spherical harmonic equations any more than of the transport equation on which they were based. But, given an initial-value problem of these equations, we wish to discuss its approximate solution by stepwise numerical methods.

To express the spherical harmonics equations in concise form, let **u** denote an $(L + 1)$-dimensional column vector defined by

$$\mathbf{u} = \mathbf{u}(r, t) = \begin{bmatrix} \psi_0(r, t) \\ \psi_1(r, t) \\ \cdots \\ \psi_L(r, t) \end{bmatrix}.$$

Then the equations are

(9.16) $$\left(\frac{1}{v}\frac{\partial}{\partial t} + M\frac{\partial}{\partial r} + \frac{1}{r}N + \sigma\right)\mathbf{u} = \sigma(1 + f)S\mathbf{u},$$

where M, N, and S are $(L + 1) \times (L + 1)$ matrices given by

$$M = \begin{bmatrix} 0 & \frac{1}{\sqrt{3}} & 0 & 0 & 0 & \cdots \\ \frac{1}{\sqrt{3}} & 0 & \frac{2}{\sqrt{3.5}} & 0 & 0 & \cdots \\ 0 & \frac{2}{\sqrt{3.5}} & 0 & \frac{3}{\sqrt{5.7}} & 0 & \cdots \\ 0 & 0 & \frac{3}{\sqrt{5.7}} & 0 & \cdot & \cdots \\ \cdot & \cdot & \cdot & \cdot & \cdot & \cdot \end{bmatrix},$$

THE TRANSPORT EQUATION

$$N = \begin{bmatrix} 0 & \dfrac{1.2}{\sqrt{1.3}} & 0 & 0 & \cdots \\ 0 & 0 & \dfrac{2.3}{\sqrt{3.5}} & 0 & \cdots \\ 0 & -\dfrac{1.2}{\sqrt{3.5}} & 0 & \dfrac{3.4}{\sqrt{5.7}} & \cdots \\ 0 & 0 & -\dfrac{2.3}{\sqrt{5.7}} & 0 & \cdots \\ \cdot & \cdot & \cdot & \cdot & \end{bmatrix},$$

and

$$S = \begin{bmatrix} p_0 & 0 & 0 & 0 & \cdots \\ 0 & \dfrac{p_1}{3} & 0 & 0 & \cdots \\ 0 & 0 & \dfrac{p_2}{5} & 0 & \cdots \\ 0 & 0 & 0 & \dfrac{p_3}{7} & \cdots \\ \cdot & \cdot & \cdot & \cdot & \end{bmatrix}.$$

The corresponding equations for the slab can be obtained from (9.16) by identifying z and r and replacing N by zero.

9.6. Slab: Difference System I for Hyperbolic Equations

The spherical harmonic equations for the slab may be written in the vector-matrix notation as

(9.17) $$\left(\frac{1}{v}\frac{\partial}{\partial t} + M\frac{\partial}{\partial z} + \sigma\right)\mathbf{u} = \sigma(1 + f)S\mathbf{u}$$

where M and S are the square matrices of order $L + 1$ given in the preceding section, and $\mathbf{u} = \mathbf{u}(z, t)$ is a vector of $L + 1$ components. We therefore have to deal with a set of $L + 1$ coupled, first-order partial differential equations with z, t as independent variables; they are linear and have

SEC. 9.6] SLAB: DIFFERENCE SYSTEM I FOR HYPERBOLIC EQUATIONS

constant coefficients, so that the methods of Chapter 4 can be used to investigate stability and other properties of the difference equations.

The equations constitute a symmetric hyperbolic system, and the results stated in this and the following sections apply to any such system.

So far, the boundary conditions for the spherical harmonic equations have not been given. For a finite slab, the true boundary conditions (9.5) are replaced by $L + 1$ conditions on the $L + 1$ functions $\psi_l(r, t)$ half of these conditions applying when $z = a$ and the remainder when $z = -a$. How such conditions can be obtained will be described below, in Section 9.12 but they are not of such a nature as to be replaceable by a periodicity condition, permitting Fourier series to be used in the analysis, hence we restrict our consideration to the case of an infinite slab in which the neutron density is a quadratically integrable function of z, permitting use of Fourier integrals. This would be the case, for example, if a thin, plane source of neutrons, embedded in a large homogeneous medium, emits neutrons for a short time and we wish to follow their subsequent motion.

The system of finite-difference equations which will be called "system I" is

$$(9.18) \quad \frac{\mathbf{u}_j^{n+1} - \mathbf{u}_j^n}{v\Delta t} + M\frac{\mathbf{u}_{j+1}^n - \mathbf{u}_{j-1}^n}{2\Delta z} + \sigma \mathbf{u}_j^n = \sigma(1 + f)S\mathbf{u}_j^n.$$

A forward time difference and divided space difference have been used. The equations are explicit and the truncation error e is $O(\Delta t) + O[(\Delta z)^2]$. The amplification matrix for the Fourier coefficients of e^{ikz} is

$$(9.19) \quad G(\Delta t, k) = (1 - v\sigma\Delta t)I + v\sigma(1 + f)\Delta t S - i\frac{v\Delta t}{\Delta z}(\sin k\Delta z)M,$$

where I is the unit matrix.

The stability considerations presuppose a prescription as to the relative rates at which Δt and Δz go to zero, and we consider two cases:

Case 1. Let $\alpha_0 = v\Delta t/\Delta z$ and suppose that α_0 is held fixed while $\Delta t, \Delta z \to 0$. It is easy to see that the equations are unstable for any value of α_0. To see this, it suffices to consider

$$(9.20) \quad G(0, k) = I - i\alpha_0(\sin k\Delta z)M.$$

M is a symmetric matrix and has real eigenvalues so that the magnitude of the largest eigenvalue of $G(0, k)$ is

$$\sqrt{1 + \alpha_0^2 \mu_0^2 \sin^2 k\Delta z},$$

where μ_0 is the largest absolute value of an eigenvalue of M. The maximum of this for all real k is

$$\sqrt{1 + \alpha_0^2 \mu_0^2} > 1,$$

independent of Δt, so that the von Neumann condition (4.18) is violated. Therefore, *system I is unstable for any fixed $\Delta t/\Delta z$ as $\Delta t, \Delta z \to 0$*.

Case 2. Let $\alpha_1 = v\Delta t/[\sigma(\Delta z)^2]$ and suppose that α_1 is held fixed while $\Delta t, \Delta z \to 0$. We prove that system I is stable for any fixed α_1. For this purpose the abbreviations

$$(1 + f)S - I = F, \qquad v\sigma\Delta t = \varepsilon,$$

are used; and we can write

$$G = I + \varepsilon F - i\sqrt{\varepsilon\alpha_1}(\sin k\Delta z)M,$$
$$G^*G = I + 2\varepsilon F + \varepsilon^2 F^2 + \varepsilon\alpha_1(\sin^2 k\Delta z)M^2$$
$$\qquad + i\sqrt{\alpha_1}\varepsilon^{3/2}(MF - FM)\sin k\Delta z,$$

where G stands for the amplification matrix $G(\Delta t, k)$. Note that one of the cross products in the formation of G^*G drops out because the unit matrix I commutes with all matrices and in particular with M. Consequently, there is no term containing just $\sqrt{\varepsilon}$ in G^*G and in fact G^*G is equal to I plus a matrix, each of whose elements is $O(\varepsilon)$. The constant implied in this O expression is a bounded function of k, because k enters G^*G only through $\sin k\Delta z$. Therefore, the maximum for all real k of the bound[5] of G is $\leq 1 + O(\varepsilon)$ and hence is $\leq 1 + O(\Delta t)$. Thus, by Section 4.11, *system I is stable for any fixed $\Delta t/(\Delta z)^2$ as $\Delta t, \Delta z \to 0$*.

9.7. A Paradox

From the point of view of the programmer performing a calculation on a computing machine, these results concerning system I seem paradoxical.

---

[5] The bound of G is the square-root of the bound of G^*G, which is also its largest eigenvalue, since G^*G is normal.

A PARADOX

He does not, of course, let Δt and Δz go to zero, but simply uses the smallest values that are consistent with the capacity and speed of his machine and the size of his budget. When he asks, "Are my equations stable?" about the only answer that can be given is that they are unstable if he regards his calculation as one member of an infinite sequence of calculations with fixed $\Delta t/\Delta z$ but that they are stable if he regards the same calculation as a member of an infinite sequence with fixed $\Delta t/(\Delta z)^2$.

This example, which applies to any hyperbolic system when approximated as in (9.19), shows that what the programmer needs is actual error estimates, not a mere assurance that the error will tend to zero at the end of an infinite sequence of calculations on an idealized computing machine. Unfortunately, actual error estimates (at least good ones) are usually difficult to obtain. However, the above stability considerations do throw some light on the question of errors. For difference system I (equation 9.19), the stability arguments show that if the programmer is for any reason dissatisfied with the accuracy he is getting, and decides to refine the mesh by reducing Δz by a factor $1/2$, then he should reduce Δt by a factor $1/4$, not merely $1/2$; otherwise he is in danger of making the error worse instead of better.

Furthermore, a practical criterion for stability can be obtained as follows: the Fourier components of the numerical solution grow as $e^{\gamma t}$, where according to case 1 the maximum value for γ is

$$\approx [\sqrt{1 + \alpha_0^2 \mu_0^2} - 1]/\Delta t \quad \text{for } k\Delta z \approx \pi/2,$$

i.e., it is attained for short wave-length components, whereas the maximum value of γ for a true solution of (9.17) cannot exceed $v\sigma f$. Since μ_0 is just slightly less than 1 and f does not generally exceed 1 by much, we take as a rough criterion

$$[\sqrt{1 + \alpha_0^2} - 1]/\Delta t < v\sigma$$

or

$$v\Delta t/\sigma(\Delta z)^2 < 1$$

(a factor $1/2$ has been dropped as insignificant because of the roughness of this criterion). In practice, this would be an uncomfortably severe restriction; hence, we search for more satisfactory finite-difference approximations to the spherical harmonic equations. (A similar practical criterion for stability, for the problem of coupled sound and heat flow, is given in the next chapter; there, also, it is found necessary to consider the finite values of Δt and Δz appearing in an actual calculation. A third example is found in Section 11.6 on the problem of a vibrating bar under tension.)

9.8. Slab: Difference System II (Friedrichs)

Friedrichs (1954) treated a class of symmetric linear hyperbolic systems, of which the spherical harmonic equations are a special case, and showed a general method for obtaining a stable finite-difference scheme. For the spherical harmonic equations his scheme is obtained from system I above by replacing $\mathbf{u}_j^n$ throughout by $\frac{1}{2}(\mathbf{u}_{j-1}^n + \mathbf{u}_{j+1}^n)$:

$$(9.21) \quad \frac{2\mathbf{u}_j^{n+1} - \mathbf{u}_{j+1}^n - \mathbf{u}_{j-1}^n}{2v\Delta t} + M\frac{\mathbf{u}_{j+1}^n - \mathbf{u}_{j-1}^n}{2\Delta z} + \frac{1}{2}\sigma(\mathbf{u}_{j+1}^n + \mathbf{u}_{j-1}^n)$$
$$= \sigma(1 + f)\frac{1}{2}S(\mathbf{u}_{j+1}^n + \mathbf{u}_{j-1}^n).$$

This is a "three-point" formula, involving only one net point at time t^{n+1} and two at time t^n.

For this system the amplification matrix is

$$G(\Delta t, k) = (I + \varepsilon F)\cos k\Delta z - i\alpha_0 M \sin k\Delta z,$$

where, as before, we let $v\sigma\Delta t = \varepsilon$, $(1 + f)S - I = F$, $\alpha_0 = (v\Delta t)/(\Delta z)$. As was shown in Sections 3.9 and 4.11, stability is unaffected by a term like $\varepsilon F \cos k\Delta z$ which is $O(\Delta t)$. The remainder of G, which is just $G(0, k)$, is normal, so that the von Neumann condition is necessary and sufficient for stability. Moreover, the eigenvalues of $G(0, k)$ have moduli $(\cos^2 k\Delta z + \alpha_0^2\mu^2 \sin^2 k\Delta z)^{1/2}$, where μ is an eigenvalue of M. If μ_0 is again the largest absolute value of an eigenvalue of M, it follows that $\alpha_0\mu_0 \leq 1$ is necessary and sufficient for stability; in other words *the Friedrichs scheme, system II for the slab, is stable if and only if*

$$(9.22) \quad \mu_0(v\Delta t/\Delta z) \leq 1.$$

This agrees, of course, with Friedrichs' more general conclusion. It will be shown below that μ_0 is just slightly less than 1, so the above restriction is just very slightly less severe than $v\Delta t \leq \Delta z$.

9.9. Implicit Schemes

A variety of implicit schemes can be constructed for the spherical harmonic equations. The simplest is a four point scheme using two adjacent space points at time t^n and the same space points at time t^{n+1}. By proper centering, the truncation error can be reduced to $O\left[(\Delta t)^2\right] + O[(\Delta z)^2]$ as contrasted with $O(\Delta t) + O[(\Delta z)^2]$ for systems I and II above, and the resulting system is unconditionally stable. Other unconditionally stable implicit schemes can be constructed in analogy with the corresponding schemes for the heat flow equation (see Table 8.I). These schemes all have the disadvantage of being implicit with respect both to the variable z and the index l that designates the components of $\mathbf{u}$ ($l = 0, \ldots, L$). The simultaneous equation system that must be solved at each time step is therefore rather complicated, and for this reason the implicit schemes have not been explored, so far as the writers are aware. It appears that system II is the only reasonable one for the spherical harmonic method and even this system is inferior to schemes available for other approximations to the transport equation, to be discussed in the remainder of the chapter.

9.10. The Wick-Chandrasekhar Method for the Slab

For slab symmetry with isotropic scattering, the transport equation takes the form (9.6), viz.,

$$\left(\frac{1}{v}\frac{\partial}{\partial t} + \mu\frac{\partial}{\partial z} + \sigma\right)\psi(z, \mu, t) = \frac{\sigma(1 + f)}{2}\int_{-1}^{1} d\mu' \psi(z, \mu', t).$$

The idea of the Wick-Chandrasekhar method[6] is to take a fixed set of values of μ, say $\mu_0, \mu_1, \ldots, \mu_L$, and construct equations for the $L + 1$ functions $\psi(z, \mu_j, t)$, evaluating the integral on the right of the transport equation as accurately as possible with these $L + 1$ values of μ'.

---

[6] G. C. Wick (1943), S. Chandrasekhar (1944b).

As is well known, the most accurate $(L + 1)$-point integration formula in a certain sense is that of Gauss,[7] namely, if $F(x)$ is any function, then

$$\tfrac{1}{2} \int_{-1}^{1} F(x)\, dx \approx \sum_{0}^{L}{}_{(j)} C_j F(\mu_j),$$

where $\mu_0, \mu_1, \ldots, \mu_L$ are the zeros of the Legendre function $P_{L+1}(\mu)$, and the coefficients C_j are given by

$$C_j = \left\{ \sum_{0}^{L}{}_{(k)} (2k + 1)[P_k(\mu_j)]^2 \right\}^{-1}.$$

With this notation, the right member of the transport equation becomes

$$\sigma(1 + f) \sum_{0}^{L}{}_{(j)} C_j \psi(z, \mu_j, t).$$

The left member of the transport equation is used unaltered for each of the $L + 1$ values of μ. The resulting system of $L + 1$ partial differential equations can best be displayed by a matrix-vector notation similar to that used for the spherical harmonic method. Let $\tilde{u} = \tilde{u}(z, t)$ denote the $(L + 1)$-component vector

$$\tilde{u} = \tilde{u}(z, t) = \begin{bmatrix} \psi(z, \mu_0, t) \\ \psi(z, \mu_1, t) \\ \cdots \\ \psi(z, \mu_L, t) \end{bmatrix}.$$

Then the equations are

(9.23)
$$\left(\frac{1}{v} \frac{\partial}{\partial t} + \tilde{M} \frac{\partial}{\partial z} + \sigma \right) \tilde{u} = \sigma(1 + f) \tilde{S} \tilde{u},$$

where $\tilde{M}$ is the diagonal matrix

$$\tilde{M} = \begin{bmatrix} \mu_0 & & & 0 \\ & \mu_1 & & \\ & & \ddots & \\ 0 & & & \mu_L \end{bmatrix},$$

---

[7] For derivation and alternative expressions for the coefficients, see Hildebrand (1956).

and $\tilde{S}$ is the matrix

$$\tilde{S} = \begin{bmatrix} C_0 & C_1 & \cdots & C_L \\ C_0 & C_1 & \cdots & C_L \\ \cdot & \cdot & \cdot & \cdot \\ C_0 & C_1 & \cdots & C_L \end{bmatrix}.$$

We shall see that these equations for the slab problem are better suited to finite-difference methods than the corresponding spherical harmonic equations. It is therefore desirable to generalize them to cover a spherical symmetry and anisotropic scattering. This can be achieved by making use of a result of J. C. Mark (unpublished) according to which there is an equivalence, for the slab with isotropic scattering, between the spherical harmonic equations and the Wick-Chandrasekhar equations. One can obtain the latter from the former by a similarity transformation of the matrices involved, and if this similarity transformation is applied to the more general spherical harmonic equations, we obtain the result we want.

Methods of this general type, containing a diagonal matrix $\tilde{M}$, and including Carlson's S_n method, described below, are called discrete-ordinate methods. In a series of papers, Keller and Wendroff (1960) have proved the convergence of these methods, as $L \to \infty$, under reasonable assumptions.

9.11. Equivalence of the Two Methods

We first state (with only a hint at the proofs) some remarkable properties of the matrix M of the spherical harmonic equations (Section 9.5).

1. *The eigenvalues of M are the zeros of the Legendre polynomial $P_{L+1}(\lambda)$.* To see this, write

$$\mathscr{P}_{L+1}(\lambda) = \frac{(2L+2)!}{2^{L+1}((L+1)!)^2} \cdot \det(\lambda I - M).$$

If the determinant is expanded with respect to its last row, there are two terms, coming from the two non-vanishing elements in the last row. These can be expressed in terms of $\mathscr{P}_L(\lambda)$ and $\mathscr{P}_{L-1}(\lambda)$ which involve principal minors of the above determinant. In this way we get a recurrence relation for the $\mathscr{P}_L(\lambda)$ which is identical with one of the standard recurrence relations of the Legendre polynomials. Furthermore, it is easily seen that $\mathscr{P}_1(\lambda) = P_1(\lambda)$ and $\mathscr{P}_2(\lambda) = P_2(\lambda)$. Therefore, $\mathscr{P}_{L+1}(\lambda) = P_{L+1}(\lambda)$ and the assertion follows. Since M is symmetric, it is now clear that there must be *an orthogonal transformation connecting M and $\tilde{M}$*, because $\tilde{M}$ is diagonal and has the same eigenvalues as M. It is

2. $U^{-1}MU = \tilde{M}$, where $U_{jk} = \sqrt{(2j+1)C_k}P_j(\mu_k)$.

To show this, one first observes that a vector $\mathbf{v}^k$ whose components are $\sqrt{2j+1}P_j(\mu_k)$ ($j = 0, \ldots, L$) satisfies the equation $(\mu_k I - M)\mathbf{v}^k = 0$, by virtue also of the recurrence relation among the $P_j(\mu)$. One thus obtains a set of eigenvectors of M, and it is only necessary to normalize them[8] to obtain the matrix U.

If there is an equivalence between the spherical harmonic method and the Wick-Chandrasekhar method, we should expect, remembering the definition of $\psi(z, t)$, that

$$\psi(z, \mu_j, t) = \sum_{0}^{L}{}_{(k)}\psi_k(z, t)\sqrt{2k+1}\,P_k(\mu_j),$$

or in matrix notation $\tilde{\mathbf{u}} = T\mathbf{u}$, $\mathbf{u} = T^{-1}\tilde{\mathbf{u}}$, where T is the matrix given by

(9.24) $$T_{jk} = \sqrt{2k+1}\,P_k(\mu_j).$$

If we apply the transformation T to the spherical harmonic equations, we find

$$\left(\frac{1}{v}\frac{\partial}{\partial t} + TMT^{-1}\frac{\partial}{\partial z} + \sigma\right)\tilde{\mathbf{u}} = \sigma(1+f)TST^{-1}\tilde{\mathbf{u}};$$

and the equivalence is established by proving that

$$TMT^{-1} = \tilde{M}, \qquad TST^{-1} = \tilde{S}.$$

To prove these relations note that the matrix T is closely related to the matrix U given above, and in fact $T = V^{-1}U^{-1}$, $T^{-1} = UV$, where V is the diagonal matrix given by $V_{jk} = \sqrt{C_j}\,\delta_{jk}$. Therefore,

$$TMT^{-1} = V^{-1}U^{-1}MUV = V^{-1}\tilde{M}V = \tilde{M},$$

---

[8] The formula given in the preceding section for the coefficient C_k is easily recognized as the normalization condition for the vector $\mathbf{v}^k$. The more usual formulas for this coefficient can be obtained therefrom by the use of the Darboux-Christoffel identity, which is discussed in Hildebrand (1956), p. 322.

SEC. 9.13]   DIFFERENCE SYSTEMS I AND II   237

the last step being justified because V and $\tilde{M}$ are both diagonal. Note further that

$$(T^{-1})_{jk} = C_k\sqrt{2j + 1}\,P_j(\mu_k)$$

and that for isotropic scattering, $p(\cos \alpha) \equiv 1$, $p_0 = 1$, $p_l = 0$ ($l > 1$), so that the only non-vanishing element of S is $S_{00} = 1$. Therefore,

$$(TST^{-1})_{jk} = T_{j0}(T^{-1})_{0k} = C_k = \tilde{S}_{jk}.$$

This completes the proof that *for the slab with isotropic scattering, the Wick-Chandrasekhar equations and the spherical harmonic equations are equivalent.* In terms of the theory of characteristics (see Section 12.7 and also Courant-Hilbert, Vol. II), the Wick-Chandrasekhar equations are the spherical harmonic equations expressed in normal or characteristic form.

9.12. Boundary Conditions

For a slab extending from $z = -a$ to $z = a$ the boundary conditions (9.5) can be taken over without essential change in the Wick-Chandrasekhar method. Let the μ_j [the zeros of $P_{L+1}(\mu)$] be so ordered that $\mu_0 > \mu_1 > \cdots > \mu_L$. (Note that μ_0 is also the largest in magnitude.) If $L + 1$ is even, we take as boundary condition:

$$\psi(-a, \mu_j, t) = 0 \quad \text{for } j = 0, 1, \ldots, (L-1)/2,$$

$$\psi(a, \mu_j, t) = 0 \quad \text{for } j = (L+1)/2, \ldots, L.$$

If $L + 1$ is odd, $\mu_{L/2} = 0$, and the boundary condition on $\psi(z, \mu_{L/2}, t)$ seems at first sight unclear. However, the equation for this function needs no boundary condition because $\partial/\partial z$ does not appear. The simplicity of these boundary conditions is an advantage of the Wick-Chandrasekhar method. One way to get boundary conditions for the spherical harmonic method is to take the above equations and translate them into the terms of the spherical harmonic method by the transformation T.

9.13. Difference Systems I and II

Finite-difference equation systems I and II are practically the same as for the spherical harmonic equations, and may be obtained formally from the equations of Sections 9.6 and 9.8 by placing tildes over the letters **u**, M, and S. They have the same properties: system I is stable for any fixed $\Delta t/(\Delta z)^2$ but unstable for any fixed $\Delta t/\Delta z$, and a practical stability condition may be taken as $v\Delta t/[\sigma(\Delta z)^2] < 1$; the system II (Friedrichs' scheme) is stable for $v\Delta t \leq \Delta z/\mu_0$. There are, moreover, some finite-difference systems which have no counterpart for the spherical harmonic method; they will now be discussed.

9.14. System III: Forward and Backward Space Differences[9]

For $\mu_j > 0$ we use a backward space difference quotient to represent $\partial \mathbf{u}/\partial z$ and for $\mu_j < 0$ we use a forward one. (For $\mu_j = 0$, which can occur if L is even, the space derivative does not enter.) To do this we split $\tilde{M}$ in two parts: $\tilde{M} = \tilde{M}_1 + \tilde{M}_2$, where

$$\tilde{M}_1 = \begin{bmatrix} \mu_0 & & & & & 0 \\ & \ddots & & & & \\ & & \mu_{l'} & & & \\ & & & 0 & & \\ & & & & 0 & \\ & & & & & \ddots \\ 0 & & & & & 0 \end{bmatrix},$$

$$\tilde{M}_2 = \begin{bmatrix} 0 & & & & & 0 \\ & 0 & & & & \\ & & \ddots & & & \\ & & & 0 & & \\ & & & & \mu_{l''} & \\ & & & & & \ddots \\ 0 & & & & & \mu_L \end{bmatrix};$$

$\mu_{l'}$ and $\mu_{l''}$ are the last positive and first negative of the μ_j. Then the difference equations are

$$(9.25) \quad \frac{\tilde{\mathbf{u}}_j^{n+1} - \tilde{\mathbf{u}}_j^n}{v\Delta t} + \tilde{M}_1 \frac{\tilde{\mathbf{u}}_j^n - \tilde{\mathbf{u}}_{j-1}^n}{\Delta z} + \tilde{M}_2 \frac{\tilde{\mathbf{u}}_{j+1}^n - \tilde{\mathbf{u}}_j^n}{\Delta z} + \sigma \tilde{\mathbf{u}}_j^n$$
$$= (1 + f)\sigma \tilde{S} \tilde{\mathbf{u}}_j^n.$$

---

[9] Compare Courant, Isaacson and Rees (1952).

The stability of these equations is readily investigated by the methods we have been using. For this system G is of the form diagonal matrix $+O(\Delta t)$. Hence, *the von Neumann condition for* (9.25) *is*

$$\mu_0 \frac{v\Delta t}{\Delta z} \leq 1,$$

and this is also sufficient for stability.

The truncation error of this system is $O(\Delta t) + O(\Delta z)$. It therefore is formally less accurate than the Friedrichs system; but in practice it is no less accurate, and perhaps a little more accurate, for the following reason. Of the two error terms of the Friedrichs system, $O(\Delta t) + O[(\Delta z)^2]$, the first greatly predominates if Δt is anywhere near the maximum allowed value $\Delta z/\mu_0 v$, whereas the two terms $O(\Delta t) + O(\Delta z)$ of system III above are comparable and opposite in sign, with consequent partial cancellation. In a sense, the proper choice of forward or backward space differences according to the sign of μ allows us to follow the characteristics more closely than in Friedrichs' scheme. The two schemes have the same stability condition.

9.15. System IV (Implicit)

In light of the experience with the heat flow equation, one might expect that the stability of the system (9.25) would be greatly increased if we use a backward time difference (or, what is almost the same thing, replace n by $n + 1$ as superscripts in the space differences). In order to follow the characteristics as closely as possible, we also interchange the roles[10] of $\tilde{M}_1$ and $\tilde{M}_2$. The resulting system is:

$$(9.26) \quad \frac{\tilde{u}_j^{n+1} - \tilde{u}_j^n}{v\Delta t} + \tilde{M}_1 \frac{\tilde{u}_{j+1}^{n+1} - \tilde{u}_j^{n+1}}{\Delta z} + \tilde{M}_2 \frac{\tilde{u}_j^{n+1} - \tilde{u}_{j-1}^{n+1}}{\Delta z} + \sigma \tilde{u}_j^n$$

$$= (1 + f)\sigma \tilde{S} \tilde{u}_j^n.$$

The amplification matrix can easily be calculated. Contrary to possible

---

[10] If the roles of $\tilde{M}_1$ and $\tilde{M}_2$ are not interchanged, equations are obtained which are unconditionally stable but which utterly fail to follow the characteristics.

expectations, this system is not unconditionally stable. In fact, *the stability condition is*

$$\mu^* \frac{v\Delta t}{\Delta z} \geq 1$$

where μ^* is the smallest positive zero of $P_{L+1}(\mu)$. Note the unusual feature of this system, namely, that for stability Δt is restricted from below rather than from above. Because of the smallness of μ^* this system is of no practical use.

9.16. System V (Carlson's Scheme)

In connection with his S_n method (described briefly below), which is somewhat similar to the Wick-Chandrasekhar method, B. Carlson devised (1953) a finite-difference scheme which is unconditionally stable. This is achieved by combining the best features of system III (equation 9.25) with those of system IV (equation 9.26). We shall give a slightly simplified version of Carlson's scheme as it applies to the Wick-Chandrasekhar method. Carlson's idea is to treat those components of $\tilde{u}$ corresponding to small μ by scheme (9.25) and those components corresponding to large μ by scheme (9.26). We suppose the quantity $v\Delta t/\Delta z$ is constant and we split the diagonal matrix $\tilde{M}$ into 4 parts: $\tilde{M} = \tilde{M}_3 + \tilde{M}_4 + \tilde{M}_5 + \tilde{M}_6$, where

$\tilde{M}_3$ contains those diagonal elements of $\tilde{M}_1$ for which $\mu > \frac{\Delta z}{v\Delta t}$,

$\tilde{M}_4$ contains those diagonal elements of $\tilde{M}_1$ for which $\mu \leq \frac{\Delta z}{v\Delta t}$,

$\tilde{M}_5$ contains those diagonal elements of $\tilde{M}_2$ for which $\mu \geq -\frac{\Delta z}{v\Delta t}$,

$\tilde{M}_6$ contains those diagonal elements of $\tilde{M}_2$ for which $\mu < -\frac{\Delta z}{v\Delta t}$,

all other elements being zero in each case.

SEC. 9.16] SYSTEM V (CARLSON'S SCHEME)

The difference equations are

$$(9.27) \quad \frac{\tilde{u}_j^{n+1} - \tilde{u}_j^n}{v\Delta t} + \tilde{M}_3 \frac{\tilde{u}_{j+1}^{n+1} - \tilde{u}_j^{n+1}}{\Delta z} + \tilde{M}_4 \frac{\tilde{u}_j^n - \tilde{u}_{j-1}^n}{\Delta z} + \tilde{M}_5 \frac{\tilde{u}_{j+1}^n - \tilde{u}_j^n}{\Delta z}$$

$$+ \tilde{M}_6 \frac{\tilde{u}_j^{n+1} - \tilde{u}_{j-1}^{n+1}}{\Delta z} + \sigma \tilde{u}_j^n = (1 + f)\sigma S \tilde{u}_j^n.$$

The significance of this scheme in terms of characteristics is indicated in Figure 9.1. Shown are six neighboring points of the net in the z, t-plane,

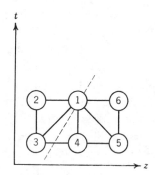

Fig. 9.1. Schematic drawing of six net points in the (z, t) plane, to illustrate Carlson's method. The dashed line represents a typical characteristic.

the top row corresponding to time t^{n+1} and the bottom row to t^n. The characteristics are $L + 1$ straight lines; the slope of the lth characteristic is $1/\mu_l v$. Depending on this slope, a characteristic through the point 1 may lie in any one of the four triangles drawn. For the lth component of the vector $\tilde{u}$, the differential operator in (9.23) is $(1/v)\partial/\partial t + \mu_l \partial/\partial z$ which is the operator for the differentiation along the lth characteristic. Carlson's scheme expresses the derivatives in terms of function values at three points: the point 1 plus one of the point pairs (2, 3) (3, 4), (4, 5), or (5, 6) according to which of the four triangles the characteristic lies in. If one of the characteristics passes through the point 3, the corresponding finite-difference operator reduces automatically to a two-point formula utilizing points 1 and 3, and is

$$(\tilde{u}_j^{n+1} - \tilde{u}_{j-1}^n)/v\Delta t.$$

In this case differentiation along the characteristic has been replaced by a difference quotient along the characteristic. This is clearly better than the less direct procedure involving partial derivatives and utilizing a third point. A similar remark applies, of course, to characteristics passing through point 4 or 5. For a characteristic not passing through one of these points, one must use (at least) three points to represent the differential operator, and it is a merit of Carlson's scheme that for each component of u it picks the three points so as to lie as closely as possible to the corresponding characteristic.

As for previous schemes, the amplification matrix G is easily written down and the stability condition readily established. The result is that the *Carlson scheme* (9.27) *is stable for any value of* $v\Delta t/\Delta z$.

Although Carlson's scheme is implicit, the numerical solution can be done easily as follows:[11] For $\mu_l < 0$, a value (namely, zero) of $\psi(z, \mu_l, t^{n+1})$ is given by the boundary condition on the right face of the slab, $z = a$. The difference equations can be solved in order of decreasing j (decreasing z) until the left face is reached. Similarly, for $\mu_1 > 0$, a boundary value is given at $z = -a$, and the equations can be solved in order of increasing j. For $\mu_1 = 0$ (which happens if L is even), the equations become explicit.

The application of Carlson's scheme to more general quasi-linear hyperbolic systems has been studied recently by H. B. Keller (1956). Because the slope of a characteristic may vary from one net point to another when the coefficients are variable, it may be necessary to change the integration scheme from one of the four procedures to another during the course of the integration. It is nevertheless possible, according to Keller, to solve the difference equations by a simple algorithm. He has also proved stability and convergence under reasonable assumptions. Keller and Wendroff have given (1957) a detailed discussion of this scheme for a wide class of discrete-ordinate methods; they have established the unconditional stability and also the convergence of the iterative procedure referred to in footnote 11.

---

[11] In the interest of accuracy, one often replaces u_j^n by $\frac{1}{2}(u_j^n + u_j^{n+1})$ in the undifferentiated terms (this has no effect on stability). When this is done, the implicit equations are not so easily solved, and one generally resorts to an iterative procedure.

9.17. Generalization of the Wick-Chandrasekhar Method

The exposition given in Section 9.10 assumes isotropic scattering and slab symmetry. If scattering is anisotropic, one can replace the matrix $\tilde{S}$ by the matrix TST^{-1}, where S is the general matrix given in Section 5 for the spherical harmonic method and T is the transformation matrix given in Section 9.11. That is, one takes $\tilde{S}$ as given by

$$(\tilde{S})_{jk} = C_k \sum_{(l)} P_l(\mu_j) p_l P_l(\mu_k)$$

to be the generalization of the matrix $\tilde{S}$ of Section 9.10. The p_l are the coefficients in the scattering law, equation (9.10). All the finite difference schemes mentioned are still available without any change of stability for no use has been made in the stability discussion of any special form of the matrix $\tilde{S}$.

To generalize the method to problems with spherical symmetry, one includes a term $(1/r)\tilde{N}\tilde{u}$ on the left side of equation (9.23), where $\tilde{N}$ is the transform of the matrix N occurring in the corresponding place in the spherical harmonic equation (see Section 9.5). That is,

$$(\tilde{N})_{jk} = (TNT^{-1})_{jk}$$
$$= C_k \sum_{(l)\,0}^{L-1} (l+1)[(l+2)P_l(\mu_j)P_{l+1}(\mu_k) - lP_l(\mu_k)P_{l+1}(\mu_j)].$$

With the addition of the term $(1/r)\tilde{N}\tilde{u}$, our stability criteria no longer hold rigorously, because we now have variable coefficients. But if one adopts the usual heuristic point of view, according to which stability is a local phenomenon, so that if the coefficients are slowly varying they can be replaced by constant values in a small neighborhood, one would expect that the addition of such a term will not affect stability, because it is easy to see that if r_0 is any constant, the addition of a term $(1/r_0)\tilde{N}\tilde{u}$ leaves the stability condition unaltered for all the finite-difference schemes discussed above. This method has not been used for time-dependent problems to the writers' knowledge, and perhaps we should withhold judgment as to what may happen at the origin because of the singularity there, but experience with the quite similar S_n equations suggests that there is not likely to be any difficulty. Carlson's scheme would again be used, integrating from the outside to the center for $\mu < 0$ and then from the center back to the outside for $\mu > 0$.

9.18. The S_n Method of Carlson (1953)

This method is similar to, but not equivalent to, the Wick-Chandrasekhar method. It differs in that the derivative with respect to μ (which occurs for spherical problems) is approximated in a more straightforward manner, and there is a certain difference in centering with respect to μ which has the effect that certain matrices, which were diagonal in the Wick-Chandrasekhar method, are here only triangular.

One divides the interval $-1 \leq \mu \leq 1$ into n subintervals (not necessarily equal):

$$-1 = \mu_0 < \mu_1 < \cdots < \mu_n = +1$$

(the quantities $\mu_0, \ldots, \mu_n$ are not generally to be identified with the quantities $\mu_0, \ldots, \mu_L$ introduced in connection with the previous methods); n partial differential equations will be written down for the n functions $N_0(r, t) = \psi(r, \mu_0, t)$, $N_1(r, t) = \psi(r, \mu_1, t)$, $\ldots, N_n(r, t) = \psi(r, \mu_n, t)$. Each except the first of these equations is centered at the midpoint of one of the subintervals.

We consider the case of a spherical system with isotropic scattering. The transport equation is (compare 9.7)

$$\left(\frac{1}{v}\frac{\partial}{\partial t} + \mu\frac{\partial}{\partial r} + \frac{1-\mu^2}{r}\frac{\partial}{\partial \mu} + \sigma\right)\psi(r, \mu, t) = \sigma\frac{1+f}{2}\int_{-1}^{1}\psi(r, \mu', t)d\mu'.$$

Let the integral be approximated as

$$\frac{1}{2}\int_{-1}^{1}\psi(r, \mu', t)d\mu' \approx \sum_{(i)0}^{n} A_i N_i(r, t),$$

where the coefficients A_i determine some integration formula, e.g., the trapezoid-rule formula. For $\mu = \pm 1$, the derivative with respect to μ

SEC. 9.18] THE S_n METHOD OF CARLSON (1953)

disappears from the transport equation and we may take as the equation for $N_0(r, t)$, which corresponds to $\mu = -1$, simply

$$(9.28) \qquad \left(\frac{1}{v}\frac{\partial}{\partial t} - \frac{\partial}{\partial r} + \sigma\right) N_0(r, t) = \sum_{0}^{n}{}_{(i)} A_i N_i(r, t).$$

For the equation that is to be centered at the midpoint of the subinterval (μ_i, μ_{i+1}), the time derivative and the term in $\sigma\psi$ are replaced by simple averages and the other terms are approximated by slightly more complicated expressions:

$$(9.29) \qquad \frac{1}{2v}\left(\frac{\partial N_i}{\partial t} + \frac{\partial N_{i+1}}{\partial t}\right) + \frac{1}{2}\left(\mu_i^+ \frac{\partial N_i}{\partial r} + \mu_{i+1}^- \frac{\partial N_{i+1}}{\partial r}\right)$$

$$+ \frac{1 - \overline{(\mu^2)}_{i+\frac{1}{2}}}{r} \frac{N_{i+1} - N_i}{\mu_{i+1} - \mu_i} + \frac{\sigma}{2}(N_i + N_{i+1})$$

$$= \sigma(1 + f) \sum_{0}^{n}{}_{(p)} A_p N_p;$$

μ_i^+ and μ_{i+1}^- are suitable points in the subinterval (μ_i, μ_{i+1}), near the left and right ends, respectively, and $\overline{(\mu^2)}_{i+\frac{1}{2}}$ is a suitable sort of average of μ^2 over the subinterval.

The S_n method is furthermore characterized by a particular choice of μ_i^+, μ_i^-, and $\overline{(\mu^2)}_{i+\frac{1}{2}}$, namely:

$$\mu_i^+ = \tfrac{1}{3}(2\mu_i + \mu_{i+1}), \qquad \mu_{i+1}^- = \tfrac{1}{3}(\mu_i + 2\mu_{i+1}),$$

$$\overline{(\mu^2)}_{i+\frac{1}{2}} = \tfrac{1}{3}(\mu_i^2 + \mu_i \mu_{i+1} + \mu_{i+1}^2).$$

Carlson's reason for this choice is partly heuristic, based on the fact that, with this choice, if we assume that ψ varies linearly with μ over the subinterval, the terms in (9.29) are precisely the averages, with respect to μ, over the subinterval of the corresponding terms of the transport equation. Keller and Wendroff (1957) have used a slightly different choice, with $\mu_i^+ = \mu_{i+1}^-$, so that the system (9.29) is then in characteristic or normal form; this facilitates the theoretical discussion of stability.

An equation similar to (9.28) can be written for $N_n(r, t)$, corresponding to $\mu = +1$. This equation is redundant, but may be used for comparison, to give an indication of the accuracy of the results.

The finite-difference scheme in r, t, used for solving these equations, is the one that we have called "Carlson's scheme" earlier in this chapter (Sections 9.16 and 9.17), except for the time-centering of undifferentiated quantities. Experience seems to indicate that it is worthwhile to replace such quantities by averages: for example, in the difference equations that carry the solution from time t^ν to time $t^{\nu+1}$, one replaces $\frac{1}{2}\sigma N_i(r, t)$ by $\frac{1}{4}\sigma[N_i(r, t^\nu) + N_i(r, t^{\nu+1})]$. There is no formal justification for this, because the truncation error of Carlson's scheme is $O(\Delta t) + O(\Delta r)$ in any case; however, because Carlson's scheme follows characteristics as closely as possible in a certain sense, there is a partial cancellation of the error terms $O(\Delta t)$ and $O(\Delta r)$ coming from the derivatives, and it is apparently worthwhile to make other terms accurate to $O[(\Delta t)^2]$.

The time-centering can be achieved without difficulty in all terms on the left side of the equation (9.29). But centering the summation on the right side has the effect of bringing all the unknown functions $N_0(r, t^{\nu+1})$, ..., $N_n(r, t^{\nu+1})$, into each equation, thereby preventing solution by the simple expedient of integrating from the surface to the center for $\mu < 0$ and then from the center back to the surface for $\mu > 0$. For this reason an iterative procedure is used: one first solves the equations with the right member uncentered to obtain provisional values of the unknowns at time $t^{\nu+1}$; these are then put into the right member (centered) and improved values of the unknowns are obtained, etc. A similar iterative procedure will be described more fully in Section 9.19 for a direct integration method of solving the transport equation. As noted above, Keller and Wendroff (1957) have proved the convergence of this iterative procedure for their variant of the S_n method.

Carlson has studied the stability of his scheme as applied to the S_n equations. His conclusions, and results of a considerable amount of numerical work in applications of the method, indicate that the scheme is unconditionally stable, just as it was for the Wick-Chandrasekhar method. With Carlson's choice of the coefficients in (9.29), the counterpart of the matrices M and $\tilde{M}$ of the earlier methods is not a normal matrix in the S_n method, so that the investigation of stability is rather complicated. There seems to be no doubt, in practice, of the stability of the method.

9.19. A Direct Integration Method

A number of finite-difference systems for integration of both stationary and time-dependent transport problems in spherical symmetry have been

SEC. 9.19]   A DIRECT INTEGRATION METHOD   247

based on the variables $x = r \cos \theta$, $y = r \sin \theta$ (instead of r, θ) suggested to one of the writers by von Neumann in 1948. One such method is described in detail in this section to illustrate some of the principles.

In terms of the straight line on which a neutron is moving at time t, y is the distance of closest approach of this line to the origin, and x is the distance measured along this line to the neutron, from the point of closest approach, in the direction of the neutron's motion (see Figure 9.2). With

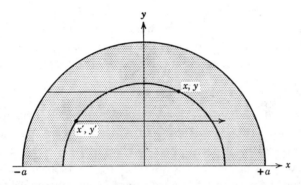

Fig. 9.2. Domain in the (x, y) plane for the transport equation in the form (9.30).

these variables, and letting $\Psi(x, y, t) = \psi(r, \cos \theta, t)$, the transport equation is

$$(9.30) \quad \left(\frac{1}{v}\frac{\partial}{\partial t} + \frac{\partial}{\partial x} + \sigma\right)\Psi(x, y, t)$$

$$= \sigma \frac{1+f}{4\pi r} \int_{-r}^{r} dx' \int_{0}^{2\pi} d\varphi\, p(\cos \alpha)\Psi(x', \sqrt{r^2 - x'^2}, t)$$

for $x^2 + y^2 \leq a^2$, $y \geq 0$, $t \geq 0$,

where now r is to be regarded as an abbreviation for $\sqrt{x^2 + y^2}$, where α is given by

$$\cos \alpha = \frac{xx' + y\sqrt{r^2 - x'^2}\cos \varphi}{r^2},$$

and the boundary condition is

$$(9.31) \quad \Psi(x, y, t) = 0 \quad \text{for } x^2 + y^2 = a^2, \quad y \geq 0, \ x \leq 0.$$

The transport equation holds in the shaded semi-circular region shown in Figure 9.2, and the path of integration in (9.30) is a semi-circular arc passing through the point (x, y). In this figure the point representing a neutron moves horizontally to the right with the constant speed v until a collision occurs, at which time the representative point jumps to some other point on the same semi-circle [say $(x'y')$ in the figure], from which it then continues to move horizontally to the right with speed v. If a neutron escapes from the system, it crosses the right-hand quarter circle of the boundary, and the boundary condition of no incident neutrons is that $\Psi = 0$ on the left-hand quarter circle of the boundary in the figure. Methods based on the use of these variables have sometimes been called direct integration methods, although they are really no more direct than, say, the S_n method.

A net of points in the x, y-plane is used which is taken sometimes as rectangular and sometimes as the intersection of a family of circles and a family of horizontal lines. In any case the points are lined up in rows parallel to the x-axis, because the derivative with respect to x appears in the equation, but not that with respect to y. Consequently the integration is along the paths which the neutrons follow between collisions.

The particular method to be described here carries this idea one step further; it integrates along the neutron trajectories in space-time and is the analogue of the method of characteristics for hydrodynamical problems. Let the three-dimensional net be as follows:

$$x = i\Delta x, \quad i = -I, -(I-1), \ldots, 0, 1, \ldots, I,$$
$$y = j\Delta y, \quad j = 0, 1, \ldots, J,$$
$$t = n\Delta t, \quad n = 0, 1, \ldots,$$

where $v\Delta t = \Delta x$ and $I\Delta x = J\Delta y = a$.

Denote $\Psi(i\Delta x, j\Delta y, n\Delta t)$ by Ψ_{ij}^n. We restrict discussion to the case of isotropic scattering and denote the right member of (9.30) by $\Phi(r, t)$, so that

(9.32) $$\Phi(r, t) = \sigma \frac{1+f}{2r} \int_{-r}^{r} dx' \, \Psi(x', \sqrt{r^2 - x'^2}, t).$$

$\Phi(r, t^n)$ will also be called $\Phi^n(r)$. An interval Δr is chosen for tabulation

SEC. 9.19] A DIRECT INTEGRATION METHOD

of Φ, and we denote $\Phi(p\Delta r, n\Delta t)$ by Φ_p^n. It is also convenient to denote $\sqrt{(i\Delta x)^2 + (j\Delta y)^2}$ by r_{ij}.

Because of the relation $v\Delta t = \Delta x$, the space time point $[(i + 1)\Delta x, j\Delta y, (n + 1)\Delta t]$ lies on the same trajectory (characteristic) as the point $(i\Delta x, j\Delta y, n\Delta t)$. We approximate the derivative terms

$$\left[\left(\frac{1}{v}\frac{\partial}{\partial t} + \frac{\partial}{\partial x}\right)\Psi\right]_{(i+\frac{1}{2})\Delta x,\ j\Delta y,\ (n+\frac{1}{2})\Delta t} = \frac{\Psi_{i+1\,j}^{n+1} - \Psi_{i\,j}^n}{\Delta x} + O[(\Delta t)^2]$$

and have, finally,

$$(9.33) \qquad \frac{\Psi_{i+1\,j}^{n+1} - \Psi_{i\,j}^n}{\Delta x} + \sigma \frac{\Psi_{i+1\,j}^{n+1} + \Psi_{i\,j}^n}{2} = \Phi^{n+\frac{1}{2}}(r_{i+\frac{1}{2}\,j}),$$

where $\Phi^{n+\frac{1}{2}}(r)$ is an approximate value of $\Phi(r, t)$ for time $(n + \frac{1}{2})\Delta t$, to be obtained by an iterative procedure described below. If the $\Phi^{n+\frac{1}{2}}$ are available, equations (9.33) give the unknowns Ψ^{n+1} explicitly at all net points, except that if $(i + 1, j)$ denotes the left most point on the jth line, inside the semi-circle, like the point "B_1" in Figure 9.3, the

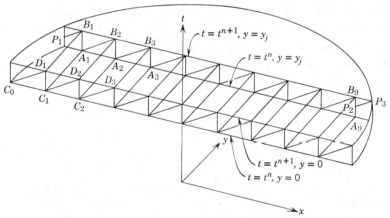

Fig. 9.3. Perspective drawing showing a portion of the three-dimensional point net for equation (9.33).

quantity Ψ_{ij}^n is not available. But the boundary condition says that Ψ vanishes at the point P_1 of space time at which the trajectory through

($i+1, j, n+1$) enters the system, at radius $r = a$. In this case, we replace Δx in (9.33) by

$$\sqrt{a^2 - (j\Delta y)^2} - |i+1|\Delta x,$$

which is the distance from the point ($i+1, j$) to the boundary along the line $y = j\Delta y$ and $t =$ constant; we also replace Ψ^n_{ij} by zero; then (9.33) gives $\Psi^{n+1}_{i+1,j}$, as for other points of the net, inside the system.

Figure 9.3 is a perspective drawing in which certain features of the three-dimensional point net are shown. The line segments A_1B_2, A_2B_3, C_1D_2 etc., are characteristics connecting net points at time t^n with those at time t^{n+1}. The right member of equation (9.33) represents the average value of $\Phi(r, t)$ along such a segment and is taken as the average of the values of $\Phi(r, t)$ at the two ends of the segment. For a segment like A_1B_2 this involves a value of $\Phi^n(r)$ at the other end. But for the special short segments P_1B_1 and P_2P_3, the equation is modified; the value of Φ at the early end is obtained by interpolating in time between times t^n and t^{n+1}. Thus, referring to the points denoted A_9, P_2, B_9 in Figure 9.3, $\Phi(P_2)$ is obtained by interpolating between $\Phi(A_9)$ and $\Phi(B_9)$.

When the Ψ^{n+1}_{ij} are known at all net-points, the Φ^{n+1}_p are obtained. Equation (9.32) shows that Φ^{n+1}_p is given by a line integral of Ψ^{n+1} around a semi-circle of radius $p\Delta r$ in the x, y-plane. Generally, this semi-circle will pass through few, if any, of the net points, and interpolation is required. A simple way to do this, and one that is easy to code for a machine, is to perform the integration with respect to x' by the trapezoid rule, using Δx or $\tfrac{1}{2}\Delta x$ for the increment of x' and obtaining values of the integrand at the required abscissas x' by linear interpolation with respect to x and y. ($\tfrac{1}{2}\Delta x$ was used as increment in the trial calculations described below.)

The iterative procedure for $\Phi^{n+\frac{1}{2}}$ is as follows: At the beginning of a complete cycle of the calculation, the Ψ^n_{ij} are known and a table of the Φ^n_p is available. The $\Phi^{n+\frac{1}{2}}(r_{i+\frac{1}{2},j})$ on the right of (9.33) is taken first as $\Phi^n(r_{i+\frac{1}{2},j})$, to be obtained by linear interpolation in the table of Φ^n_p at $r = r_{i+\frac{1}{2},j}$. The equations (9.33) are then solved, to give preliminary values of Ψ^{n+1}_{ij}. From these, provisional values of Φ^{n+1}_p are obtained, as described above. A second calculation (iteration) is then performed, in which $\Phi^{n+\frac{1}{2}}$ is taken as the mean of Φ^n and Φ^{n+1}. This iteration results

SEC. 9.19] A DIRECT INTEGRATION METHOD

in improved values of Ψ_{ij}^{n+1} from which improved values of Φ_p^{n+1} are obtained.

Experience indicates that generally further iterations are not worthwhile—using a finer net pays off more rapidly, whereas the first iteration generally results in greater improvement than would result from devoting the same additional effort to refining the net. This is presumably the case because the first iteration decreases the truncation error formally from $O(\Delta t)$ to $O[(\Delta t)^2]$, after which there is no further change in the formal order of magnitude of the error.

The method is illustrated by trial calculations performed on the Univac. The problem is a simplified one of possible astrophysical interest but chosen mainly to test the method. It applies to photons rather than neutrons, hence the speed is called c instead of v. A short burst of light (as from a variable star) is emitted at time $t = 0$ from a point source situated in a large homogeneous spherical cloud of purely scattering material ($f = 0$), which is assumed to scatter isotropically and without polarization effects. Required is a curve of intensity vs. time for the light emitted from the surface of the cloud.

Several calculations were made for a sphere of two mean free paths radius and one calculation for a sphere of four mean free paths radius. Some of the results are given graphically in Figures 9.4a, b, 9.5, 9.6, 9.7. Choosing the mean free path and mean free time as units of length and time, we have:

case I $\quad \sigma = 1, \quad v = c = 1, \quad a = 2,$

case II $\quad \sigma = 1, \quad v = c = 1, \quad a = 4.$

In the first two attempts to calculate case I ($I = J = 10$, 162 interior net points; and $I = J = 15$, about 350 interior net points), the problem was treated in a straightforward way as an initial-value problem: at $n = 0$ ($t = 0$), Ψ_{ij}^n was set equal to zero except at the central net point $i = j = 0$, where it was arbitrarily set equal to 0.1. The photons were thus initially concentrated into the immediate neighborhood of the point source and then allowed to move in accordance with the transport equation. This treatment was unsuccessful, except in a rough, qualitative way, because of the violently discontinuous nature of the distribution at early times. Spreading out the initial distribution over 3 or 4 net points was also tried, but produced little improvement.

The procedure was then modified as follows: Ψ_{ij}^n was taken identically zero for $n = 0$, and a source term was introduced into the transport equation during the interval $0 \leq t \leq a/c$ while the initial pulse was traveling outward through the sphere, the source being located at position $r = ct$ at time t and representing photons emerging from their first scattering collision. In other words, the first collision was treated analytically and subsequent ones by the numerical calculation. The initial pulse travels out as an expanding spherical shell with a number of photons per unit area of the shell proportional to $e^{-\sigma r}/r^2$. To allow for a shell source proportional to this, it is only necessary to add the quantity $Ae^{-\sigma p \Delta r}/p^2$ to Φ_p^{n+1} after the cycle for which $n + 1 = p$. Before the first cycle, Φ_0^0 was set equal to a value chosen to represent approximately the photons injected into the system by the shell source during the time interval $(0, \frac{1}{2}\Delta t)$ (the other Φ_p^0 were zero). The constant A was taken proportional to I^3 so that when a problem was rerun with a finer net the total number of photons introduced during the interval $(0, a/c)$ should be the same.

With this modification the functions Φ and Ψ which one is trying to compute still have discontinuities during the time interval $(0, a/c)$, but at least they are bounded and do not require Dirac delta-functions for their representation.

The emergent flux F at $r = a$ was obtained by trapezoid-rule evaluation of the integral in the equation

$$F = 2\pi \int_0^{\pi/2} \sin \theta \, d\theta \, \Psi(a \cos \theta, a \sin \theta, t).$$

The flux F is shown plotted on a logarithmic scale as function of t for case I in Figure 9.4a and for case II in Figure 9.4b. In both cases F is practically zero (as it should be) until the instant $t = a/c$ when the original pulse emerges. Thereafter the cloud shines with an intensity which varies with time as shown. At first the curve of F versus t shows a transient behavior, but later it settles down to a very nearly exponential decline.

The spatial density $\Phi = \Phi(r, t)$ is shown as function of r at various times t for case II in Figure 9.5. In the three earliest curves the discontinuous

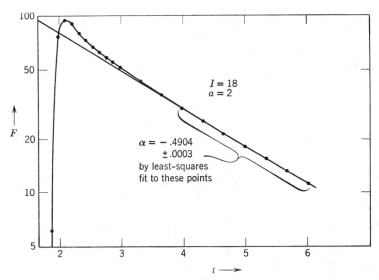

Fig. 9.4a. Emergent light flux vs. time for a cloud of radius equal to 2 mean free paths.

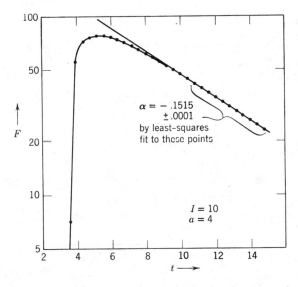

Fig. 9.4b. Same as Figure 9.4a for a cloud of radius equal to 4 mean free paths.

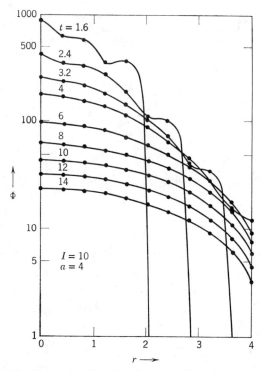

Fig. 9.5. Photon density in the cloud vs. radius at various times.

radiation front at $r = ct$ is clearly visible. The oscillations behind the front are presumably spurious, having been somehow introduced by the truncation error of the finite-difference equations. After the instant $t = a/c$, the distribution is quite smooth and settles down to a steady shape.

We have no exact information on which to compare these results but one can get an idea of the accuracy attained in various indirect ways. In the first place there are several internal checks which can be applied to the calculation. For example, the total number N of photons in the cloud at an instant t can be obtained either as the space integral of the space density

$$N = N_1 \propto \int_0^a r^2 \, dr \, \Phi(r),$$

or as the phase space integral of the phase space density

SEC. 9.19] A DIRECT INTEGRATION METHOD

$$N = N_2 \propto \tfrac{1}{2} \iint dx\, dy\, y\Psi(x, y),$$

the last integral being extended over the semi-circular region: $x^2 + y^2 \leq a^2, y \geq 0$. N_1 and N_2 should of course be equal. After each cycle the machine evaluated N_1 and N_2 (essentially by the trapezoid rule) and printed them out. During the interval $0 \leq t \leq a/c$, while the discontinuous front was moving outward, there was generally a quite large discrepancy between N_1 and N_2 but later the agreement was better. For the calculation shown in Figure 9.4a, the difference between N_1 and N_2 was approximately

15% at $t = \tfrac{1}{2} a/c$,

3% at $t = a/c$,

1.6% (average) for $a/c < t < 2a/c$.

Considering the small number of net points used (18 on the radius), this discrepancy is perhaps not surprising. Furthermore there is some evidence that even this discrepancy was due in part to poor scaling (loss of significant figures) rather than truncation errors, for the portion $4 \leq t \leq 6$ of the calculation was performed after rescaling, and the average discrepancy between N_1 and N_2 was then only 0.4% for $4 \leq t \leq 6$.

As a second internal check, the quantity

$$T = N + \int F\, dt$$

should be constant, by conservation of photons, for $t > a/c$, i.e., after the source term is no longer present in the equation. If N_2 is used for N in the above equation, it is found that T is indeed quite constant; the average deviation of T from its mean value for $a/c < t < 2a/c$ was 0.11% for the calculation shown in Figure 9.4a. This suggests, incidentally, that N_2 (integration over phase space) provides a better estimate of N than does N_1 (integration over space).

A second type of check on the accuracy resulted from varying the net spacing. The calculation was also performed with a coarser net than used for Figure 9.4a: $I = 12$ instead of $I = 18$. The results agreed well with the ones shown, except in the immediate vicinity of $t = a/c$, where the

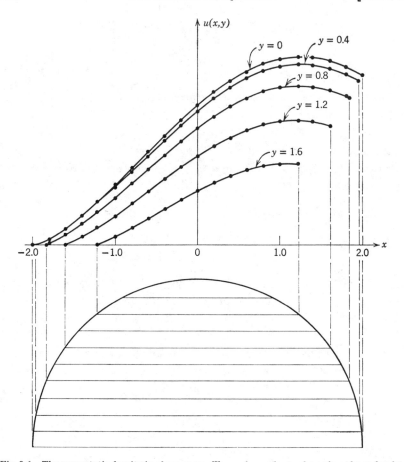

Fig. 9.6. The asymptotic density in phase space: Ψ vs. x for various values of y at large fixed t.

sudden onset of the luminosity was slightly less abrupt than for $I = 18$. For $t > a/c$, the two curves have the same shape to an accuracy of about 0.5%, but the one for $I = 12$ is consistently about 2.5% lower than that for $I = 18$.

Further evidence of the accuracy of the method comes from the asymptotic behavior of the solution for large values of t. If we look for a solution (9.30) having the form $\Psi(x, y, t) = \Psi_0(x, y)e^{\alpha t}$ (such a solution is called a *normal mode* solution), it can be shown that the corresponding density function

SEC. 9.19] A DIRECT INTEGRATION METHOD

$$\Phi_0(r) = \alpha \frac{1+f}{2r} \int_{-r}^{r} \Psi_0(x, \sqrt{r^2 - x^2})\, dx$$

(compare equation (9.32)) is a solution of an integral equation well known in transport theory. The fundamental solution of this equation was obtained by standard numerical methods of such accuracy that the results may be regarded as exact for purposes of the present discussion.

It has often been surmised that the general solution of the transport equation (9.30) has the form of a linear superposition of discrete normal modes. As t increases, the fundamental mode (i.e., the one with the highest value of α) predominates, and the asymptotic form of the solution is, therefore[12]

$$\Psi(x, y, t) \sim \Psi_0(x, y) e^{\alpha t}$$

and correspondingly,

$$\Phi(r, t) \sim \Phi_0(r) e^{\alpha t}.$$

Examination of the numerical solution depicted in Figure 9.4a shows that for $4 < t < 6$ the decay is quite accurately exponential and the value of the decay constant is $\alpha = -0.4904$. We therefore suppose that the asymptotic form of the solution has been fairly well established by $t = 4$ and we compare this form with the correct one as given by solution of (9.34). The correct value of α, according to integral equation, is $\alpha = -0.4925$, which agrees with the above value to a fraction of a percent. In Figure 9.7, the radial dependence of $\Phi(r, t)$ for $t = 5$, for one of the calculations is compared with the fundamental solution $\Phi_0(r)$ of the integral equation, suitably normalized. The discrepancy is within 1%, except for the point on the surface, $r = 2$, where it is 5%.

It seems that our solution has quite accurately the required asymptotic form, considering the coarseness of the net used.

The asymptotic distribution in phase space obtained from the numerical calculation is shown in Figure 9.6, where $\Psi_0(x, y)$ is plotted versus x for various values of y.

---

[12] The work of Lehner and Wing (1956) showed that for the similar problem of the slab the surmise made above is wrong, but the conclusion drawn from it nevertheless correct. Lehner and Wing found that the general solution has the form of a linear superposition of finitely many discrete normal modes plus a residuum which decays faster, as t increases, than any of the normal modes. Van Norton (1962) showed that the sphere has infinitely many normal modes.

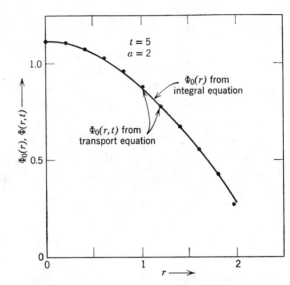

Fig. 9.7. The asymptotic density: Φ vs. r for large fixed t. The curve shows the accurate values as obtained from the integral equation (9.34).

In summary, the main features of the method are:

1. In the coordinates x, y, a neutron's free path in space is given by $y = $ const.
2. A rectangular net is used in the x, y, t-space.
3. Δt is taken $= \Delta x/v$, and the equations are differenced along characteristics (trajectories in space-time).
4. The fractional net intervals in x near the surface of the sphere are specially treated so as to retain the second-order accuracy.
5. The flux term Φ is centered in time by an iteration.
6. If the source has singularities, one should calculate at least the first collision analytically so as to obtain a smoother source distribution.

CHAPTER 10

Sound Waves

10.1. Physical Basis

In this chapter we deal with problems of the infinitesimal vibrations of fluids. Viscosity is neglected and so is thermal conduction until the last section of the chapter. Body forces, if present, are assumed balanced by static pressure gradients.

Let p_0, ρ_0 be the static pressure and density and let $p = f(\rho)$ be the equation of state, for adiabatic compression and rarefaction. We assume $f'(\rho) > 0$ and call $f'(\rho_0) = c^2$; p_0, ρ_0, c^2 are thought of as possibly depending on x, y, z but not on t. If $\mathbf{g} = \mathbf{g}(x, y, z)$ is the body force per unit volume, we assume $\mathbf{g} + \nabla p_0 = 0$ (static equilibrium). If the pressure and density are $p_0 + p$ and $\rho_0 + \rho$ where $p \ll p_0$ and $\rho \ll \rho_0$, then, to first order,

$$p = c^2 \rho.$$

If $\mathbf{v} = \mathbf{v}(x, y, z)$ is the fluid velocity, the equation of motion is

$$\rho_0 \frac{\partial \mathbf{v}}{\partial t} = -\nabla p$$

and the equation of continuity, to first order of small quantities, is

$$\frac{\partial \rho}{\partial t} = -\rho_0 \nabla \cdot \mathbf{v}.$$

Now,

$$\frac{\partial p}{\partial t} = c^2 \frac{\partial \rho}{\partial t} = -\rho_0 c^2 \nabla \cdot \mathbf{v},$$

and therefore,

(10.1) $$\frac{\partial^2 p}{\partial t^2} = \rho_0 c^2 \nabla \cdot \frac{1}{\rho_0} \nabla p.$$

Equation (10.1) is the general linear wave equation, to which the finite-difference methods of this chapter apply. In the examples and stability discussions however, we shall restrict consideration to the one-dimensional form of the equation with constant ρ_0 and c^2. In this approximation the fluid displacement, velocity, acceleration, density, etc., all satisfy equations of the same form. For example, if $y = y(x, t)$ denotes the displacement (this is the usual choice of dependent variable), we have

(10.2) $$\frac{\partial^2 y}{\partial t^2} = c^2 \frac{\partial^2 y}{\partial x^2}, \quad c^2 = \text{constant} > 0.$$

An equation of this form is called the *wave equation*. We assume the initial conditions

$$y(x, 0) = f_1(x),$$

$$\frac{\partial y}{\partial t}(x, 0) = f_2(x)$$

and a periodicity condition

$$y(x, t) = y(x + L, t)$$

in lieu of boundary conditions. It will be evident to the reader how to generalize the stability discussion to the multi-dimensional cases, if the constancy of ρ_0 and c^2 and the periodicity are retained, and how to generalize the finite-difference equations to the general case of variable ρ_0 and c^2.

10.2. The Usual Finite-Difference Equation

This equation is

(10.3) $$y_j^{n+1} - 2y_j^n + y_j^{n-1} = \left(\frac{c \Delta t}{\Delta x}\right)^2 (y_{j+1}^n - 2y_j^n + y_{j-1}^n)$$

and to fit it into the general scheme of Chapter 4 one might be tempted to take as a second variable $v = \partial y / \partial t$, whereupon (10.2) becomes

$$\frac{\partial v}{\partial t} = c^2 \frac{\partial^2 y}{\partial x^2},$$

$$\frac{\partial y}{\partial t} = v.$$

SEC. 10.2] THE USUAL FINITE-DIFFERENCE EQUATION 261

However, this is unsuitable if we regard y, v as the components of a vector $\mathbf{u}(x, t)$ and take the norm of the corresponding element u of $\mathcal{B}$ as that given in Chapter 4, namely,

$$\|u\| = \left[\int_0^L (|y|^2 + |v|^2)dx\right]^{\frac{1}{2}},$$

because in this case the operator $E_0(t)$ is unbounded and the initial-value problem is not properly posed. This can be seen by considering the simple standing wave

$$f_1 = \sin(2\pi n x/L), \quad f_2 = 0,$$

for which the solution $u(t)$ is easily obtained; we find that

$$\frac{\|u(t)\|}{\|u(0)\|} = \sqrt{\cos^2 \frac{2\pi n c t}{L} + \left(\frac{2\pi n c}{L}\right)^2 \sin^2 \frac{2\pi n c t}{L}},$$

which can be made to exceed any given bound by suitable choice of n and t. This example shows merely that a choice of norm is implied in the concept of a properly posed initial-value problem, and that we have chosen the norm inappropriately.

A better procedure is to introduce an additional variable $w = c\partial y/\partial x$, whereupon (10.2) becomes

(10.4)
$$\frac{\partial v}{\partial t} = c\frac{\partial w}{\partial x},$$
$$\frac{\partial w}{\partial t} = c\frac{\partial v}{\partial x},$$

and the norm is taken to be

$$\|u\| = \left[\int_0^L (|w|^2 + |v|^2)dx\right]^{\frac{1}{2}}.$$

Now $E_0(t)$ is bounded—in fact the bound is 1 because $\|u(t)\|$ is constant for any solution, in consequence of the conservation of energy (the square of the norm is proportional to the energy of the wave motion in the region $0 \leq x \leq L$).

A further parenthetical remark is that the rather obvious finite-difference scheme

$$(v_j^{n+1} - v_j^n)/\Delta t = c(w_{j+1}^n - w_{j-1}^n)/2\Delta x,$$
$$(w_j^{n+1} - w_j^n)/\Delta t = c(v_{j+1}^n - v_{j-1}^n)/2\Delta x$$

is unstable unless $\Delta t/(\Delta x)^2$ is bounded as $\Delta t, \Delta x \to 0$. This would be an impractical restriction, and we abandon this scheme.

To avoid difference quotients over the double space interval $2\Delta x$, we shall henceforth associate values of w with the midpoints of the interval and write $w_{j+\frac{1}{2}}^n$, etc. This is purely a matter of notation. One sometimes also treats the time intervals in this way by introduction of fractional superscripts—this results in a more symmetrical appearance and an obviously correct centering—but we shall avoid this in order to adhere more closely to the notation of Chapter 4.

The scheme

(10.5)
$$(v_j^{n+1} - v_j^n)/\Delta t = c(w_{j+\frac{1}{2}}^n - w_{j-\frac{1}{2}}^n)/\Delta x,$$
$$(w_{j-\frac{1}{2}}^{n+1} - w_{j-\frac{1}{2}}^n)/\Delta t = c(v_j^{n+1} - v_{j-1}^{n+1})/\Delta x$$

is equivalent to the usual scheme (10.3) if one identifies

$$v_j^n \text{ with } (y_j^n - y_j^{n-1})/\Delta t,$$

and

$$w_{j-\frac{1}{2}}^n \text{ with } c(y_j^n - y_{j-1}^n)/\Delta x.$$

This scheme was one of the subjects studied by Courant, Friedrichs and Lewy (1928) in the paper in which the concept of stability was first introduced.

The amplification matrix for the system (10.5) is

$$G = G(\Delta t, k) = \begin{bmatrix} 1 & ia \\ ia & 1 - a^2 \end{bmatrix},$$

where a is an abbreviation for $(2c\Delta t/\Delta x) \sin(k\Delta x/2)$. The characteristic equation of G is

$$\lambda^2 - \lambda(2 - a^2) + 1 = 0.$$

If $a^2 \leq 4$, both roots have absolute value 1. Clearly, we may suppose that

$$c\Delta t/\Delta x = \text{constant as } \Delta t, \Delta x \to 0,$$

and then, from the definition of a, we see that the von Neumann condition is
$$c\Delta t/\Delta x \leq 1.$$

A sufficient condition for stability was obtained in Section 4.11. If $\mathbf{v}_1$ and $\mathbf{v}_2$ denote the normalized eigenvectors of G, we found that $|\det(\mathbf{v}_1, \mathbf{v}_2)|$ is bounded away from zero for all real k, and hence the equations are stable by the theorem given there, if $c\Delta t/\Delta x < 1$. For $c\Delta t/\Delta x = 1$, the system (10.5) is unstable; to show this it suffices to exhibit an unbounded solution, namely,
$$v_j^n = (-1)^{n+j}(1 - 2n),$$
$$w_{j+\frac{1}{2}}^n = (-1)^{n+j} 2n.$$

Although such a slow growth of errors would probably be harmless in practice, perturbations can make the growth more rapid, as was pointed out in Section 5.2. However, exact equality of $c\Delta t = \Delta x$ would be meaningless in a real calculation, where c may be variable and is in any case known to only a finite number of decimal places.

10.3. An Implicit System

The system

(10.6)
$$\frac{v_j^{n+1} - v_j^n}{\Delta t} = c \frac{w_{j+\frac{1}{2}}^n - w_{j-\frac{1}{2}}^n + w_{j+\frac{1}{2}}^{n+1} - w_{j-\frac{1}{2}}^{n+1}}{2\Delta x},$$

$$\frac{w_{j-\frac{1}{2}}^{n+1} - w_{j-\frac{1}{2}}^n}{\Delta t} = c \frac{v_j^{n+1} - v_{j-1}^{n+1} + v_j^n - v_{j-1}^n}{2\Delta x},$$

which is equivalent to the second-order system

(10.7) $$\frac{y_j^{n+1} - 2y_j^n + y_j^{n-1}}{(\Delta t)^2} = c^2 \frac{(\delta^2 y)_j^{n+1} + 2(\delta^2 y)_j^n + (\delta^2 y)_j^{n-1}}{4(\Delta x)^2},$$

has the amplification matrix

$$G = \begin{bmatrix} \dfrac{1 - a^2/4}{1 + a^2/4} & \dfrac{ia}{1 + a^2/4} \\ \dfrac{ia}{1 + a^2/4} & \dfrac{1 - a^2/4}{1 + a^2/4} \end{bmatrix}.$$

Both eigenvalues of G have absolute value 1, and G^*G is the unit matrix, so that *the system* (10.6) *is unconditionally stable*.

From (10.7) it is clear that the solution of this implicit system can be achieved by the algorithm given in Chapter 8 for implicit systems for the diffusion equation. Therefore, equations (10.6) provide a practically satisfactory system for the wave equation.

10.4. Coupled Sound and Heat Flow

In the flow of a compressible fluid there are often considerable differences of temperature from one point to another, and the transfer of energy by thermal conduction may have a significant effect on the motion. The parabolic equation of heat flow is then coupled to the hyperbolic equations of fluid dynamics and the two phenomena must be calculated concurrently. This effect occurs also for infinitesimal or acoustic vibrations and is responsible for absorption of ultrasonic waves.

Let the pressure, specific volume and specific internal energy be $p_0 + p$, $V_0 + V$ and $\mathscr{E}_0 + \mathscr{E}$, where p_0, V_0 and $\mathscr{E}_0$ are the ambient values and where $p \ll p_0$, $V \ll V_0$, $\mathscr{E} \ll \mathscr{E}_0$; and let the material velocity be u. p, V, $\mathscr{E}$, u are functions of x and t. The quantity $c = \sqrt{p_0 V_0}$ is the isothermal sound speed. We take the equation of state to be

$$\mathscr{E}_0 + \mathscr{E} = (p_0 + p)(V_0 + V)/(\gamma - 1)$$

and denote by σ the ratio of thermal conductivity to specific heat at constant volume.

In terms of auxiliary dependent variables defined by $w = cV/V_0$ and $e = \mathscr{E}/c$, the differential equations, to first order of small quantities, are

(10.8)
$$\frac{\partial u}{\partial t} = c\frac{\partial}{\partial x}(w - (\gamma - 1)e),$$

$$\frac{\partial w}{\partial t} = c\frac{\partial u}{\partial x},$$

$$\frac{\partial e}{\partial t} = \sigma\frac{\partial^2 e}{\partial x^2} - c\frac{\partial u}{\partial x}.$$

SEC. 10.4] COUPLED SOUND AND HEAT FLOW

Finite-difference equations for this system can be constructed in various ways. A simple explicit system is

(10.9)
$$\frac{u_j^{n+1} - u_j^n}{\Delta t} = c\,\frac{w_{j+\frac{1}{2}}^n - w_{j-\frac{1}{2}}^n - (\gamma - 1)(e_{j+\frac{1}{2}}^n - e_{j-\frac{1}{2}}^n)}{\Delta x},$$

$$\frac{w_{j+\frac{1}{2}}^{n+1} - w_{j+\frac{1}{2}}^n}{\Delta t} = c\,\frac{u_{j+1}^{n+1} - u_j^{n+1}}{\Delta x},$$

$$\frac{e_{j+\frac{1}{2}}^{n+1} - e_{j+\frac{1}{2}}^n}{\Delta t} = \sigma\,\frac{e_{j+\frac{3}{2}}^n - 2e_{j+\frac{1}{2}}^n + e_{j-\frac{1}{2}}^n}{(\Delta x)^2} - c\,\frac{u_{j+1}^{n+1} - u_j^{n+1}}{\Delta x}.$$

The advanced values of the velocity u have been used in the second and third equations just as in the second equation (10.5) of the system without heat flow (it will be recalled that that was necessary in order to achieve a reasonable stability condition for the sound wave problem). Nevertheless, equations (10.9) are effectively explicit because the first equation can be solved first to obtain the values of u_j^{n+1} and u_{j+1}^{n+1} needed in the other two.

This system has been found satisfactory. If the sound waves and heat flow were uncoupled, the respective stability conditions would be

(10.10)
$$\sqrt{\gamma}\,c\Delta t/\Delta x < 1,$$

$$\sigma \Delta t/(\Delta x)^2 < \tfrac{1}{2}.$$

Surely these conditions are necessary. In the limit, as Δt and $\Delta x \to 0$, the second of these conditions always implies the first, and it is generally conjectured that it is the stability condition. In an actual calculation, one should of course choose Δx and Δt so as to satisfy the first condition as well as the second.

To avoid the small time increment required by condition (10.10) one often uses an implicit treatment of the third differential equation, for example,

(10.11)
$$\frac{e_{j+\frac{1}{2}}^{n+1} - e_{j+\frac{1}{2}}^n}{\Delta t} = \sigma\,\frac{e_{j+\frac{3}{2}}^{n+1} - 2e_{j+\frac{1}{2}}^{n+1} + e_{j-\frac{1}{2}}^{n+1}}{(\Delta x)^2} - c\,\frac{u_{j+1}^{n+1} - u_j^{n+1}}{\Delta x}.$$

The first two equations (10.9) are retained.

In the first edition of this book it was conjectured that the stability condition is $c\Delta t/\Delta x < 1$ for this system. The isothermal sound speed c

was used because instabilities generally involve very short wavelength disturbances, and for these the thermal equilibration time from crest to trough is less than the period of the wave. The equilibration time is proportional to the square of the wavelength λ, and the compressions and rarefactions in the wave motion are very nearly isothermal rather than adiabatic if $\lambda \ll \sigma/c$.

The correctness of this conjecture was confirmed in 1962 by Morimoto. In the limit $\sigma \to 0$, however, the condition is the more stringent inequality $\sqrt{\gamma} c\Delta t/\Delta x < 1$, and this suggests that, unless $\Delta x \ll \sigma/c$, there should be a practical stability condition, for a finite net, involving the two dimensionless constants

(10.12) $$\nu = c\Delta t/\Delta x, \qquad \mu = \sigma\Delta t/(\Delta x)^2,$$

and also γ; it would presumably be intermediate between the two conditions above, namely, it would be of the form $\nu < \nu_0$, where ν_0 is some function of γ and μ whose values lie between $1/\sqrt{\gamma}$ and 1.

To find the practical stability condition, we adopt the principle, already stated in Section 9.7, that no Fourier component of the approximation should be allowed to grow more rapidly than the most rapid possible growth of the exact solution. Since sound waves do not grow in amplitude (in fact, they are damped out, in the present problem, by the action of the heat flow, but by a negligible amount of damping for long wavelengths), we must require the amplification factors g to be ≤ 1 in modulus, not merely $\leq 1 + O(\Delta t)$. The g satisfy the equation

$$\begin{vmatrix} g-1 & -2i\nu \sin\alpha & 2i\nu(\gamma-1)\sin\alpha \\ -2ig\nu \sin\alpha & g-1 & 0 \\ 2ig\nu \sin\alpha & 0 & g-1+2\mu(1-\cos 2\alpha)g \end{vmatrix} = 0,$$

where $\alpha = \tfrac{1}{2} k\Delta x$. The requirement that $|g| \leq 1$ for all α leads to the condition

(10.13) $$\nu < \sqrt{(1+2\mu)/(\gamma+2\mu)},$$

which is the practical stability condition; this condition was shown in Section 6.4 to be also sufficient for stability in the sense that a certain positive definite quadratic form in u, w, e, and $\Delta_+ e_j = e_{j+1} - e_j$ is bounded for all time if (10.13) is satisfied.

SEC. 10.4] COUPLED SOUND AND HEAT FLOW

Equation (10.13) shows that, as was to be expected, stability is determined by the adiabatic sound speed for $\mu \ll 1$, by the isothermal sound speed for $\mu \gg 1$, and by an intermediate speed for $\mu \sim 1$.

In a series of numerical tests, the difference equations (the first two of (10.9) and equation (10.11)) were used to calculate the propagation of a wave which starts out as a simple jump or infinitesimal shock (infinitesimal in the sense that the linearized equations are being used). Such a wave propagates with speed $\sqrt{\gamma}c$, the profile becoming gradually rounded off so as to resemble the normal distribution curve

$$\int_{-\infty}^{x} \exp[-y^2/2]dy$$

with a width that increases as $\sqrt{t}$. An approximate (but rather good) analytic solution is known.

Figure 10.1 shows the calculated profile in a stable calculation after 165 time steps, and Figure 10.2 shows the values obtained in an unstable calculation after only five time steps; in both calculations the ordinary stability condition $c\Delta t/\Delta x < 1$ was satisfied. Figure 10.3 summarizes the series; each dot indicates the values of μ and γ of a run that was found to

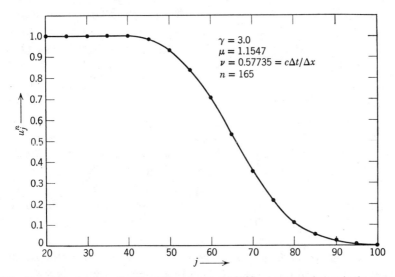

Fig. 10.1. Calculated profile after 165 cycles of an initially sharp sound wave in the presence of heat conduction.

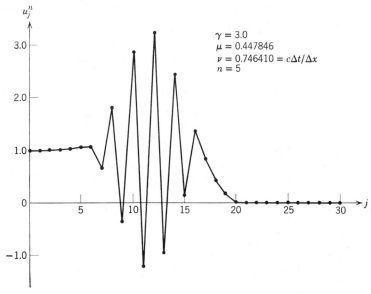

Fig. 10.2. Calculated profile after 5 cycles, in a run in which the practical criterion (10.13) was violated, although $c\Delta t/\Delta x < 1$.

be stable, and each cross indicates the values of a run that was found to be unstable; the curve gives the maximum permissible ν as a function of ν/μ, according to (10.13). It is seen that the observed stability condition is in agreement with the theoretical one. Two runs for which the inequality (10.13) was replaced by the equality turned out to be stable.

To understand the nature of this kind of instability, consider a calculation with $\nu = c\Delta t/\Delta x = 0.9$ and $\gamma = 3$. According to the usual criterion, this is a stable scheme, and convergence follows, for $\Delta t \to 0$. However, if Δt is such that μ is also equal to 0.9 (this case was chosen just for illustration), then, according to the determinantal equation above, $g \approx -2.6$ for the Fourier component with $k\Delta x = \pi$; therefore, this component becomes amplified a thousandfold in less than 10 time steps. To avoid this, if one wishes to keep $c\Delta t/\Delta x = 0.9$, both Δx and Δt have to be reduced by a factor ≈ 0.25, so as to satisfy (10.13).

An alternative approach is to retain the explicit character of the difference equations but to use a smaller time step, say $\Delta t/K$, where K is an integer, for the third equation than for the first two. Then each cycle

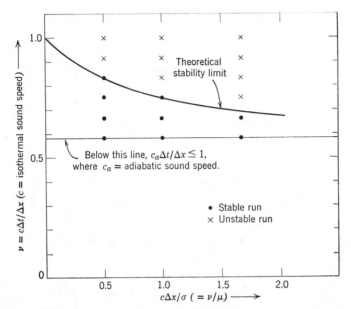

Fig. 10.3. Stability diagram for coupled sound and heat flow with $\gamma = 3$. In the numerical tests, a run was counted as stable if the profile was still smooth, as in Figure 10.1, after about 100 cycles and as unstable if the value of $|u_j^n|$ exceeded 3.0 at any point (when this happened, it happened in $\leq$ 5 cycles in every case).

of the fluid dynamic calculation is accompanied by K steps of the heat flow calculation. The procedure can be described by use of fractional superscripts:

$$\frac{e_{j+\frac{1}{2}}^{n+(m+1)/K} - e_{j+\frac{1}{2}}^{n+m/K}}{\Delta t/K} = \sigma \frac{e_{j+\frac{3}{2}}^{n+m/K} - 2e_{j+\frac{1}{2}}^{n+m/K} + e_{j-\frac{1}{2}}^{n+m/K}}{(\Delta x)^2} - c\frac{u_{j+1}^{n+1} - u_j^{n+1}}{\Delta x},$$
$$m = 0, 1, \cdots, K-1.$$

It is conjectured that in this case the stability condition is

$$\frac{c\Delta t}{\Delta x} < 1 \quad \text{and} \quad \frac{\sigma \Delta t/K}{(\Delta x)^2} < \tfrac{1}{2}.$$

10.5. A Practical Stability Criterion

The above example and those of Sections 9.7 and 11.6 show that stability, as usually defined, is not always adequate for the practical utility of a method. Although stability implies convergence in the limit $\Delta t \to 0$ under the conditions of the Lax equivalence theorem, it can happen that for the finite Δt used in practice the errors are nevertheless unacceptably amplified. To ensure that no Fourier component can grow faster than the most rapid possible growth of the exact solution, we must impose some restriction on the constant implied by the O symbol of the von Neumann condition $|g| \leq 1 + O(\Delta t)$.

According to Section 4.3, a Fourier component of the exact solution varies as

(10.14) $$\hat{u}(\mathbf{k}, t) = e^{tP(i\mathbf{k})}u(\mathbf{k}, 0),$$

where $P = P(\mathbf{q}) = P(q_1, \ldots, q_d)$ is the matrix introduced there, whose elements are polynomials in the $q_i = \partial/\partial x_i$, and which gives the differential operator A of the equation system $du/dt = Au$. Therefore, if we call

$$m = \text{Max}\,[\text{Re}\,\lambda \mid \lambda = \text{eigenvalue of } P(i\mathbf{k})],$$

maximized by considering all real vectors $\mathbf{k}$ and all eigenvalues for each $\mathbf{k}$, then the solution cannot grow faster than e^{mt}, possibly multiplied by a polynomial in t. We therefore take as a *modified von Neumann condition* the restriction

(10.15) $$|g| = |g(\Delta t, \mathbf{k})| \leq e^{m\Delta t}$$

on the eigenvalues of G.

One may wonder whether (10.15) might be too severe and might exclude all possible schemes, for some problems, but this seems unlikely in view of Kreiss' theorem on dissipative schemes for symmetric hyperbolic systems (Section 5.4) where $m = 0$ and where we actually require that $|g| < 1$ (not $= 1$) except for $k = 0$. Indeed, it may well be that a useful requirement for acceptability of a scheme of accuracy $2r - 1$ is

$$|g| \leq e^{m\Delta t}(1 - \delta|\boldsymbol{\xi}|^{2r})$$

for some $\delta > 0$, for all $\boldsymbol{\xi}$ whose components $(k_1 \Delta x_1, \ldots, k_d \Delta x_d)$ all lie between $-\pi$ and π. In a sense, the trouble with the usual theory is that it does not specify the values of either m or δ. We have now specified m. Probably δ should always be of the order $\frac{1}{2}(\sqrt{d}\pi)^{-2r}$.

CHAPTER 11

Elastic Vibrations

11.1. Vibrations of a Thin Beam

Problems in dynamic elasticity present some interesting questions for numerical analysis which we shall illustrate by simple one-dimensional problems. Suppose that a straight thin beam or rod or wire of length L is supported or clamped at the ends and set into vibration. We assume that the deformation is everywhere small, so that Hooke's law holds. This requires that the radius of curvature of the curve into which the rod is bent be much larger than the radius of the rod. It does not require that the wire be nearly straight: a piece of piano wire can be tied into a loose knot without exceeding the elastic limit anywhere. To make our problem linear, however, we further assume that the rod remains nearly straight and that the amplitude of vibration is small. We also assume that the vibration is in one transverse plane, and we denote by $y = y(x, t)$ the displacement from the equilibrium position at distance x along the rod from one end. In these terms the assumption leading to linearity are $y(x, t) \ll L$, $\partial y(x, t)/\partial x \ll 1$.

If we divide the rod into two parts by an imaginary cut, as by the dotted line in the drawing, Figure 11.1, the portion on the left exerts a

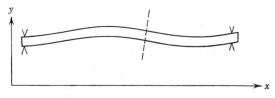

Fig. 11.1. A deformed elastic rod, showing the end supports and the imaginary cut discussed in the text.

certain force F and a certain turning moment M on the portion on the right. For this problem we assume that the rod is under no tension or compression lengthwise; then the force F is transverse. If we take the y direction as positive for F and the counter-clockwise direction as positive for M, we may summarize the elastic properties of the rod in the equations

$$M = -\varkappa \frac{\partial^2 y}{\partial x^2},$$

$$F = \frac{\partial}{\partial x} \varkappa \frac{\partial^2 y}{\partial x^2},$$

where $\varkappa$ is a positive quantity depending on the material and cross-section of the rod. The movement and force exerted on the left portion of the rod by the right are $-M$ and $-F$. The net force exerted on a small segment of length δl is therefore $-\delta l \partial F/\partial x$; consequently, if $\mu = \mu(x)$ is the mass per unit length, the equation of motion is

$$\mu \frac{\partial^2 y}{\partial t^2} = -\frac{\partial^2}{\partial x^2} \varkappa \frac{\partial^2 y}{\partial x^2}.$$

We simplify the problem further by supposing $\varkappa$ and μ to be constant. The equation is then

(11.1) $$\frac{\partial^2 y}{\partial t^2} = -a^2 \frac{\partial^4 y}{\partial x^4}, \qquad a = \sqrt{\frac{\varkappa}{\mu}} = \text{const.}$$

A typical boundary condition is the condition of "simple support," according to which the supports are supposed to prevent lateral motion but to exert no turning moments. That is,

(11.2) $$y = 0, \quad \frac{\partial^2 y}{\partial x^2} = 0 \quad \text{at } x = 0 \text{ and } x = L.$$

To rewrite (11.1) as two equations of first order in time, call $\partial y/\partial t = v$, $a\partial^2 y/\partial x^2 = w$; then

(11.3)
$$\frac{\partial v}{\partial t} = -a \frac{\partial^2 w}{\partial x^2},$$

$$\frac{\partial w}{\partial t} = a \frac{\partial^2 v}{\partial x^2}.$$

The difference equations which we give may be interpreted as applying either to (11.1) or to the system (11.3).

We note in passing that these boundary conditions (11.2) are of such a nature that they can be replaced by periodicity and oddness conditions

(11.2a) $\qquad y(x, t) = y(x + 2L, t) = -y(-x, t),$

for functions that are now defined for all real x. The proof is as follows: The differential equations (11.3) are invariant under the transformations $x + 2L \to x$ and $-x \to x$. Therefore, any pair of odd trigonometric polynomials with period $2L$ may be taken as initial data $v(x, 0)$, $w(x, 0)$, and the resulting genuine solution is also invariant under these transformations. Thus a unique solution $y(x, t)$ of (11.1) satisfying (11.2a) is determined via $a\partial^2 y/\partial x^2 = w$. This y also satisfies (11.2) by symmetry. In other words, the two initial-value problems (with boundary condition (11.2) and with boundary condition (11.2a)) have genuine solution operators $E_0(t)$ which agree for the odd trigonometrical polynomials. Now both initial-value problems are properly posed (we omit the proof) for the Banach space $\mathscr{B}$ consisting of function pairs v, w on $0 \leq x \leq L$ with norm

$$\left[\int_0^L (|v|^2 + |w|^2) dx \right]^{1/2},$$

so $E_0(t)$ is bounded for both problems. Furthermore the odd trigonometrical polynomials provide a set which is dense in $\mathscr{B}$ and therefore the generalized solution operators $E(t)$ of the two problems are identical, which is what we wanted to prove.

The properties of the simple finite-difference systems are so similar to those for the wave equation given in the preceding chapter, that we shall summarize them rather briefly.

11.2. Explicit Difference Equations

If forward time differences are used in both equations (11.3) the resulting system is unstable unless $\Delta t \to 0$ as $(\Delta x)^3$ and is therefore unsatisfactory.

If one forward and one backward time difference are used, we get the system

(11.4)
$$\frac{v_j^{n+1} - v_j^n}{\Delta t} = -a \frac{(\delta^2 w)_j^n}{(\Delta x)^2},$$
$$\frac{w_j^{n+1} - w_j^n}{\Delta t} = a \frac{(\delta^2 v)_j^{n+1}}{(\Delta x)^2},$$

which is equivalent to the approximation

(11.5) $$\frac{y_j^{n+1} - 2y_j^n + y_j^{n-1}}{(\Delta t)^2} = -a^2 \frac{(\delta^4 y)_j^n}{(\Delta x)^4}$$

to equation (11.1) and is effectively explicit. The truncation error of (11.5) is $O[(\Delta t)^2] + O[(\Delta x)^2]$.

If we abbreviate:

$$\frac{4a\Delta t}{(\Delta x)^2} \sin^2 \frac{k\Delta x}{2} = \omega,$$

the amplification matrix of the system (11.4) is

$$G = G(\Delta t, k) = \begin{bmatrix} 1 & \omega \\ -\omega & 1 - \omega^2 \end{bmatrix}.$$

The eigenvalues of G have magnitude unity if $|\omega| \leq 2$, hence if we assume that $\Delta t \to 0$ as $(\Delta x)^2$, the von Neumann *necessary condition for stability* is

$$\frac{a\Delta t}{(\Delta x)^2} \leq \tfrac{1}{2}.$$

A sufficient condition for stability can be obtained from the theorem of Section 4.11, just as for the corresponding system for the wave equation. (In fact, the above matrix G is equivalent under a unitary transformation to the matrix given in equation (4.20), except for notation.) If $\mathbf{v}_1$ and $\mathbf{v}_2$ denote the normalized eigenvectors of G, we easily find that $|\det (\mathbf{v}_1 \, \mathbf{v}_2)|^2 = 1 - (\omega^2/4)$; this is bounded away from zero for all real k if

$$\frac{a\Delta t}{(\Delta x)^2} < \tfrac{1}{2};$$

this inequality is therefore sufficient for stability.

11.3. An Implicit System

Of the many conceivable implicit systems, we shall consider one that is similar to (10.6) for the wave equation, namely,

(11.6) $$\frac{v_j^{n+1} - v_j^n}{\Delta t} = -a \frac{(\delta^2 w)_j^{n+1} + (\delta^2 w)_j^n}{2(\Delta x)^2},$$

$$\frac{w_j^{n+1} - w_j^n}{\Delta t} = a \frac{(\delta^2 v)_j^{n+1} + (\delta^2 v)_j^n}{2(\Delta x)^2}.$$

SEC. 11.5]   SOLUTION OF IMPLICIT EQUATIONS

The amplification matrix is

$$G = \begin{bmatrix} \dfrac{1 - \frac{1}{4}\omega^2}{1 + \frac{1}{4}\omega^2} & \dfrac{\omega}{1 + \frac{1}{4}\omega^2} \\ \dfrac{-\omega}{1 + \frac{1}{4}\omega^2} & \dfrac{1 - \frac{1}{4}\omega^2}{1 + \frac{1}{4}\omega^2} \end{bmatrix}$$

which is easily seen to be a unitary matrix. Therefore, (1) the eigenvalues of G lie on the unit circle, and the von Neumann condition is always satisfied; (2) G is normal and the von Neumann condition is sufficient for stability. Consequently *the system* (11.6) *is always stable.* The truncation error of the equivalent second order (in time) equation for y is $O[(\Delta t)^2] + O[(\Delta x)^2]$, just as for the explicit system discussed in the preceding section.

11.4. Virtue of the Implicit System

The vibrating beam problem is an example of a problem in which the use of the implicit system really pays off. The truncation error is of the same order in Δt as in Δx for system (11.6) and also for (11.4). Therefore as increasing accuracy is required, one would only have to decrease Δt at the same rate as Δx from the point of view of accuracy; but the stability condition of the explicit system requires that Δt be decreased as $(\Delta x)^2$, and this eventually leads to the use of many more time steps than required by consideration of accuracy alone, with the further danger that the increments which must be added to v^n and w^n to give v^{n+1} and w^{n+1} may become so small as to be disturbed by round-off errors. The implicit system is free of this defect.

11.5. Solution of Implicit Equations of Arbitrary Order

The algorithm for solving the simultaneous equations of the implicit system (11.6) is similar to that for the implicit systems encountered in the heat flow and sound wave problems, but there are enough new features to warrant a discussion. In terms of the equivalent system for the y_j^{n+1}, the matrix of the set of simultaneous equations now has five non-vanishing diagonals instead of three.

The typical equation of the simultaneous set may be written as a five-term recurrence for the y_j^{n+1} or two three-term recurrence relations for the v_j^{n+1} and w_j^{n+1}. We choose the latter representation, and let $\mathbf{u}_j$ stand for the two-component vector

$$\mathbf{u}_j = \begin{bmatrix} v_j^{n+1} \\ w_j^{n+1} \end{bmatrix}.$$

Then it is seen from (11.6) that the recurrence relation (after multiplication through by Δt) is

(11.7) $\quad -A_j \mathbf{u}_{j+1} + B_j \mathbf{u}_j - C_j \mathbf{u}_{j-1} = \mathbf{d}_j, \quad j = 1, 2, \ldots, J-1,$

(note the similarity with equation (8.15) for the diffusion equation discussion), where A_j, B_j, and C_j are the 2×2 matrices

$$B = \begin{bmatrix} 1 & -\dfrac{a\Delta t}{(\Delta x)^2} \\ \dfrac{a\Delta t}{(\Delta x)^2} & 1 \end{bmatrix},$$

$$A_j = C_j = \begin{bmatrix} 0 & -\dfrac{1}{2}\dfrac{a\Delta t}{(\Delta x)^2} \\ \dfrac{1}{2}\dfrac{a\Delta t}{(\Delta x)^2} & 0 \end{bmatrix},$$

and $\mathbf{d}_j$ is a known vector. Note that in the present problem A_j, B_j, C_j are independent of j and that $A = C$, $B = I + A + C$. In order to have the algorithm applicable to more general problems we make no use of these special properties of the matrices. The algorithm can also be applied to problems in which $\mathbf{u}_j$ has any number of components. We do assume that A_j, B_j, C_j are non-singular.

The boundary condition of simple support for the bar gives $\mathbf{u}_0 = \mathbf{u}_J = 0$, but let us suppose more general ones given, corresponding to conditions relating $\mathbf{u}$ and $\partial \mathbf{u}/\partial x$ at each end:

(11.8) $\quad \mathbf{u}_0 = H\mathbf{u}_1 + \mathbf{l},$

(11.9) $\quad \mathbf{u}_{J-1} = M\mathbf{u}_J + \mathbf{n},$

where H, M are given matrices and $\mathbf{l}$, $\mathbf{n}$ are given vectors. Still more general boundary conditions will be considered below.

SEC. 11.5]   SOLUTION OF IMPLICIT EQUATIONS   277

Now (11.7) and (11.8) determine a two-parameter family[1] of solutions of (11.7). Given a set of non-singular matrices E_j and vectors $\mathbf{f}_j$, the equations

(11.10) $\qquad \mathbf{u}_j = E_j \mathbf{u}_{j+1} + \mathbf{f}_j, \quad j = 0, 1, \ldots,$

also have a two-parameter family of solutions; and those two families are identical if

$$E_0 = H, \quad \mathbf{f}_0 = \mathbf{l},$$

(11.11) $\qquad E_j = (B_j - C_j E_{j-1})^{-1} A_j$

$\qquad \mathbf{f}_j = (B_j - C_j E_{j-1})^{-1} (\mathbf{d}_j - C_j \mathbf{f}_{j-1}) \qquad j \geq 1.$

The last two equations are obtained by substituting $E_{j-1} \mathbf{u}_j + \mathbf{f}_{j-1}$ for $\mathbf{u}_{j-1}$ in (11.7), solving the result for $\mathbf{u}_j$, and then identifying the coefficients with those of (11.10).

As for the diffusion equation problem, equations (11.11) serve to determine the E_j and $\mathbf{f}_j$ inductively ($j = 0, 1, 2, \ldots$). The induction can be continued (so long as the matrix $(B_{j+1} - C_{j+1} E_j)$ is non-singular, which we assume to be the case) until all the E_j and $\mathbf{f}_j$ have been calculated, for $j = 1, 2, \ldots, J - 1$. Note that the E_j so obtained are non-singular. We still have to invert a 2×2 matrix; one can either work out the formulas in advance or provide an inversion subroutine in the machine—if the number of components is greater than two, the latter is probably preferable.

The $\mathbf{u}_j$ are now obtained from an induction on decreasing j. First, $\mathbf{u}_J$ is determined from the equations

$$\mathbf{u}_{J-1} = E_{J-1} \mathbf{u}_J + \mathbf{f}_{J-1},$$
$$\mathbf{u}_{J-1} = M \mathbf{u}_J + \mathbf{n},$$

by elimination of $\mathbf{u}_{J-1}$. We must assume that the matrix $E_{J-1} - M$ is non-singular. We make this assumption because otherwise the solution either would not exist or would not be unique, indicating that the boundary conditions had been improperly formulated in the finite-difference form (11.8), (11.9). Then equations (11.10) determine the solution by an induction in order of decreasing j ($j = J - 1, J - 2, \ldots 0$).

---

[1] More generally, a k-parameter family if we are dealing with k-component vectors and $k \times k$ matrices.

For this to be a reasonable scheme, the elements of E_j and $\mathbf{f}_j$ must remain reasonably in scale. We shall show, first, that under certain assumptions, the matrices E_j are uniformly bounded. Noting that for matrices the reciprocal of a product is the product of the reciprocals in inverse order, we can write (11.11) as

$$E_j = [B_j B_j^{-1}(B_j - C_j E_{j-1})]^{-1} A_j$$
$$= [B_j^{-1}(B_j - C_j E_{j-1})]^{-1} B_j^{-1} A_j$$
$$= [I - B_j^{-1} C_j E_{j-1}]^{-1} B_j^{-1} A_j.$$

Letting $\|M\|$ denote the norm of a matrix M, we use the fact[2] that if $\|M\| < 1$, $\|(I - M)^{-1}\| \leq (1 - \|M\|)^{-1}$ and that $\|MN\| \leq \|M\| \cdot \|N\|$. Applying these relations to the above recurrence formula for E_j, and calling $\|B_j^{-1} A_j\| = \gamma_j$, $\|B_j^{-1} C_j\| = \delta_j$, we find

(11.12) $$\|E_j\| \leq \frac{\gamma_j}{1 - \delta_j \|E_{j-1}\|}.$$

provided that $\delta_j \|E_{j-1}\| < 1$.

This suggests that we consider the sequence $\xi_0, \xi_1, \ldots$ given by

(11.13) $$\xi_0 = \|E_0\| = \|H\|,$$
$$\xi_j = \gamma_j / (1 - \delta_j \xi_{j-1}),$$

for then it follows by induction that if

(11.14) $$\delta_j \xi_{j-1} < 1, \quad j = 1, 2, \ldots,$$

then

(11.15) $$\|E_j\| \leq \xi_j, \quad j = 0, 1, \ldots.$$

We are therefore interested in finding conditions on ξ_0 and the sequences (γ_j), (δ_j) such that (1) the sequence (ξ_j) is bounded, (2) $\delta_j \xi_{j-1} < 1$ for all j.

---

[2] To see this, note first that $I - M$ has an inverse, because if $I - M$ were singular, we would have $\det(I - M) = 0$, and an eigenvalue of M would be 1, which contradicts $\|M\| < 1$. Second, if $v = (I - M)w$,

$$\text{Max} \frac{|(I - M)^{-1} v|}{|v|} = \text{Max} \frac{|w|}{|(I - M)w|} \leq \frac{|w|}{|w| - |Mw|}$$
$$\leq \frac{|w|}{|w| - \|M\| \cdot |w|} = \frac{1}{1 - \|M\|},$$

the next to the last step because $|Mw| \leq \|M\| \cdot |w| < |w|$.

SEC. 11.5] SOLUTION OF IMPLICIT EQUATIONS

We shall consider the simple case in which γ_j and δ_j are independent of j, as they are for problems of differential equations with constant coefficients. We write $\gamma_j = \gamma$, $\delta_j = \delta$, and note that $\gamma > 0$, $\delta > 0$.

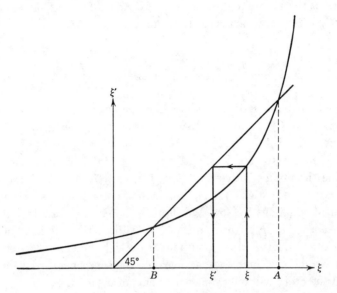

Fig. 11.2. Construction for investigating the boundedness of the matrices E_j given by equations (11.11) (see equation (11.13)).

In Figure 11.2 is plotted a graph of the quantity

$$\xi' = \gamma/(1 - \delta\xi)$$

against ξ. The transformation

$$\gamma/(1 - \delta\xi) \to \xi$$

is effected by following the arrows in the figure. If, as assumed in drawing the figure, the curve of ξ' vs. ξ intersects the 45° line, then any value of ξ to the left of A in the figure is moved closer to B by the transformation. The condition for intersection is $\gamma\delta < 1/4$, and then $A = (1 + \sqrt{1 - 4\gamma\delta})/2\delta$. Clearly, if $\xi < A$, $\delta\xi < 1$. Therefore, all conditions are satisfied if

(11.16)
$$\gamma\delta < 1/4,$$
$$\xi_0 < (1 + \sqrt{1 - 4\gamma\delta})/2\delta.$$

Taking matrices A, B, and C, as given above for the problem of the uniform, simply-supported bar, conditions (11.16) are satisfied. There $A = C$, and therefore $\gamma = \delta$; specifically, $\gamma = \delta = \|B^{-1}A\|$. The bound of $B^{-1}A$ is the square root of the absolute value of the largest eigenvalue of $(B^{-1}A)^*(B^{-1}A)$ which is easily found to be

$$(11.17) \qquad \gamma = \delta = \frac{1}{2} \frac{a\Delta t/(\Delta x)^2}{\sqrt{1 + [a\Delta t/(\Delta x)^2]^2}} < \frac{1}{2};$$

furthermore, the boundary condition for simple support is $0 = H = E_0$, therefore $\xi_0 = 0$. It follows that *the matrices E_j remain bounded as j increases indefinitely.*

It is possible to handle slightly more general boundary conditions. Let p be the number of components of **u**, so that A_j, B_j, C_j are $p \times p$ matrices. In equations (11.7), the number of unknowns exceeds the number of equations by $2p$. We suppose that half the additional equations needed (i.e., p of them) are linear conditions at the left boundary and the other half are linear conditions at the right boundary. By this we mean that half the conditions involve only the components of $\mathbf{u}_0, \mathbf{u}_1, \ldots, \mathbf{u}_{j_0}$ and the other half only the components of $\mathbf{u}_J, \mathbf{u}_{J-1}, \ldots, \mathbf{u}_{J-j_0}$, where j_0 is a (generally small) integer which remains fixed as $\Delta x \to 0$ and hence as $J \to \infty$, $\Delta x = L/J$. Then in the limit, these conditions become boundary conditions in the sense that they involve only values of the functions and certain derivatives at $x = 0$ or at $x = L$. We suppose $J > 2j_0$.

The boundary conditions are supposed independent; the equations (11.7) together with the left boundary conditions have a general solution depending linearly on p parameters, which we call $\lambda_1, \ldots, \lambda_p$. In particular, the components of $\mathbf{u}_j$ for any j, that is, the quantities u_{ji} ($i = 1, 2, \ldots, p$) are linear functions of these parameters. Call U_j the matrix given by

$$(U_j)_{ik} = \frac{\partial u_{ji}}{\partial \lambda_k},$$

so that

$$(11.18) \qquad \mathbf{u}_j = U_j \boldsymbol{\lambda} + \mathbf{g}_j, \quad j = 0, 1, \ldots, J,$$

SEC. 11.5] SOLUTION OF IMPLICIT EQUATIONS

where $\boldsymbol{\lambda}$ is the vector

$$\boldsymbol{\lambda} = \begin{bmatrix} \lambda_1 \\ \vdots \\ \lambda_p \end{bmatrix}$$

and $\mathbf{g}_j$ is a vector whose components do not depend on the parameters.

To see the interpretation of this formalism in special cases, note that if the left boundary condition is of the form (11.8), as assumed in the preceding discussion, we may take the parameters $\lambda_1, \ldots, \lambda_p$ as the components of $\mathbf{u}_1$, so that U_1 is the unit matrix, which is (among other things) non-singular. However, the boundary condition for a clamped bar, which is $y = 0$, $\partial y/\partial x = 0$ at the boundary, we may suppose translated into finite-difference terms as $v_0 = v_1 = 0$, so that w_0 and w_1 may be taken as the parameters. In this case, the first three of the matrices U_j are

$$U_0 = \begin{bmatrix} 0 & 0 \\ 1 & 0 \end{bmatrix}, \quad U_1 = \begin{bmatrix} 0 & 0 \\ 0 & 1 \end{bmatrix}, \quad U_2 = \begin{bmatrix} 0 & 2(\Delta x)^2/a\Delta t \\ -1 & 2 \end{bmatrix},$$

of which the first non-singular one is U_2. (To obtain U_2 one uses equation (11.7) with $j = 1$.) The boundary condition for the free end of a bar, i.e. the condition of no force and no turning moment, or $\partial^2 y/\partial x^2 = 0$, $\partial^3 y/\partial x^3 = 0$, takes the form $w_0 = w_1 = 0$. The result is the same as above, except for interchange of the roles of v and w.

Now let U_{j_1} be the first non-singular matrix of the sequence $U_1, U_2, \ldots$ (note that U_0 is not included). Then

$$\boldsymbol{\lambda} = U_{j_1}^{-1}(\mathbf{u}_{j_1} - \mathbf{g}_{j_1})$$

$$\therefore \mathbf{u}_{j_1-1} = U_{j_1-1}U_{j_1}^{-1}(\mathbf{u}_{j_1} - \mathbf{g}_{j_1}) + \mathbf{g}_{j_1-1}.$$

This may be written as

(11.19) $$\mathbf{u}_{j_1-1} = H\mathbf{u}_{j_1} + \mathbf{l}$$

and is therefore of the same form as (11.8) except that it applies to the point $j = j_1$ instead of $j = 1$. The inductive calculation of E_j and $\mathbf{f}_j$ is as before but starts with $j = j_1$. The right-hand boundary condition can similarly be given the same form as (11.9) at say $j = J - j_2$, and the descending inductive calculation of the $\mathbf{u}_j$ started at $j = J - j_2$. This leaves a few of the $\mathbf{u}_j$ to be calculated near each boundary—they can be obtained from (11.7).

In practice, H, $\mathbf{l}$, M, $\mathbf{n}$ are usually quite simple expressions in terms of the primary boundary conditions and perhaps one or two of the coefficients of (11.7). It is probably usually best to obtain these expressions by hand and code them directly into the machine program.

Another observation is that in practice one will not in general have the idealized situation in which γ_j and δ_j are independent of j; instead, the points A and B of Figure 11.2 may shift slightly from one iteration of the transformation

$$(1 - \delta_j \xi)/\gamma_j \to \xi$$

to the next. But the point B is a point of positive stability in the sense that any value of ξ to the left of A is moved definitely closer to B with each iteration, and it is intuitively clear that if the initial value $\xi = \xi_0$ is well to the left of point A, and if the shift of the points A and B is sufficiently small for each iteration, and if the points A and B remain well separated, ξ will always remain to the left of A. Whether these conditions are fulfilled in a given case may be difficult to determine in advance by analytic methods, and one may have to rely, in running a problem, either on the overflow detection circuits of the machine or on a programmed sub-routine which would automatically take appropriate action if the matrix E_j gets out of scale or the matrix $B_j - C_j E_{j-1}$ threatens to become singular.

11.6. Vibration of a Bar Under Tension

The introduction of a tension term into the preceding discussion introduces two new points worthy of note.

If T denotes the tension, the equation of motion is

$$\mu \frac{\partial^2 y}{\partial t^2} = -\varkappa \frac{\partial^4 y}{\partial x^4} + T \frac{\partial^2 y}{\partial x^2},$$

and in place of (11.3) we write

(11.20)
$$\frac{\partial v}{\partial t} = -a \frac{\partial^2 w}{\partial x^2} + b \frac{\partial w}{\partial x},$$
$$\frac{\partial w}{\partial t} = +a \frac{\partial^2 v}{\partial x^2} + b \frac{\partial v}{\partial x},$$

SEC. 11.6] VIBRATION OF A BAR UNDER TENSION

where $v = \partial y/\partial t$ and $w = a\partial^2 y/\partial x^2 + b\partial y/\partial x$, and where $a = \sqrt{\varkappa/\mu}$, $b = \sqrt{T/\mu}$.

The implicit difference system

$$\frac{v_j^{n+1} - v_j^n}{\Delta t} = -a \frac{(\delta^2 w)_j^{n+1} + (\delta^2 w)_j^n}{2(\Delta x)^2} + b \frac{w_{j+1}^{n+1} - w_{j-1}^{n+1} + w_{j+1}^n - w_{j-1}^n}{4\Delta x},$$

(11.21)

$$\frac{w_j^{n+1} - w_j^n}{\Delta t} = +a \frac{(\delta^2 v)_j^{n+1} + (\delta^2 v)_j^n}{2(\Delta x)^2} + b \frac{v_{j+1}^{n+1} - v_{j-1}^{n+1} + v_{j+1}^n - v_{j-1}^n}{4\Delta x},$$

analogous to (11.6), is unconditionally stable.

However, in this case there may be a difficulty with the algorithm given in the preceding section for solving the implicit equations. The matrices A_j, B_j, and C_j of equation (11.7) are modified by additional terms in such a way that we now have

$$\gamma = \delta = \frac{1}{2} \frac{a\Delta t/(\Delta x)^2 + b\Delta t/2\Delta x}{\sqrt{1 + [a\Delta t/(\Delta x)^2]^2}}$$

in place of (11.17), and the condition $\gamma\delta < 1/4$ (see (11.16)) may or may not be fulfilled, depending on the values of a, b, Δt, Δx. If $\gamma\delta > 1/4$, there is no intersection in Figure 11.2, the sequence ξ_j is unbounded, and the matrices E_j are likely to get out of scale. In this case *the stability of the difference equations has been achieved at the expense of the stability of this particular algorithm for solving them.*

This difficulty is most severe if the vibration is governed mainly by tension, rather than stiffness, as in the case of a piano wire, because then $b = \sqrt{T/\mu}$ is large while $a = \sqrt{\varkappa/\mu}$ is small. Then the problem is qualitatively similar to that of the vibrating string, for which the equation of motion is the one-dimensional equation. This suggests that we consider explicit difference systems similar to those that have been found useful for the wave equation. One of these systems is

(11.22)

$$\frac{v_j^{n+1} - v_j^n}{\Delta t} = -a \frac{(\delta^2 w)_j^n}{(\Delta x)^2} + b \frac{w_{j+1}^n - w_{j-1}^n}{2\Delta x},$$

$$\frac{w_j^{n+1} - w_j^n}{\Delta t} = a \frac{(\delta^2 v)_j^{n+1}}{(\Delta x)^2} + b \frac{v_{j+1}^{n+1} - v_{j-1}^{n+1}}{2\Delta x}.$$

It has a forward time difference in the first equation and a backward time difference in the second one, and is analogous to system (10.5) for the wave equation and system (11.4) for the vibrating bar without tension.

The stability condition for the system (11.22) is readily found to be

(11.23) $$a\Delta t/(\Delta x)^2 < \tfrac{1}{2},$$

and this not very serious in practice; because of the small value of a, Δt is not severely restricted by this inequality for any reasonable value of Δx. In fact there is a pitfall here, in that Δt may not be restricted enough by (11.23), which shows a basic weakness of the concept of stability, similar to the weakness mentioned in Section 9.7 on the transport equation and in Section 10.4 on coupled sound and heat flow. To see the pitfall, note that if the terms in a were absent completely from (11.20), we should have for (11.22) the stability condition

(11.24) $$b\Delta t/2\Delta x < 1.$$

In the limit as $\Delta t \to 0$, $\Delta x \to 0$, condition (11.23) of course implies (11.24), no matter what values a and b have, if $a \neq 0$. But if the vibration is controlled primarily by tension rather than stiffness, so that b is large and a small, the actual point net used in the calculation may be quite fine and satisfy (11.23) but nevertheless fail to satisfy (11.24) by a wide margin. It is intuitively clear that then very violent oscillations may develop in the solution, just as when the stability condition for the wave equation is violated. Just as for the paradox mentioned in Chapter 9, the programmer wants to know not merely what happens in the limit $\Delta t, \Delta x \to 0$, but what happens for the point net he is actually using. Stability in the limit is not enough.

To get a practical criterion for the finite difference scheme (11.22), we appeal (as we did in Chapters 9 and 10) to the principle that the rate of growth of the amplitude of any Fourier component should be not appreciably greater than the maximum rate of growth expected (and presumed roughly known) in the true solution. Now for the simply supported bar, under tension or not, there is a conservation-of-energy principle according to which the amplitude of each component remains constant in time. Let us therefore try to require that the amplitudes do not grow in time for the finite-difference equations either. (This is not

SEC. 11.6]   VIBRATION OF A BAR UNDER TENSION

always possible, even for conservative systems, but happens to be possible here.) That is, whereas the von Neumann condition in general permits $\lambda \leq 1 + O(\Delta t)$, we shall here require $\lambda \leq 1$.

For the system (11.22), the amplification matrix is

$$G = G(\Delta t, k) = \begin{bmatrix} 1 & \beta \\ -\bar{\beta} & 1-|\beta|^2 \end{bmatrix},$$

where β is an abbreviation for

$$\beta = \frac{4a\Delta t}{(\Delta x)^2} \sin^2 \frac{k\Delta x}{2} + i\frac{b\Delta t}{\Delta x} \sin k\Delta x.$$

The characteristic equation for G is

$$\lambda^2 - \lambda(2 - |\beta|^2) + 1 = 0.$$

We wish the roots of this equation to lie in or on the unit circle, hence we must have $|\beta|^2 \leq 4$, not merely $|\beta|^2 \leq 4 + O(\Delta t)$, as previously. We therefore require

(11.25) $$\left(\frac{4a\Delta t}{(\Delta x)^2}\right)^2 \sin^4 \frac{k\Delta x}{2} + \left(\frac{b\Delta t}{\Delta x}\right)^2 \sin^2 k\Delta x \leq 4$$

for all real k. Observing that the left member of this inequality is majorized if we replace both sine functions by 1, we obtain as a simple sufficient stability condition for (11.22), which is good enough for most practical purposes,

(11.26) $$\left(\frac{2a\Delta t}{(\Delta x)^2}\right)^2 + \left(\frac{b\Delta t}{2\Delta x}\right)^2 < 1;$$

we note that this implies both (11.23) and (11.24).

It is interesting to compare this with the stability conditions obtainable for this system from the energy methods of Chapter 6, for this is one of the simplest examples in which the energy method fails to give, in a practical way, the whole stability range as determined by Fourier methods. If we denote $2a\Delta t/(\Delta x)^2$ by μ and $b\Delta t/2\Delta x$ by ν, equations (11.22) may be written in the form

$$v^{n+1} - v^n = -\tfrac{1}{2}\mu\delta^2 w^n + 2\nu\Delta_0 w^n$$

$$w^{n+1} - w^n = \tfrac{1}{2}\mu\delta^2 v^{n+1} + 2\nu\Delta_0 v^{n+1},$$

where $\Delta_0 u_j = \frac{1}{2}(u_{j+1} - u_{j-1})$. Multiplying the first by $v^{n+1} + v^n$, the second by $w^{n+1} + w^n$, adding them together, and summing over all the mesh points yields

$$\|v^{n+1}\|^2 + \|w^{n+1}\|^2 - \|v^n\|^2 - \|w^n\|^2$$
$$= \tfrac{1}{2}\mu[(w^{n+1} + w^n, \delta^2 v^{n+1}) - (v^{n+1} + v^n, \delta^2 w^n)]$$
$$+ 2\nu[(w^{n+1} + w^n, \Delta_0 v^{n+1}) + (v^{n+1} + v^n, \Delta_0 w^n)].$$

By the summation-by-parts formulas (6.17) to (6.19), $(w^{n+1} + w^n, \delta^2 v^{n+1}) = -(\Delta_+(w^{n+1} + w^n), \Delta_+ v^{n+1})$ and similarly for the second term on the right also, $(w^{n+1} + w^n, \Delta_0 v^{n+1}) = -(\Delta_0(w^{n+1} + w^n), v^{n+1})$. Collecting terms together we then find that the equation above reduces to $S_{n+1} = S_n$, where

$$S_n = \|v^n\|^2 + \|w^n\|^2 + \tfrac{1}{2}\mu(\Delta_+ v^n, \Delta_+ w^n) + 2\nu(v^n, \Delta_0 w^n).$$

Hence, stability follows if $S_n \geq$ const. $(\|v^n\|^2 + \|w^n\|^2)$. The crudest treatment of the inner products in S_n yields

$$|(\Delta_+ v^n, \Delta_+ w^n)| \leq 2(\|v^n\|^2 + \|w^n\|^2)$$

and

$$|(v^n, \Delta_0 w^n)| \leq \tfrac{1}{2}(\|v^n\|^2 + \|w^n\|^2),$$

resulting in the sufficient stability condition

(11.27) $$\mu + \nu < 1.$$

This is to be compared with $\mu^2 + \nu^2 \leq 1$ from (11.26)—see Figure 11.3. However, by a more careful treatment of the inner products above, we have, after putting $(\Delta_+ v^n, \Delta_+ w^n) = -(v^n, \delta^2 w^n)$,

$$|-\tfrac{1}{2}\mu(v^n, \delta^2 w^n) + 2\nu(v^n, \Delta_0 w^n)|$$
$$= \left|\sum_{(j)} v_j^n[(\nu - \tfrac{1}{2}\mu)w_{j+1}^n + \mu w_j^n - (\nu + \tfrac{1}{2}\mu)w_{j-1}]\right|$$
$$\leq \tfrac{1}{2}[|\nu - \tfrac{1}{2}\mu| + \mu + (\nu + \tfrac{1}{2}\mu)](\|v^n\|^2 + \|w^n\|^2).$$

The resulting stability condition, which appears to be the best practically obtainable by the energy method,[3] is

(11.28) $$\tfrac{1}{2}\mu + \mathrm{Max}(\tfrac{1}{2}\mu, \nu) < 1.$$

---

[3] The bounds given above cannot be attained for any mesh functions v_j^n and w_j^n when $\nu > \tfrac{1}{2}\mu$. But to take advantage of this and so weaken (11.28), one needs to consider the cancellation of cross-product terms arising from successive mesh points. This leads naturally back to a Fourier analysis, by which means we know by Kreiss' matrix theorem (see Chapter 4) that the full stability range can be found.

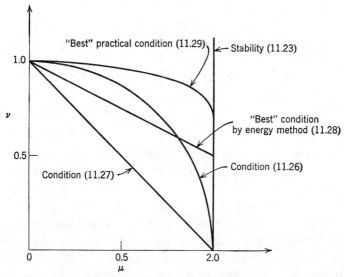

Fig. 11.3. Stability conditions for the difference scheme (11.22) as obtained by different methods.

One can see from Figure 11.3 that this is sometimes less restrictive than even (11.26). This is because (11.26) is not the best result obtainable from (11.25): if we put $z = \sin^2 \tfrac{1}{2}k\Delta x$, (11.25) becomes

$$\mu^2 z^2 + 4\nu^2(z - z^2) \leq 1, \quad \text{for } 0 \leq z \leq 1.$$

The maximum of the left number is attained at $z = 1$ for $2\nu^2 \leq \mu^2$ and then equals μ^2; otherwise, the maximum equals $4\nu^4/(4\nu^2 - \mu^2)$. Thus, *the necessary and sufficient condition that the scheme* (11.22) *not only be stable but also that no Fourier component grow is*

(11.29) $\quad\quad\begin{aligned}&\text{if } \mu^2 \geq 2\nu^2, \quad \mu < 1; \\ &\text{otherwise,} \quad (\mu/2\nu)^2 + \nu^2 \leq 1.\end{aligned}$

Thus, either the full range of μ is allowable or it is increased by a factor of 2ν compared with (11.26).

CHAPTER 12

Fluid Dynamics in One Space Variable

12.1. Introduction

Finite difference methods for the calculation of time-dependent flows have mostly been based on either the Eulerian or the Lagrangean form of the equations.[1] These are equivalent, except that the Lagrangean form gives more information: it tells where each bit of fluid came from originally; this considerably facilitates calculations of flows when two or more fluids, with different equations of state, are present. Furthermore, the only stable Eulerian difference schemes that were known until fairly recently were less accurate than the Lagrangean ones, which were therefore preferred. This was subsequently changed by the development of the Lax-Wendroff second-order scheme, which is especially effective for the Eulerian formulation.

Fluid flows are frequently characterized by internal discontinuities, such as shocks, on which special boundary conditions (jump conditions) are required. These conditions are provided by the well-known Rankine-Hugoniot equations, but their application in practice is encumbered by the difficulties that the surfaces on which the conditions are to be applied are in motion through the fluid and that their motion is not known in advance but is determined by the equations themselves (the differential equations plus the jump conditions) so that in numerical work a highly implicit situation prevails. We shall start by describing the differential equations which apply to the smooth part of the flow, as for problems without

---

[1] In the Eulerian form, the independent space variables refer to a coordinate system fixed in space and through which the fluid is thought of as moving, the flow being characterized by a time-dependent velocity field which is to be found by solving the initial value problem; in the Lagrangean form, the independent space variables refer to a coordinate system fixed in the fluid and undergoing all the motion and distortion of the fluid, so that particles of the fluid are permanently identified by their Lagrangean variables, while their actual positions in space are among the dependent quantities to be solved for. For further details see Courant and Friedrichs (1948), or Bethe *et al.* (1944).

12.2. The Eulerian Equations

Let $\mathbf{X} = (X, Y, Z)$ denote the position vector in a cartesian coordinate system. The properties of the fluid medium are characterized by the density $\rho = \rho(\mathbf{X}, t)$, the pressure $p = p(\mathbf{X}, t)$, the internal energy per gram $\mathscr{E} = \mathscr{E}(\mathbf{X}, t)$, and the fluid velocity $\mathbf{u} = \mathbf{u}(\mathbf{X}, t)$. We assume given an equation of state,

$$\text{(12.1)} \qquad \mathscr{E} = F(p, \rho).$$

We also assume that viscous forces, body forces (like gravity), heat conduction, and energy sources are absent, although body forces and energy sources could be included without essential modification. Later we shall introduce certain artificial pseudo-viscous forces in connection with shocks, and we shall leave the door open for heat conduction by leaving the energy equation in the system rather than eliminating it by use of the constancy of entropy along the world lines of fluid particles.

The conservation of matter is expressed by the following equation (sometimes called the "equation of continuity"):[2]

$$\text{(12.2)} \qquad \left(\frac{\partial}{\partial t} + \mathbf{u} \cdot \nabla\right)\rho = -\rho \nabla \cdot \mathbf{u};$$

the equation of motion is

$$\text{(12.3)} \qquad \rho\left(\frac{\partial}{\partial t} + \mathbf{u} \cdot \nabla\right)\mathbf{u} = -\nabla p,$$

and the equation of energy is

$$\text{(12.4)} \qquad \rho\left(\frac{\partial}{\partial t} + \mathbf{u} \cdot \nabla\right)\mathscr{E} = -p \nabla \cdot \mathbf{u}.$$

---

[2] For derivation of equations such as these and a general discussion of fluid-dynamical phenomena, see Courant and Friedrichs (1948).

For slab symmetry, where nothing depends on Y or Z, we have

(12.5)
$$\left(\frac{\partial}{\partial t} + u\frac{\partial}{\partial X}\right)\rho = -\rho\frac{\partial u}{\partial X},$$
$$\rho\left(\frac{\partial}{\partial t} + u\frac{\partial}{\partial X}\right)u = -\frac{\partial p}{\partial X},$$
$$\rho\left(\frac{\partial}{\partial t} + u\frac{\partial}{\partial X}\right)\mathscr{E} = -p\frac{\partial u}{\partial X},$$

(in place of (12.2), (12.3), (12.4)) together with the equation of state, (12.1).

For spherical symmetry, the equations are slightly different, and it is convenient to combine the two systems by writing

(12.6)
$$\left(\frac{\partial}{\partial t} + u\frac{\partial}{\partial R}\right)\rho = -\rho\frac{1}{R^{\alpha-1}}\frac{\partial}{\partial R}R^{\alpha-1}u,$$
$$\rho\left(\frac{\partial}{\partial t} + u\frac{\partial}{\partial R}\right)u = -\frac{\partial p}{\partial R},$$
$$\rho\left(\frac{\partial}{\partial t} + u\frac{\partial}{\partial R}\right)\mathscr{E} = -p\frac{1}{R^{\alpha-1}}\frac{\partial}{\partial R}R^{\alpha-1}u,$$

where, for spherical symmetry, $R = \sqrt{X^2 + Y^2 + Z^2}$, $\alpha = 3$, while for slab symmetry, $R = X$, $\alpha = 1$. The case of cylindrical symmetry (in the sense of a flow invariant both under translations parallel to an axis and under rotations about the axis) is given by $R = \sqrt{X^2 + Y^2}$, $\alpha = 2$.

12.3. Difference Equations, Eulerian

Finite-difference equations may be constructed in various ways; their choice depends on many factors, including stability, accuracy, and the reflection of important physical considerations, e.g., conservation laws. The method of Courant, Isaacson, and Rees[3] (1952) should be a very good one, but has not been much used, presumably because of the necessity of reduction of the equations to normal, or characteristic form. In this method, time differences are forward, and each equation of the normal

---

[3] See Chapter 9 for an application of this method to the transport equation, where it is known as the method of "forward and backward space differences."

form is given a forward or backward space difference according to the sign of the slope of the corresponding characteristic. Stability and convergence have been proved, in a suitably restricted domain, provided $\Delta t/\Delta X$ nowhere exceeds the slope of any characteristic, which would give the stability condition

$$(|u| + c) \Delta t/\Delta X \leq 1,$$

where $c = c(X, t)$ is the local adiabatic sound speed (a more accurate adjective would be "isentropic"). For isentropic variations, we have $d\mathscr{E} + p\,d(1/\rho) = 0$ or, using (12.1),

$$\frac{\partial F}{\partial p} dp + \frac{\partial F}{\partial \rho} d\rho - \frac{p}{\rho^2} d\rho = 0.$$

The sound speed c is the square root of $dp/d\rho$ along an adiabat; therefore,

$$(12.7) \qquad c^2 = \frac{p/\rho^2 - \partial F/\partial \rho}{\partial F/\partial p}.$$

The procedure of Courant, Isaacson, and Rees may be interpreted as follows: Along each characteristic, a certain quantity is almost constant (exactly constant in the isentropic case; it is then the corresponding Riemann invariant). To obtain the quantity at net-point $n + 1, j$, one follows the characteristic backward from this point until it intersects the line $t = n\Delta t$; one then obtains the quantity by linear interpolation between n, j and $n, j + 1$ or between $n, j - 1$ and n, j. Ansorge (1963) has constructed higher-order methods of this sort for semi-linear hyperbolic systems by analogy with the familiar Adams procedure for ordinary differential equations. Multi-dimensional generalizations have been devised and studied by Bruhn and Haack (1958) and by Törnig (1963). A similar multi-dimensional method using co-characteristics has been devised and studied by Sauer. These multi-dimensional methods will be described briefly in Chapter 13.

Lax and Keller (1951) proposed a scheme using a staggered net of points in the X, t-plane. This scheme is only of first order accuracy, but it has been useful for proofs of existence and other properties of solutions of the differential equations. An ingenious method similar to that of Lax and Keller, but designed to handle discontinuities as well as smooth flow, was devised by Godunov (1959); it will be described briefly in Section 12.15.

The most commonly used schemes (for both the Lagrangean and the Eulerian equations) have been patterned after the scheme (10.5) for the wave equation (the original scheme of Courant, Friedrichs, and Lewy). The schemes have in common that the middle equation of (12.5) or of (12.6) is used first to advance u from time t to $t + \Delta t$,[4] and then the advanced value of u is used in the other two equations.

For the Eulerian equations, such schemes are unstable unless care is used in the manner of treating the terms containing $u\partial/\partial X$.[5] For example, it is not satisfactory to approximate $u\partial\rho/\partial X$ by

$$u_j^{n+1} \frac{\rho_{j+1}^n - \rho_{j-1}^n}{2\Delta X}$$

and the other terms of this type similarly. This is easily seen by writing down the equations of first variation of the difference equations and computing their amplification matrix for the limit $\Delta t = \Delta X = 0$, i.e., the matrix $G(0, k)$. One of the eigenvalues of this matrix is $1 + (u\Delta t/\Delta X) 2i \sin k\Delta X$; the maximum magnitude of this is $(1 + (u\Delta t/\Delta X)^2)^{1/2}$ which exceeds 1 (unless $u \equiv 0$), so the von Neumann condition cannot be satisfied for any fixed $\Delta t/\Delta X$. Lelevier's remedy is to use a forward or backward space difference in such terms according to whether $u < 0$ or $u > 0$, that is, to replace the above expression by

$$u_j^{n+1} \frac{\rho_{j+1}^n - \rho_j^n}{\Delta X} \quad \text{if } u_j^{n+1} < 0$$

$$u_j^{n+1} \frac{\rho_j^n - \rho_{j-1}^n}{\Delta X} \quad \text{if } u_j^{n+1} \geqq 0$$

and similarly for the other terms of this type in equations (12.5) or (12.6). In a first sweep through the net, the second of the equations gives u_j^{n+1} for all j; these values, which are appropriate to time $t + \frac{1}{2}\Delta t$,[4] are then used in the first and third equation, to advance ρ and $\mathscr{E}$ from time t to time $t + \Delta t$.

The resulting scheme (we omit details) is stable if $(|u| + c)\Delta t/\Delta X < 1$.

---

[4] From the point of view of centering, the old and new values of u should really be referred to times $t - \frac{1}{2}\Delta t$ and $t + \frac{1}{2}\Delta t$, respectively, whereas other quantities are referred to times t and $t + \Delta t$, but we prefer to retain our present notation.

[5] R. Lelevier, private communication (1953).

THE LAGRANGEAN EQUATIONS

However, the scheme has only first-order accuracy, which is inadequate for most applications. The Lax-Wendroff equations, which are now preferred because of their second-order accuracy, are described in Section 12.7.

12.4. The Lagrangean Equations

For slab symmetry, let x denote the X-coordinate of a small fluid element at time $t = 0$ and let $X = X(x, t)$ denote the X-coordinate of the same fluid element at time t, so that x and t are regarded as independent variables. Each bit of fluid is labeled with a certain value of x called the Lagrangean coordinate and retains that label as it moves about. Clearly the label x can be chosen in many other ways than the one given above. For example, it could be defined as the X-coordinate at some other fixed time $t = t_0$. Furthermore, if $g(x)$ is any function with a continuous positive derivative, the variable $x' = g(x)$ would also be suitable as a label; in other words, if $h(x)$ is continuous, > 0, we may take

$$x' = \int h(x)dx,$$

in place of x, as Lagrangean coordinate. A particular choice which is convenient, and will be used henceforth is

$$x' = \frac{1}{\rho_0} \int \rho(x)dx$$

where $\rho(x)$ is the density at $t = 0$, and ρ_0 is a constant having the dimensions of a density. x' can be interpreted as the X-coordinate in a *reference configuration* of the system in which the various layers of the fluid have been shifted and expanded or contracted until they are all at the common density ρ_0. This we shall call the *standard reference configuration*. Henceforth we drop the prime, and write just x for x'. Similarly, for the sphere, let r denote the radial coordinate of a fluid element in the standard reference configuration in which the various spherical shells of fluid have been moved in or out until they all have density ρ_0 and let $R = R(r, t)$ be the radial coordinate of that same fluid element at time t in the motion. r and R can be similarly defined for cylindrical symmetry (by which we again mean invariance under both rotations about and translations along an axis).

As for the Eulerian equations, we combine the formulas by writing $\alpha = 1, 2, 3$ for slab, cylindrical, or spherical symmetry, and letting r and R stand for x and X in the case of slab symmetry. We have the three principal equations

(12.8)
$$\frac{\partial R}{\partial t} = u,$$
$$\frac{\partial u}{\partial t} = -V_0\left(\frac{R(r,t)}{r}\right)^{\alpha-1}\frac{\partial p}{\partial r},$$
$$\frac{\partial \mathscr{E}}{\partial t} = -p\frac{\partial V}{\partial t},$$

where $V_0 = 1/\rho_0$, and the two auxiliary equations

(12.9)
$$V = V_0\left(\frac{R(r,t)}{r}\right)^{\alpha-1}\frac{\partial R}{\partial r},$$
$$\mathscr{E} = F_1(p, V).$$

Here, V is the specific volume $1/\rho$; u, p, and $\mathscr{E}$ are again the fluid velocity, the pressure, and the internal energy per unit mass, as for the Eulerian equations, but are now to be thought of as functions of r and t.

We note that for the stability analysis (given below) the primary dependent variables are to be thought of as V, u, and p (or $\mathscr{E}$); R is merely a subsidiary variable, from which u and V are to be obtained via the first equation of (12.8) and the first equation of (12.9). To see this, note that for small deviations from a constant state (sound waves), in the case of slab symmetry ($\alpha = 1$), the system reduces to

$$\frac{\partial R}{\partial t} = u, \quad \frac{\partial u}{\partial t} = (cV_0/V)^2\frac{\partial^2 R}{\partial r^2},$$

which is precisely of the form we advised against at the beginning of Section 10.2 of Chapter 10; hence, if R and u were used to construct a Banach-space norm, as described in that section, the problem would not be properly posed. This means in practice that the numerical values of R have to be retained with more relative accuracy (more decimal or

SEC. 12.5] DIFFERENCE EQUATIONS, LAGRANGEAN

binary places) than the other variables, since it is only the differences of R (with respect to time and space) that have primary physical significance.

It is sometimes more convenient (as also for the Eulerian equations) to have the equation of state in a form solved for p, namely, $p = f(\mathscr{E}, V)$. This equation and the first equation (12.9) are then to be thought of as abbreviations, and their substitution into (12.8) gives us three partial differential equations for R, u, and $\mathscr{E}$.

12.5. Difference Equations, Lagrangean

As for the Eulerian formulation, the system is hyperbolic, and the difference equations might be taken as those of Courant, Isaacson, and Rees (1952) or those of Lax and Keller (1951), but in practice they usually have been patterned after system (10.5), for the wave equation, or have been of the Lax-Wendroff type. Implicit schemes are seldom justifiable on the ground of increased stability, because for most problems in fluid dynamics the system changes significantly in a time $\Delta t = \Delta x/c$, so that there is no motivation for using longer time intervals. However, they have been found useful for certain astrophysical problems, in which the expected motion is a slow succession of near-equilibrium states, in problems where one part of the system has a much larger sound speed than the other parts (hence, is always nearly in equilibrium), and in applications to meteorology, where the important effects are propagated by mechanisms considerably slower than acoustic waves. See Turner and Wendroff (1964).

The usual system is

(12.10)
$$\frac{R_j^{n+1} - R_j^n}{\Delta t} = u_j^{n+1},$$

$$\frac{u_j^{n+1} - u_j^n}{\Delta t} = -V_0 \left(\frac{R_j^n}{r_j}\right)^{\alpha-1} \frac{p_{j+\frac{1}{2}}^n - p_{j-\frac{1}{2}}^n}{\Delta r},$$

$$\mathscr{E}_{j+\frac{1}{2}}^{n+1} - \mathscr{E}_{j+\frac{1}{2}}^n = -p_{j+\frac{1}{2}}^n (V_{j+\frac{1}{2}}^{n+1} - V_{j+\frac{1}{2}}^n),$$

$$V_{j+\frac{1}{2}}^{n+1} = V_0 \frac{(R_{j+1}^{n+1})^\alpha - (R_j^{n+1})^\alpha}{(r_{j+1})^\alpha - (r_j)^\alpha},$$

$$p_{j+\frac{1}{2}}^{n+1} = f(\mathscr{E}_{j+\frac{1}{2}}^{n+1}, V_{j+\frac{1}{2}}^{n+1}).$$

We have used r_j to denote the jth net point of the Lagrangean net. If we wish to make the system as a whole correct to $O[(\Delta t)^2] + O[(\Delta r)^2]$, the third equation must be centered by replacing $p_{j+\frac{1}{2}}^n$ by some better approximation to p at $t = t^{n+\frac{1}{2}}$; this can be done in various ways, but at the expense of having two unknowns, $p_{j+\frac{1}{2}}^{n+1}$ and $\mathscr{E}_{j+\frac{1}{2}}^{n+1}$, appear in the third equation; then an iterative procedure is generally required to solve this equation together with the fifth one. The other equations of (12.10) are properly centered in fact, though not in appearance, because u_j^{n+1} and u_j^n are really to be associated with times $t^{n+\frac{1}{2}}$ and $t^{n-\frac{1}{2}}$.

Typical boundary conditions at point $j = J$ are:

rigid wall, $u_J^n = 0$ for all n;

free surface, $p_{J+\frac{1}{2}}^n = -p_{J-\frac{1}{2}}^n$ for all n.

The second of these has the effect of making the interpolated value of p vanish at $j = J$. Boundary conditions can also be applied at the midpoint of an interval, for example, at $j = J + \frac{1}{2}$, as follows:

rigid wall, $u_{J+1}^n = -u_J^n$ for all n;

free surface, $p_{J+\frac{1}{2}}^n = 0$ for all n.

Special treatment of interfaces will be discussed below.

The system (12.10) is effectively explicit, provided the quantities are solved for in the proper order. If all quantities are known at $t = t^n$, and if there are boundaries at $j = 0$ and $j = J$, the order might be:

	Quantities	Equation
u_j^{n+1}	$(j = 1, 2, \ldots, J-1)$	second of (12.10)
u_0^{n+1}, u_J^{n+1}		boundary conditions
R_j^{n+1}	$(j = 0, 1, \ldots, J)$	first of (12.10)
$V_{j+\frac{1}{2}}^{n+1}$	$(j = 0, 1, \ldots, J-1)$	fourth of (12.10)
$\mathscr{E}_{j+\frac{1}{2}}^{n+1}$	$(j = 0, 1, \ldots, J-1)$	third of (12.10)
$p_{j+\frac{1}{2}}^{n+1}$	$(j = 0, 1, \ldots, J-1)$	fifth of (12.10)

As described, this order would require five sweeps through the net; in practice, these are generally interleaved and combined into a single sweep.

To investigate the stability of these equations, consider first the case of small perturbations of a constant state in slab symmetry ($\alpha = 1$), and

SEC. 12.5]   DIFFERENCE EQUATIONS, LAGRANGEAN   297

write Δx for Δr. By differencing the fourth equation of (12.10) with respect to time and substituting into it from the first equation, R can be eliminated:

(12.11) $$V_{j+\frac{1}{2}}^{n+1} - V_{j+\frac{1}{2}}^{n} = V_0 \frac{\Delta t}{\Delta x}(u_{j+1}^{n+1} - u_j^{n+1}).$$

Linearizing the third equation of (12.10) and using

$$\Delta \mathscr{E} = \frac{\partial F_1}{\partial p}\Delta p + \frac{\partial F_1}{\partial V}\Delta V,$$

we find

$$\frac{\partial F_1}{\partial p}(p_{j+\frac{1}{2}}^{n+1} - p_{j+\frac{1}{2}}^n) + \left(p + \frac{\partial F_1}{\partial V}\right)(V_{j+\frac{1}{2}}^{n+1} - V_{j+\frac{1}{2}}^n) = 0,$$

which relates the time increments of p to the time increments of V. Because the basic state is constant, the same relation governs the space increments; therefore the second equation of (12.10) gives

(12.12) $$u_j^{n+1} - u_j^n = \frac{c^2}{V^2} V_0 \frac{\Delta t}{\Delta x}(V_{j+\frac{1}{2}}^n - V_{j-\frac{1}{2}}^n),$$

where c^2 is given by the equation

$$c^2 = \frac{V^2(p + \partial F_1/\partial V)}{\partial F_1/\partial p},$$

which is equivalent to (12.7). (The equation of state is now in the form of (12.9) rather than (12.1).) Equations (12.11) and (12.12) are of the same form as the system (10.5) for the wave equation; hence, the stability condition is

$$\frac{V_0}{V}\frac{c\Delta t}{\Delta x} \leq 1.$$

The extension to the spherical and cylindrical cases is only slightly more complicated, and leads to the condition:

$$\frac{V_0}{V}\left(\frac{R}{r}\right)^{\alpha-1}\frac{c\Delta t}{\Delta r} \leq 1.$$

This may be written in a more natural way, if we define

$$\Delta R = \frac{V}{V_0}\left(\frac{r}{R}\right)^{\alpha-1}\Delta r,$$

and note that, to the accuracy of the difference formulas, ΔR is the difference of the values of $R(r, t)$ for adjacent net points r_j and r_{j+1}, hence the actual distance between fluid particles bearing the Lagrangean labels r_j and r_{j+1}. We therefore have the condition

(12.13) $$\frac{c\Delta t}{R_{j+1}^n - R_j^n} \leq 1 \quad \text{for all } n, j$$

as stability condition. It is customary, in machine calculations, to have the machine automatically test the inequality (12.13) at all, or selected values of n and j and stop the calculation if the test fails. The sound speed c of course also depends on n and j and must be calculated from the equation of state; if the equation of state is complicated, one may use an approximate formula for c and then allow a margin of safety by replacing the right member of the inequality by a number somewhat less than 1.

12.6. Treatment of Interfaces in the Lagrangean Formulation

For an interface between two fluids at an index $j = J$, the equations (12.10) may be used just as they are, remembering only to use the appropriate equation of state on each side of the interface. However, this may lead to inaccuracies, and it is better to give special treatment to the interfaces. The inaccuracies result in part from the fact that if the densities of the two fluids are very different (as for a gas and a liquid) one may wish to use quite different increments Δr on the two sides of the interface, with consequent bad centering of the formulas (12.10), and in part from the fact that, although the pressure gradient $\partial p/\partial r$ varies continuously across the interface (this is a consequence of the particular definition of the Lagrangean variable adopted here—see below), the *derivative* of the pressure gradient may not, so that $\partial p/\partial r$ may not be given to second order accuracy by the centered difference quotient. We shall derive an improved formula for the treatment of interfaces.

First, to show the continuity of $\partial p/\partial r$, we assume that the initial data

SEC. 12.6]   TREATMENT OF INTERFACES

are smooth (they are usually analytic), except possibly at a finite number of points where different sets of conditions meet, and we assume that the equation of state is smooth. Then the solution is smooth, except at a finite number of lines or curves in the x, t-plane, representing shocks, rarefaction heads, contact discontinuities, interfaces, and the like. (We make no attempt to prove this statement or even to be very precise about what it means, but we note that similar questions have been considered by Lax (1953) concerning the analyticity of the solution.) If we exclude those instants of time at which a shock or a rarefaction wave or the like crosses the interface, we assert that $\partial p/\partial r$ is continuous across the interface, for the following reasons: in a small neighborhood of the interface, the solution is smooth on both sides; clearly, R and u must be continuous in r across the interface (a violation of this would represent a cavity or the penetration of one fluid into the other); their common values, say $R_0(t)$ and $u_0(t)$, give the position and velocity of the interface itself. From the smoothness on either side, which we take to include uniform one-sided limits of the functions and their derivatives on both sides, we see that $\partial u/\partial t$ approaches du_0/dt from either side and, hence, is continuous. From the equation of motion, second equation of (12.8), it then immediately follows that $\partial p/\partial r$ is continuous.

We shall use the continuity of $\partial p/\partial r$ to obtain an approximate formula for the acceleration of the interface. Let the interface be at $j = J$ and let $\Delta_1 r$ and $\Delta_2 r$ be the increments of r used on the left and right sides, respectively. The pressures at the midpoints of the intervals are denoted as usual by $p^n_{J-3/2}$, $p^n_{J-1/2}$, $p^n_{J+1/2}$, $p^n_{J+3/2}$, etc. In addition, let p^n_J denote the pressure at the interface (p is of course continuous). We don't know the value of p^n_J, but it will be eliminated from the final formula. On the left side of the interface, we have

$$\frac{p^n_{J-1/2} - p^n_{J-3/2}}{\Delta_1 r} \approx \frac{\partial p}{\partial r} \text{ at } r = r_{J-1},$$

$$\frac{p^n_J - p^n_{J-1/2}}{\tfrac{1}{2}\Delta_1 r} \approx \frac{\partial p}{\partial r} \text{ at } r = r_{J-1/4},$$

and, therefore, by extrapolation,

$$\frac{4}{3}\frac{p^n_J - p^n_{J-1/2}}{\tfrac{1}{2}\Delta_1 r} - \frac{1}{3}\frac{p^n_{J-1/2} - p^n_{J-3/2}}{\Delta_1 r} \approx \frac{\partial p}{\partial r} \text{ at } r = r_J.$$

Similarly, from the right side, we have

$$\frac{4}{3}\frac{p_{J+\frac{1}{2}}^n - p_J^n}{\frac{1}{2}\Delta_2 r} - \frac{1}{3}\frac{p_{J+\frac{3}{2}}^n - p_{J+\frac{1}{2}}^n}{\Delta_2 r} \approx \frac{\partial p}{\partial r} \text{ at } r = r_J.$$

By elimination of p_J^n, we find

(12.14) $$\left.\frac{\partial p}{\partial r}\right|_{r=r_J} \approx \frac{3(p_{J+\frac{1}{2}}^n - p_{J-\frac{1}{2}}^n) - \frac{1}{3}(p_{J+\frac{3}{2}}^n - p_{J-\frac{3}{2}}^n)}{\Delta_1 r + \Delta_2 r}.$$

This expression is to be used, for $j = J$, in place of

$$\frac{p_{J+\frac{1}{2}}^n - p_{J-\frac{1}{2}}^n}{\Delta r}$$

in the equation of motion—the second equation of (12.10).

The formula (12.14) can be used even if the two fluids are identical, so that the interface is not a real one but only a point at which the increment Δr is changed. But in such a case one can also use a three-point formula involving only $p_{J-\frac{3}{2}}^n$, $p_{J-\frac{1}{2}}^n$, and $p_{J+\frac{1}{2}}^n$ (or $p_{J-\frac{1}{2}}^n$, $p_{J+\frac{1}{2}}^n$, and $p_{J+\frac{3}{2}}^n$), by taking advantage of the continuity of $\partial^2 p/\partial r^2$: one simply fits a quadratic function of r to the three values of p and finds its derivative at $r = r_J$, taking the variable spacing properly into account. The three-point formula is often more convenient from the point of view of the logistics of a machine computation; for a real interface, (12.14) should be used.

12.7. Conservation-Law Form and the Lax-Wendroff Equations

The difference scheme described by Lax and Wendroff (1960) is well suited for solving the Eulerian equations (12.6) and the Lagrangean equations (12.8) in the special case of slab symmetry. A variant of it, called the *two-step* Lax-Wendroff scheme, is suitable for multi-dimensional problems in the Eulerian formulation, as will be shown in the next chapter.

The equations are first put into conservation-law form. In the Eulerian formulation, new dependent variables m and e are defined by

(12.15) $$m = \rho u, \quad e = \rho(\mathscr{E} + \tfrac{1}{2}u^2);$$

ρ, m, and e are the mass, momentum, and energy, respectively, per unit volume. The equations then take the form

$$(12.16) \qquad \frac{\partial}{\partial t}\mathbf{U} + \frac{\partial}{\partial x}\mathbf{F}(\mathbf{U}) = 0,$$

where $\mathbf{U}$ and $\mathbf{F}(\mathbf{U})$ are vectors (in the mathematical sense), defined as

$$(12.17) \qquad \mathbf{U} = \begin{bmatrix} \rho \\ m \\ e \end{bmatrix}, \quad \mathbf{F}(\mathbf{U}) = \begin{bmatrix} m \\ (m^2/\rho) + p \\ (e + p)m/\rho \end{bmatrix}.$$

The pressure is given by the equation

$$(12.18) \qquad p = f(\mathscr{E}, V) = f\left(\frac{e}{\rho} - \frac{m^2}{2\rho^2}, \frac{1}{\rho}\right),$$

where $f(\cdot, \cdot)$ is the function introduced in Section 12.4 to express the equation of state. By virtue of (12.18), $\mathbf{F}(\mathbf{U})$ is a function of $\mathbf{U}$, as the notation indicates. The system (12.16) is called the *conservation-law form* of the equations of fluid dynamics. The physical interpretation is the following: each component of $\mathbf{U}$ is a quantity per unit volume; if a slab extends from x_1 to x_2, the total amount of this quantity per unit area of the slab is the integral from x_1 to x_2 of this component; but (12.16) shows that

$$(12.19) \qquad \frac{d}{dt}\int_{x_1}^{x_2} \mathbf{U}\,dx = -\mathbf{F}(\mathbf{U})|_{x=x_2} + \mathbf{F}(\mathbf{U})|_{x=x_1};$$

hence, the corresponding component of $\mathbf{F}(\mathbf{U})|_x$ is the rate of flux of this quantity, in the $+x$ direction, per unit area of the slab, at point x. The laws of conservation of mass, momentum, and energy will appear again in Section 12.8, where the Rankine-Hugoniot jump conditions at a shock will be deduced from them.

In the Lagrangean formulation, R is first eliminated from (12.8) and (12.9) (slab symmetry $\alpha = 1$ is assumed), to give

$$(12.20) \qquad \frac{\partial V}{\partial t} = V_0 \frac{\partial u}{\partial x}$$

as one of the equations. The total energy per unit mass $\mathscr{E} + u^2/2$ is called E. Then, with

$$(12.21) \qquad \mathbf{U} = \begin{bmatrix} V \\ u \\ E \end{bmatrix}, \qquad \mathbf{F}(\mathbf{U}) = V_0 \begin{bmatrix} -u \\ p \\ pu \end{bmatrix},$$

the equations assume the conservation-law form (12.16). Here, p stands for $f(E - u^2/2, V)$, $f(\cdot, \cdot)$ being again the function introduced in Section 12.4. As in the Eulerian formulation, the three equations express conservation of mass, momentum, and energy.

The Lax-Wendroff equations start from a Taylor's series in t, viz.,

$$\mathbf{U}_j^{n+1} = \mathbf{U}_j^n + \Delta t \left(\frac{\partial \mathbf{U}}{\partial t}\right)_j^n + \frac{\Delta t^2}{2} \left(\frac{\partial^2 \mathbf{U}}{\partial t^2}\right)_j^n + \cdots.$$

The t derivatives indicated are replaced by x-derivatives by means of the differential equation (12.16) and the further equation

$$\frac{\partial^2 \mathbf{U}}{\partial t^2} = -\frac{\partial}{\partial t}\frac{\partial \mathbf{F}}{\partial x} = -\frac{\partial}{\partial x}\frac{\partial \mathbf{F}}{\partial t} = -\frac{\partial}{\partial x}\left(A\frac{\partial \mathbf{U}}{\partial t}\right) = \frac{\partial}{\partial x}\left(A\frac{\partial \mathbf{F}}{\partial x}\right),$$

where the matrix $A = A(\mathbf{U})$ is the Jacobian of $\mathbf{F}(\mathbf{U})$ with respect to $\mathbf{U}$, that is, $A_{ij} = \partial F_i/\partial U_j$. The x-derivatives are then approximated by difference quotients, to give the Lax-Wendroff system,

$$(12.22) \quad \mathbf{U}_j^{n+1} = \mathbf{U}_j^n - \frac{1}{2}\frac{\Delta t}{\Delta x}(\mathbf{F}_{j+1}^n - \mathbf{F}_{j-1}^n)$$

$$+ \frac{1}{2}\left(\frac{\Delta t}{\Delta x}\right)^2 [A_{j+\frac{1}{2}}^n(\mathbf{F}_{j+1}^n - \mathbf{F}_j^n) - A_{j-\frac{1}{2}}^n(\mathbf{F}_j^n - \mathbf{F}_{j-1}^n)],$$

where $A_{j+\frac{1}{2}}^n$ denotes $A(\frac{1}{2}\mathbf{U}_{j+1}^n + \frac{1}{2}\mathbf{U}_j^n)$. If A were a constant matrix the system would be

$$(12.23) \quad \mathbf{U}_j^{n+1} = \mathbf{U}_j^n - \frac{1}{2}A\frac{\Delta t}{\Delta x}(\mathbf{U}_{j+1}^n - \mathbf{U}_{j-1}^n)$$

$$+ \frac{1}{2}\left(A\frac{\Delta t}{\Delta x}\right)^2(\mathbf{U}_{j+1}^n - 2\mathbf{U}_j^n + \mathbf{U}_{j-1}^n).$$

We shall present a procedure, the two-step Lax-Wendroff procedure, which also has second-order accuracy and which reduces to (12.23) when the coefficients are constant. It is slightly simpler to use, because the matrix A does not appear. First, provisional values are calculated at the centers of the rectangular meshes of the net in the x, t-plane:

(12.24a) $$\mathbf{U}_{j+\frac{1}{2}}^{n+\frac{1}{2}} = \tfrac{1}{2}(\mathbf{U}_{j+1}^n + \mathbf{U}_j^n) - \frac{\Delta t}{2\Delta x}(\mathbf{F}_{j+1}^n - \mathbf{F}_j^n),$$

where $\mathbf{F}_j^n$ is an abbreviation for $\mathbf{F}(\mathbf{U}_j^n)$ etc.; the final values are then obtained from the equation

(12.24b) $$\mathbf{U}_j^{n+1} = \mathbf{U}_j^n - \frac{\Delta t}{\Delta x}(\mathbf{F}_{j+\frac{1}{2}}^{n+\frac{1}{2}} - \mathbf{F}_{j-\frac{1}{2}}^{n+\frac{1}{2}}).$$

From its structure, equation (12.24b) is seen to be of second-order accuracy (the error term is $O[(\Delta t)^3]$ for fixed ratio $\Delta t/\Delta x$); the provisional values have only first-order accuracy, but the difference on the right of (12.24b) has second-order accuracy, owing to the subtraction of quantities whose difference is $O(\Delta x)$.

It is easily verified that, in the special case where $\mathbf{F}(\mathbf{U}) = A\mathbf{U}$, substitution of (12.24a) into (12.24b) yields (12.23), as claimed above.

In the case of constant coefficients, the stability of the Lax-Wendroff system is easily established. The amplification matrix of (12.23) is

(12.25) $$G = I - i\frac{\Delta t}{\Delta x} A \sin \alpha - \left(\frac{\Delta t}{\Delta x} A\right)^2 (1 - \cos \alpha),$$

where $\alpha = k\Delta x$. The matrix A has real eigenvalues and a complete set of linearly independent eigenvectors, for this is the defining property of hyperbolicity of the system $\partial \mathbf{U}/\partial t + A\partial \mathbf{U}/\partial x = 0$, and the equations of fluid dynamics are hyperbolic. Since G is a rational function of the single matrix A, its eigenvectors are the same as those of A; hence, the Gram determinant of these eigenvectors is independent of α and of $\Delta t/\Delta x$, and by Condition 1 of Section 4.11 of Chapter 4, the von Neumann condition is sufficient as well as necessary for stability. If λ is an eigenvalue of A, the corresponding eigenvalue of G is

$$g = 1 - i\frac{\lambda \Delta t}{\Delta x} \sin \alpha - \left(\frac{\lambda \Delta t}{\Delta x}\right)^2 (1 - \cos \alpha).$$

As α varies from 0 to 2π, g traces out an ellipse in the complex plane, and a simple calculation shows that this ellipse lies inside the unit circle if

$|\lambda|\Delta t/\Delta x < 1$; for stability, this inequality must be satisfied for each eigenvalue λ of A. We note that the ellipse has second-order contact with the unit circle at the point $g = 1$, i.e., it has the same curvature there. This reflects the second-order accuracy of the Lax-Wendroff system and the non-dissipativity of the differential equation system, which exactly conserves the amplitude of each Fourier component of the solution, for constant A. On the other hand, the above formula for g yields $|g|^2 = 1 - 4\mu^2(1 - \mu^2)\sin^4 \frac{1}{2}\alpha$, where $\mu = |\lambda|\Delta t/\Delta x$, so that the difference scheme is dissipative of fourth order in the sense of Kreiss—see Section 5.4—when $0 < \mu < 1$. In the Eulerian formulation, which will be considered first, the matrix A obtained from (12.17) is rather cumbersome, but its eigenvalues can be found indirectly: The original Eulerian equations (12.5) are also of the form $\partial \mathbf{U}/\partial t + A\partial \mathbf{U}/\partial x = 0$, with a different matrix A, but the eigenvalues of these two matrices are the same, since the reciprocals of these eigenvalues are the slopes of the characteristics in the x, t-plane. To find the eigenvalues, expression (12.1) for $\mathscr{E}$ is substituted into the third equation of (12.5), and the derivative of ρ is eliminated from this equation by use of the first equation of (12.5); this gives

$$\frac{\partial p}{\partial t} + u\frac{\partial p}{\partial X} + \rho c^2 \frac{\partial u}{\partial X} = 0.$$

The system then takes the desired form, with A and $\mathbf{U}$ given by

(12.26) $\qquad A = \begin{bmatrix} u & \rho & 0 \\ 0 & u & 1/\rho \\ 0 & \rho c^2 & u \end{bmatrix}, \quad \mathbf{U} = \begin{bmatrix} \rho \\ u \\ p \end{bmatrix}.$

The eigenvalues are u, $u + c$, $u - c$; hence, the stability condition of the Lax-Wendroff equations is

(12.27) $\qquad (|u| + c)\frac{\Delta t}{\Delta X} < 1,$

as for the other Eulerian difference schemes.

Similarly in the Lagrangean formulation, the matrix A as obtained directly from (12.21) is rather cumbersome, but the third equation can be rewritten as

$$\frac{\partial p}{\partial t} + \frac{c^2}{V^2}V_0\frac{\partial u}{\partial x};$$

then, taking V, u, p as the components of $\mathbf{U}$, the matrix A is

$$(12.28) \qquad A = \begin{bmatrix} 0 & -V_0 & 0 \\ 0 & 0 & V_0 \\ 0 & V_0(c/V)^2 & 0 \end{bmatrix};$$

its eigenvalues are 0, $\pm cV_0/V$, so that the stability condition is

$$(12.29) \qquad \frac{V_0}{V} \frac{c\Delta t}{\Delta x} < 1,$$

as for the previous Lagrangean scheme.

The matrix (12.26) or (12.28) can be reduced to diagonal form by a similarity transformation PAP^{-1} (for example, the columns of P^{-1} can be taken as the normalized eigenvectors of A). If the equation system (12.22) is multiplied through by the matrix P, the equations become transformed to the so-called *characteristic form*—see Courant-Hilbert, vol. II (1962), or Courant-Friedrichs (1948). In the Eulerian formulation, the equations in characteristic form are

$$(12.30) \qquad \begin{aligned} \left(\frac{\partial}{\partial t} + u \frac{\partial}{\partial X}\right)p - c^2\left(\frac{\partial}{\partial t} + u \frac{\partial}{\partial X}\right)\rho &= 0, \\ \left(\frac{\partial}{\partial t} + (u+c) \frac{\partial}{\partial X}\right)p + \rho c\left(\frac{\partial}{\partial t} + (u+c) \frac{\partial}{\partial X}\right)u &= 0, \\ \left(\frac{\partial}{\partial t} + (u-c) \frac{\partial}{\partial X}\right)p - \rho c\left(\frac{\partial}{\partial t} + (u-c) \frac{\partial}{\partial X}\right)u &= 0; \end{aligned}$$

in the Lagrangean formulation, they are

$$\frac{\partial p}{\partial t} + \frac{c^2}{V^2} \frac{\partial V}{\partial t} = 0,$$

$$(12.31) \qquad \begin{aligned} \left(\frac{\partial}{\partial t} + c\frac{V_0}{V} \frac{\partial}{\partial x}\right)p + \frac{c}{V}\left(\frac{\partial}{\partial t} + c\frac{V_0}{V} \frac{\partial}{\partial x}\right)u &= 0, \\ \left(\frac{\partial}{\partial t} - c\frac{V_0}{V} \frac{\partial}{\partial x}\right)p - \frac{c}{V}\left(\frac{\partial}{\partial t} - c\frac{V_0}{V} \frac{\partial}{\partial x}\right)u &= 0. \end{aligned}$$

An advantage of the two-step procedure is that, since the difference equations (12.24a) and (12.24b) are also in conservation form, the quantities that ought to be conserved, according to the differential equations

(12.16), are exactly conserved in the difference form. In place of the integral in (12.19), we have a sum

$$S^n = \sum_{j=j_1}^{j_2} \mathbf{U}_j^n \Delta x,$$

and in place of (12.19) we have the equation

(12.32) $$\frac{S^{n+1} - S^n}{\Delta t} = -\mathbf{F}_{j_2+\frac{1}{2}}^{n+\frac{1}{2}} + \mathbf{F}_{j_1-\frac{1}{2}}^{n+\frac{1}{2}},$$

which follows directly from summing (12.24b).

The further advantage that the matrix A and its derivatives do not appear explicitly (this simplifies the calculation) is even more important in more dimensions (see Chapter 13). The calculational program needs a sub-routine for computing the components of $\mathbf{F(U)}$ from those of $\mathbf{U}$, but this is generally a small sub-routine.

A class of difference schemes for the conservation equation (12.16), obtained as iterative approximations to the implicit equation

$$\mathbf{U}_j^{n+1} - \mathbf{U}_j^n + \frac{\Delta t}{4\Delta x}(\mathbf{F}_{j+1}^{n+1} + \mathbf{F}_{j+1}^n - \mathbf{F}_{j-1}^{n+1} - \mathbf{F}_{j-1}^n) = 0,$$

was considered by J. Gary (1962b).

12.8. The Jump Conditions at a Shock

As is well known—see Courant and Friedrichs (1948), for example—the motion of an ideal fluid is often characterized by lines or curves in the x, t-plane (more generally surfaces in the $\mathbf{X}, t$-space) on which certain of the dependent quantities are discontinuous, but possess one-sided limits on both sides. (Even if the flow is smooth in the beginning, discontinuities may develop after a finite time, after which no smooth, single-valued solution of the system (12.8) exists.) Examples are an interface, a contact discontinuity (where ρ and $\mathscr{E}$ are discontinuous, but p and u are continuous), a shock (where ρ, $\mathscr{E}$, p, u are all discontinuous) and the head of a rarefaction wave (where only certain derivatives are discontinuous). At such discontinuities, the differential equations (which now involve one-sided derivatives only) must be supplemented by jump conditions, which serve

SEC. 12.8]   THE JUMP CONDITIONS AT A SHOCK   307

as internal boundary conditions and are needed to make the solution unique.

For an interface, a contact discontinuity, or a rarefaction head, the only jump conditions needed are the continuity of certain variables, and no modification of the finite-difference system is required, although for an interface, as noted, modification may be desirable for increased accuracy. But when shocks are present, the unmodified difference equations (12.10) do not give even approximately the right answers, except in certain limiting cases. These equations were studied by J. von Neumann in experimental calculations during the war.[6]

We state the jump conditions for slab symmetry, and we let the Lagrangean position of the shock be $x = \xi(t)$. The mass of fluid swept over by the shock in unit time per unit area is

$$(12.33) \qquad M = \rho_0 \, d\xi/dt.$$

Let $V = 1/\rho$ denote specific volume (it is customary to express the jump conditions in terms of V rather than ρ); and let subscripts 1, 2 denote left and right-hand limits as $x \to \xi$ so that $u_1, V_1, p_1, \mathscr{E}_1$ describes conditions immediately behind the shock and $u_2, V_2, p_2, \mathscr{E}_2$ conditions immediately ahead, if the shock is moving in the $+x$-direction, and conversely if it is moving in the $-x$-direction. *Conservation of mass* is expressed by the continuity of $X(x, t)$ across the shock, so that the time derivative of X along the line $x = \xi(t)$, namely,

$$\frac{d\xi}{dt}\frac{\partial X}{\partial x} + \frac{\partial X}{\partial t},$$

must be the same on both sides; but $\partial X/\partial x$ is $\rho_0 V$ and $\partial X/\partial t$ is u, so

$$(12.34) \qquad MV_1 + u_1 = MV_2 + u_2.$$

---

[6] We must not give the impression that von Neumann expected them to give the right answers. He pointed out that the unmodified equations do not allow for the entropy increase known to be produced by a shock, and he was investigating the size of the error caused by this. Nevertheless, ignoring the shocks completely in a calculation is known in some circles as the "von Neumann method of calculating shocks"! As is shown in section 12.14, the Lax-Wendroff equations can sometimes be used in unmodified form for shocks because they contain a built-in dissipative mechanism that can increase the entropy where a shock occurs.

Conservation of momentum says that the increase of momentum of the fluid swept over in unit time is equal to the pressure difference, or

(12.35) $$M(u_1 - u_2) = p_1 - p_2.$$

Conservation of energy says that the increase in energy of this fluid is the net work done in unit time, or

(12.36) $$M\left(\mathscr{E}_1 + \frac{u_1^2}{2} - \mathscr{E}_2 - \frac{u_2^2}{2}\right) = p_1 u_1 - p_2 u_2.$$

Equations (12.33) to (12.36) can be combined in various ways, of which one is:

(12.37)
$$\rho_0 \frac{d\xi}{dt} = \sqrt{\frac{p_1 - p_2}{V_2 - V_1}},$$
$$u_1 - u_2 = \sqrt{(p_1 - p_2)(V_2 - V_1)},$$
$$\mathscr{E}_1 - \mathscr{E}_2 = \frac{p_1 + p_2}{2}(V_2 - V_1);$$

these are known as the Rankine-Hugoniot equations or simply the Hugoniot equations.[7] They hold also for spherical and cylindrical shocks, where $r = \xi(t)$ is the Lagrangean coordinate of the shock front. Furthermore, they hold locally for a general smoothly curved shock surface with $\rho_0 d\xi/dt$ interpreted as the mass swept over per unit area per unit time, and u_1 and u_2 interpreted as the components of fluid velocity normal to the shock front; the parallel components are continuous across the shock.

12.9. Shock Fitting

The Hugoniot equations, together with the continuity of $X(x, t)$ or $R(r, t)$ across the shock, just suffice to provide the needed boundary conditions for the differential equations on each side and also to determine the motion of the shock front itself. They have been used directly in numerical calculations—unpublished work, Los Alamos, also L. H. Thomas (1954), J. Gary (1962a), D. Quarles (1964). We shall outline the

---

[7] The derivation of (12.37) from the preceding equations is ambiguous with respect to the signs of the square roots. The plus signs should be taken, as we have done, for a shock moving to the right, and the minus signs, for a shock moving to the left.

SEC. 12.9] SHOCK FITTING

procedure for the special case where the fluid ahead of the shock is at rest and in equilibrium; then u_2, V_2, p_2, $\mathscr{E}_2$ are known constants ($u_2 = 0$), while u_1, V_1, p_1, and $\mathscr{E}_1$ are functions of t to be determined by the computation. Among the quantities to be calculated for each n are the position $\xi^n = \xi(t^n)$ of the shock (this does not generally fall at a net point) and the values of u_1, V_1, p_1, $\mathscr{E}_1$ at the shock, i.e., at $\xi = \xi^n$, for $t = t^n$; we denote these by u_1^n, V_1^n, p_1^n, $\mathscr{E}_1^n$ and ask the reader to avoid confusion with u_j^n for $j = 1$, etc.

When all quantities are known at $t = t^n$, the Rankine-Hugoniot equations (12.37), in which each quantity is imagined to carry a superscript $n + 1$, together with an equation like

$$(12.38) \qquad \xi^{n+1} - \xi^n = \frac{\Delta t}{2}(\dot\xi^n + \dot\xi^{n+1}),$$

where $\dot\xi = d\xi/dt$, and with the equation of state,

$$(12.39) \qquad p_1^{n+1} = f(\mathscr{E}_1^{n+1}, V_1^{n+1}),$$

constitute a system of five equations for the six unknowns ξ^{n+1}, $\dot\xi^{n+1}$, u_1^{n+1}, V_1^{n+1}, $\mathscr{E}_1^{n+1}$, p_1^{n+1}. An additional equation is needed (obviously to express the influence on the shock of the motion of the material just behind it); for this purpose, one of the partial differential equations, or some combination of them, is written in difference form, the space differences being taken between values at $x = \xi^{n+1}$ and values at nearby net-points behind the shock, and the time differences being taken between values at ξ^{n+1}, t^{n+1} and ξ^n, t^n, with due allowance for the displacement of the shock between times t^n and t^{n+1}.

In order to decide which of the differential equations, or what combination of them, to take to augment the jump conditions in this way, appeal is made to the theory of characteristics given at the end of Section 12.7.[8] The curves in the x, t-plane whose inverse slopes are the eigenvalues of A are called *characteristics*: through each point there are three charac-

---

[8] In the early work it was found empirically that the continuity equation could be used; that choice may possibly still be the best. The analysis in the text is intended to throw some light on the behavior of possible choices, with generalization to the much more difficult multidimensional problems (see Chapter 13) kept very much in mind.

teristics, one following forward sound signals, one following backward sound signals, and one in between following a particle path.

In a sense, information flows along the characteristics. For example, the entropy S is constant along the middle characteristic, by the third equation of (12.8), so that a knowledge of S at any point yields information on S at any later point on the same characteristic. Furthermore, when S is constant throughout the flow, so that V and c can be expressed in terms of p, equations (12.31) show that the quantities $\int (V/c)dp \pm u$ (they are called the *Riemann invariants*) are constant along the forward and backward characteristics, respectively. Even when S is not constant, except along the particle paths, quantities similar to the Riemann invariants are nearly constant, for short times, along the forward and backward characteristics; for example, one possible difference approximation to the second equation of (12.31) says that, along the forward characteristic, the quantity $p^{n+1} + (c/V)^{n+\frac{1}{2}} u^{n+1}$ is approximately equal to $p^n + (c/V)^{n+\frac{1}{2}} u^n$.

In the shock-fitting, the incomplete information provided by the jump conditions must be supplemented by additional information coming to the shock at time t^{n+1} from the fluid behind it. Only one of the three characteristics is in a position to do this, namely, the one representing a forward-moving sound signal, which in time overtakes the shock; the other two characteristics, if followed backward from a point on the shock, lie in the region ahead so that, for example, the middle characteristic tells us the value of the entropy just ahead of the shock, not just behind, where we need it. Therefore, of the three equations (12.31), only the second (for a shock moving to the right) is suitable, in difference form, for supplementing the jump conditions. (The question remains open, of course, whether some other combination of the original equations, in non-characteristic form, may be as good or better). The resulting equations that must be solved for the six unknowns at ξ^{n+1}, t^{n+1} are non-linear and rather complicated, so that an iterative method must be used for solving them. The equations must be modified when the shock passes a net point, that is, when ξ^n and ξ^{n+1} lie in different intervals of the r-net.

It is probably clear from this sketchy outline that shock fitting is rather uncomfortably complicated. It has been found to be feasible to use shock fitting for shocks running into previously undisturbed fluid, so that $u_2, p_2, V_2, \mathscr{E}_2$ are known in advance, but hardly feasible for the general case. For complicated problems, there is an additional difficulty because shocks may develop spontaneously, within the fluid—see, for example, J. W. Calkin (1942) or Courant and Friedrichs (1948) pages 115, 116—and there is no way of telling when and where shock fitting should be used, except possibly by constantly monitoring the results of the calculation with a very practiced eye.

In order to get around these difficulties, von Neumann and Richtmyer (1950) devised an approximate method of calculation for fluid dynamical problems, discussed in the next three sections, in which shocks are taken care of automatically, whenever and wherever they arise. The method is based on certain physical notions concerning real (and therefore non-ideal) fluids rather than on a direct attempt to incorporate the Hugoniot jump-conditions into the calculations. Specifically, dissipative mechanisms, like viscosity and heat conduction, have a smoothing effect on a shock, so that the surface of discontinuity is replaced by a thin transition layer in which quantities change rapidly, but not discontinuously. The differential equations, with the dissipative mechanism included, apply in this layer as well as elsewhere, so that no boundary conditions are needed. At the same time, the basic conservation laws on which the Hugoniot conditions were based are retained, and the jump conditions still hold across the transition layer, in the approximation in which this layer is regarded as thin in comparison with other dimensions occurring in the problem. A similar method is discussed in Ludford, Polachek and Seeger (1953). The method was improved by Lax and Wendroff (1959); their work is described in Section 12.14.

12.10. Effect of Dissipation

The effect of dissipation on shocks has been studied by various investigators, including Lord Rayleigh, R. Becker (1922), and L. H. Thomas (1944). Becker studied the effect of heat conduction and viscosity. He showed that, when heat conduction is allowed for, the temperature varies smoothly through the shock layer and that for shocks of strength[9] less than a critical value, the pressure and density vary smoothly also;

---

[9] A convenient measure of the strength of a shock is the ratio of the final pressure to the initial pressure.

whereas, for stronger shocks, the transition of pressure and density from their initial to their final values is partly by a smooth variation and partly by a discontinuous jump. When viscosity is allowed for, however, all quantities vary smoothly through the shock region for a shock of any strength. In either case the thickness of the transition zone is proportional to the coefficient of the dissipative mechanism, so that in the limit of no heat conduction and no viscosity the variations approach the discontinuous ones of the Hugoniot theory.

The idea of von Neumann and Richtmyer was to introduce a purely artificial dissipative mechanism of such form and strength that the shock transition would be a smooth one extending over a small number (say three or four) of intervals Δr of the space variable, and then to construct finite-difference equations with the dissipation included, but without any necessity for shock fitting. In a calculation with such equations, the shocks show up automatically as near discontinuities, across which quantities have very nearly the correct jump, and which travel at very nearly the correct speed through the fluid.

It is clear that a mechanism like viscosity is more suitable for our purpose than one like heat conduction, because, with viscosity, all quantities vary smoothly. However, Becker showed that with ordinary viscosity, in which the stress is proportional to the rate of shear, and which is therefore represented by linear terms in the differential equations, the thickness of the transition layer varies with the shock strength, approaching zero for a very strong shock and infinity for a very weak one. But we wish the thickness to be about the same—namely, about $(3-4)\Delta r$—for all shocks, and we therefore add *quadratic* terms to the differential equation; this is equivalent to using a small viscosity coefficient for weak shocks and a large one for strong shocks. It will be shown below that we achieve a thickness independent of the shock strength.[10] Other features of the

---

[10] Possibly von Neumann and Richtmyer were unduly concerned about this point. The method of Lax (1954) has the effect of introducing a linear viscosity term. Lax finds indeed that the thickness varies with shock strength, but not enough to interfere with practical calculations. However, the von Neumann-Richtmyer method has the advantage that the pseudo-viscous terms, being quadratic in the strain rate, are very small indeed in the smooth part of the flow between shocks, where one wishes the behavior to approximate that of an ideal fluid. One wishes this because generally the true viscosity of real fluids is smaller by orders of magnitude than the viscosity required to give shocks the thickness needed for the present purpose.

SEC. 12.10]   EFFECT OF DISSIPATION

added terms ensure that the thickness is also independent of the pressure and density of the material into which the shock is running.

The equations of the pseudo-viscosity method will be written first for slab symmetry, and first of all on the differential equation level. A Lagrangean formulation is used, and the equations are

(12.40)
$$\frac{\partial X}{\partial t} = u,$$
$$\frac{\partial u}{\partial t} = -\frac{1}{\rho_0}\frac{\partial(p+q)}{\partial x},$$
$$\frac{\partial \mathscr{E}}{\partial t} = -(p+q)\frac{\partial V}{\partial t},$$
$$V = \frac{1}{\rho_0}\frac{\partial X}{\partial x},$$
$$\mathscr{E} = F(p, \rho),$$

where $V = 1/\rho$ is the specific volume; and q, the pseudo-viscous pressure, is given by[11]

(12.41)
$$q = \begin{cases} (l^2/V)(\partial u/\partial x)^2 & \text{if } \partial u/\partial x < 0 \\ 0 & \text{if } \partial u/\partial x \geq 0, \end{cases}$$

where l is a constant having the dimensions of a length.

To give an idea of the effect of the added terms, we analyze the structure of a plane, steady-state shock, and therefore look for solutions of the differential equations in which p, q, u, $\mathscr{E}$, and V depend on x and t only through the combination $w = x - st$, where s is the speed of the shock in the Lagrangean coordinate system; that is, we are looking for running-wave solutions. We furthermore assume the simple equation of state:

$$\mathscr{E} = pV/(\gamma - 1)$$

---

[11] This expression for q incorporates the suggestion of M. Rosenbluth (unpublished work, Los Alamos) that q should be $= 0$ when the fluid is undergoing expansion. We had previously set

$$q = -\frac{l^2}{V}\left|\frac{\partial u}{\partial x}\right|\frac{\partial u}{\partial x}$$

which agrees with (12.41) for compression and is negative for expansion. For most problems either formula is satisfactory, because q is exceedingly small except in a shock; but when extreme expansion (as of a free surface) can take place, Rosenbluth's formula is superior.

of an ideal gas. In the special case, considered here, of slab symmetry, the first and fourth equations of (12.40) can be combined into the single equation

$$\frac{\partial V}{\partial t} = \frac{1}{\rho_0}\frac{\partial u}{\partial x},$$

from which X has been eliminated. Substituting the running-wave solution into this and the other equations, we find

$$\rho_0 s \frac{dV}{dw} + \frac{du}{dw} = 0,$$

$$\rho_0 s \frac{du}{dw} - \frac{d(p+q)}{dw} = 0,$$

$$\frac{d\mathscr{E}}{dw} + (p+q)\frac{dV}{dw} = 0.$$

Letting $M = \rho_0 s$ denote the mass swept over per unit area per unit time, we solve the above equations by writing

(12.42)
$$MV + u = C_1,$$
$$M^2 V + p + q = C_2,$$
$$\tfrac{1}{2}M^2 V^2 + \mathscr{E} + (p+q)V = C_3,$$

where C_1, C_2, and C_3 are integration constants.

If we now consider a solution[12] for which u, V, p, q approach asymptotic values (denoted by subscript 1 and 2, respectively) as $x \to -\infty$ and $+\infty$ and for which the derivatives of these quantities approach 0 as $x \to \pm\infty$, so that in particular $q \to 0$, then equations (12.42) require that

$$M(V_1 - V_2) + u_1 - u_2 = 0,$$
$$M^2(V_1 - V_2) + p_1 - p_2 = 0,$$
$$\tfrac{1}{2}M^2(V_1^2 - V_2^2) + \mathscr{E}_1 - \mathscr{E}_2 + p_1 V_1 - p_2 V_2 = 0,$$

from which the Hugoniot equations (12.37) follow by a slight rearrangement.

From (12.42),

$$\mathscr{E} = \tfrac{1}{2}M^2 V^2 - C_2 V + C_3 = pV/(\gamma - 1).$$

---
[12] The general solution of this kind will be exhibited below.

Therefore, from the second equation of (12.42),

$$qV = -\tfrac{1}{2}(\gamma + 1) M^2 V^2 + C_4 V + C_5,$$

where C_4 and C_5 are constants. But since $q = 0$ for $V = V_1$ and for $V = V_2$, the quadratic function of V on the right of the last equation is uniquely determined, viz.,

(12.43) $$qV = \tfrac{1}{2}(\gamma + 1) M^2 (V_2 - V)(V - V_1).$$

We can now conclude that for the running wave solution $du/dw \leq 0$ everywhere; for, in an interval where $du/dw \geq 0$, we have $q = 0$ by (12.41) and this implies, by (12.43), that in such an interval V must have either the constant value V_2 or the constant value V_1. But V = constant implies u = constant since $\rho_0 \partial V / \partial t = \partial u / \partial x$, therefore $du/dw \geq 0$ implies $du/dw = 0$. *The running wave represents a compression of the fluid, never a rarefaction.* The Hugoniot equations (12.37) admit negative shocks (sudden decompressions) as well as positive ones, whereas other but perfectly clear physical grounds[13] prohibit them, and it is gratifying to see that the pseudo-viscosity method admits only positive shocks.

Since $du/dw \leq 0$, (12.41) gives for the running wave

$$q = \frac{l^2}{V}\left(\frac{du}{dw}\right)^2 = \frac{l^2 M^2}{V}\left(\frac{dV}{dw}\right)^2,$$

and, therefore,

$$\left(l\frac{dV}{dw}\right)^2 = \tfrac{1}{2}(\gamma + 1)(V_2 - V)(V - V_1),$$

whose general solution is

$$V = \frac{V_2 + V_1}{2} + \frac{V_2 - V_1}{2} \sin \sqrt{\frac{\gamma + 1}{2}} \frac{w - w_0}{l},$$

w_0 being a constant. But because of the condition $du/dw \leq 0$ or $dV/dw \geq 0$, we can use only a half sine-wave of this solution, which must be pieced

---

[13] A negative shock would result in a decrease of entropy without any corresponding increase elsewhere and thus violate the second law of thermodynamics. Stability arguments (von Neumann) show that a negative shock would disintegrate immediately into a rarefaction fan in which the variations are continuous and no entropy change occurs.

together continuously with the two other solutions, $V \equiv V_1$ and $V \equiv V_2$. Taking the origin so that $w_0 = 0$ for convenience, we get finally the running-wave solution

(12.44) $$V = \begin{cases} \dfrac{V_2 + V_1}{2} + \dfrac{V_2 - V_1}{2} \sin \sqrt{\dfrac{\gamma + 1}{2}} \dfrac{x - st}{l} \\ \qquad\qquad\qquad \text{for } |x - st| \leq \dfrac{\pi\sqrt{2}l}{2\sqrt{\gamma + 1}} \\ V_1 \qquad\qquad\quad \text{for } x - st \leq -\dfrac{\pi\sqrt{2}l}{2\sqrt{\gamma + 1}} \\ V_2 \qquad\qquad\quad \text{for } x - st \geq +\dfrac{\pi\sqrt{2}l}{2\sqrt{\gamma + 1}} \end{cases}$$

Similarly, the total pressure and the pseudo-viscous pressure are given in the shock region, that is for $|x - st| \leq \pi\sqrt{2}l/2\sqrt{\gamma + 1}$, by

$$p + q = \dfrac{p_1 + p_2}{2} - \dfrac{p_1 - p_2}{2} \sin \sqrt{\dfrac{\gamma + 1}{2}} \dfrac{x - st}{l},$$

$$q = \dfrac{p_1 - p_2}{2} \dfrac{\gamma + 1}{2} \dfrac{\left[\cos \sqrt{\dfrac{\gamma + 1}{2}} \dfrac{x - st}{l}\right]^2}{\dfrac{V_2 + V_1}{V_2 - V_1} + \sin \sqrt{\dfrac{\gamma + 1}{2}} \dfrac{x - st}{l}};$$

they assume constant values in front of the shock region ($p = p_2$, $q = 0$) and behind it ($p = p_1$, $q = 0$). Figure 12.1 shows V, $p + q$ and q graphically for a typical case.

The fact that the Hugoniot relations are satisfied exactly in this solution means that this "smeared out" shock travels with exactly the same speed as a discontinuous one would, for the given driving pressure p_1, and produces exactly the same entropy increase. (This should occasion no surprise, because the Hugoniot relations are derived from the conservation laws of mass, momentum, and energy, and the conservation laws follow directly from (12.40) independently of any assumption about the form of q. Of course, it is essential for this that the same total pressure, $p + q$, appear in the energy equation as in the equation of motion.) Conse-

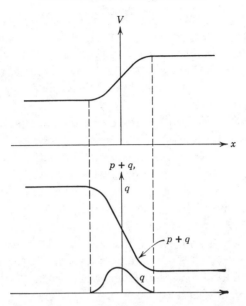

Fig. 12.1. Properties of a steady shock under the influence of the pseudo-viscosity. A typical solution given by the equation (12.44) and ff: profiles of specific volume, total pressure, and pseudo-viscous pressure.

quently, this solution is an acceptable approximation to the true shock if its thickness, which is approximately 2 to 3 times l, is small enough.[14]

12.11. Finite-Difference Equations

Finite-difference systems may be obtained in many ways, for example, by modification of (12.10) to include the pseudo-viscous pressure terms. In the pseudo-viscous terms, we replace l by $a\Delta x$, where a is a dimensionless constant around 1.5 to 2.0, thus giving the shocks a thickness around 3 to 5 times Δx. If the mesh size is variable this will give the shocks a greater thickness in the coarse part of the net than in the fine part.

---

[14] The actual thickness resulting in practice is effectively slightly smaller than the distance ($\sqrt{2/(\gamma+1)}\,\pi l$) between the dotted lines in Figure 12.1; the points of join between the half sine wave and the constant parts are of course not visible in a numerical solution, and if one defines the effective thickness as the total change in V divided by the maximum slope, dV/dw, one gets $\sqrt{2/(\gamma+1)}\,2l$ for the analytic solution (12.44); the observed width is roughly the same in numerical solutions.

P. Lax has pointed out (1954, p. 165) that a fundamental inconsistency is implied, or at least suggested, by setting $l = a\Delta x$. The limiting process of letting Δt and $\Delta x \to 0$ presumably yields an exact solution of the differential equations (12.40) for fixed l; and the solution of (12.40) presumably approaches the exact solution of the original fluid-dynamics equations (including the Hugoniot conditions) in the limit as $l \to 0$; but there is no apparent reason for believing that we will get the same result if we let Δt, Δx and l all approach zero simultaneously, which seems to be suggested by writing $l = a\Delta x$. Of course, in an actual calculation one does not carry out the limiting process at all, but is satisfied with a single set of values of a, Δt, Δx; nevertheless, the question raised by Lax has relevance, because if, in the limiting process, it is really necessary to let Δt, $\Delta x \to 0$ first with fixed l, one would conclude that the numerical solution can be accurate only if $a \gg 1$; available evidence (See Section 12.13) indicates that the solution is, however, quite accurate for values of a as low as 1 or 2; this must be counted as in some sense a lucky coincidence.

The most used difference equations are:

$$\frac{R_j^{n+1} - R_j^n}{\Delta t} = u_j^{n+1},$$

$$\frac{u_j^{n+1} - u_j^n}{\Delta t} = -\frac{1}{\rho_0} \frac{(\delta p)_j^n + (\delta q)_j^n}{\Delta r} \left(\frac{R_j^n}{r_j}\right)^{\alpha-1},$$

$$V_{j+\frac{1}{2}}^{n+1} = \frac{1}{\rho_0} \frac{(R_{j+1}^{n+1})^\alpha - (R_j^{n+1})^\alpha}{(r_{j+1})^\alpha - (r_j)^\alpha},$$

(12.45) $\quad \mathscr{E}_{j+\frac{1}{2}}^{n+1} - \mathscr{E}_{j+\frac{1}{2}}^n + \left(\frac{p_{j+\frac{1}{2}}^{n+1} + p_{j+\frac{1}{2}}^n}{2} + q_{j+\frac{1}{2}}^{n+1}\right)(V_{j+\frac{1}{2}}^{n+1} - V_{j+\frac{1}{2}}^n) = 0,$

$$p_{j+\frac{1}{2}}^{n+1} = f(\mathscr{E}_{j+\frac{1}{2}}^{n+1}, V_{j+\frac{1}{2}}^{n+1}),$$

$$q_{j+\frac{1}{2}}^n = \begin{cases} \dfrac{2a^2}{V_{j+\frac{1}{2}}^n + V_{j+\frac{1}{2}}^{n-1}} [(\delta u)_{j+\frac{1}{2}}^n]^2 & \text{if } (\delta u)_{j+\frac{1}{2}}^n < 0 \\ 0 & \text{if } (\delta u)_{j+\frac{1}{2}}^n \geq 0, \end{cases}$$

where $\alpha = 1, 2, 3$, for slab, cylinder, and sphere, respectively.

It is to be noticed that $q_{j+\frac{1}{2}}^n$, like u_j^n, is really centered, timewise, at $t = t^{n-\frac{1}{2}}$, so that the term in q of the fourth equation of (12.45) is correctly centered while that in the second equation is not; it has not been found worthwhile to rewrite the second equation to center it, and in fact doing so may result in instabilities unless it is done in a particular, and not very convenient, way. Perhaps one may justify the lack of centering by noting that since q is not a physically relevant quantity, it need not be treated quite so accurately as other quantities in the difference equations.

It is sometimes preferred to express q in terms of $\partial V/\partial t$ rather than $\partial u/\partial r$. The formula is

$$q = \frac{(\rho_0 l)^2}{V}\left(\frac{\partial V}{\partial t}\right)^2,$$

which is exactly equivalent to (12.41) for slab symmetry, and has the same dominant term as (12.41) for cylindrical and spherical symmetry. But now, in place of $l = a\Delta r$ we must write $l = a\Delta r(r/R)^{\alpha-1}$, for the following reasons: we have no analytic solution to guide us for the cylindrical and spherical cases but, presumably, for a shock whose thickness is small compared to its distance from the origin the plane theory should be approximately correct. The actual thickness of the shock is of course the interval δR of the dependent variable R covered by it, not the interval δr of the Lagrangean coordinate. The plane theory gave δr proportional to l, so δR is proportional to $l\partial R/\partial r = l\rho_0/\rho$ in the plane case. Therefore, δR should be approximately proportional to $l\rho_0/\rho$ in the non-plane cases, too. What is required however, is not that all shocks have the same thickness δR but that they cover the same number of net-points (namely about 3 to 5), hence we want $\delta r = a\Delta r$, where a is a dimensionless quantity as before. That is, we want

$$a\Delta r = \delta r = \frac{\delta R}{\partial R/\partial r} = \frac{l\rho_0/\rho}{(r/R)^{\alpha-1}(\rho_0/\rho)}$$

which gives the result stated for l. We therefore replace the last line of (12.45) by

(12.46) $\quad q_{j+\frac{1}{2}}^n = \begin{cases} \dfrac{2(a\rho_0\Delta r)^2}{V_{j+\frac{1}{2}}^n + V_{j+\frac{1}{2}}^{n-1}}\left(\dfrac{V_{j+\frac{1}{2}}^n - V_{j+\frac{1}{2}}^{n-1}}{\Delta t}\right)^2 \left(\dfrac{r_{j+1}}{R_{j+1}^n}\right)^{2\alpha-2} \\ \qquad\qquad\qquad\qquad\qquad\qquad \text{if } V_{j+\frac{1}{2}}^n - V_{j+\frac{1}{2}}^{n-1} < 0 \\ 0 \qquad\qquad\qquad\qquad\qquad\qquad \text{otherwise.} \end{cases}$

The factor $(r_{j+1}/R^n_{j+1})^{2\alpha-2}$ can become quite important in multi-region problems. If preferred, the factor may be time-centered as

$$\left(\frac{2r_{j+1}}{R^n_{j+1} + R^{n-1}_{j+1}}\right)^{2\alpha-2};$$

there is no point to centering it in space because it is already correct to second order if r and R are taken at the same place. We have written $j+1$ rather than say j to avoid an indeterminacy at the origin where $r_0 = R^n_0 = 0$.

That no similar factor belongs in the last equation of (12.45) can be seen by noting that, with the old formula (12.45), the conversion of $(\delta R)^2$ to $(\delta r)^2$ is just compensated by the conversion of $(\partial u/\partial R)^2$ to $(\partial u/\partial r)^2$.

If the Eulerian formulation is used, it is appropriate to take

$$q = \begin{cases} l^2 \rho \left(\dfrac{\partial u}{\partial R}\right)^2 & \text{for } \dfrac{\partial u}{\partial R} < 0 \\ 0 & \text{otherwise,} \end{cases}$$

with $l = a\Delta R$, so that in the difference equations,

(12.47) $\quad q^n_{j+1/2} = \begin{cases} \tfrac{1}{2}a^2(\rho^n_{j+1/2} + \rho^{n-1}_{j+1/2})(u^n_{j+1} - u^n_j)^2 & \text{if } u^n_{j+1} - u^n_j < 0 \\ 0 & \text{otherwise.} \end{cases}$

This essentially has the same form as the last equation of (12.45) but has a different consequence, because of the different significance of the subscripts: it is now the true shock thickness δR that is rendered constant, not δr.

Alternative forms of the viscosity term have been proposed and tested by numerous investigators, but we shall not describe them because we feel that they are all superseded by the method of Lax and Wendroff, described in Section 12.14.

12.12. Stability of the Finite-Difference Equations

Stability of (12.45) is investigated in the usual intuitive fashion: the equations of first variation of the difference equations are written down, the coefficients are regarded as (almost) constant (in any small region), and the theory of Chapter 4 is applied. For slab symmetry and the simple equation of state $\mathscr{E} = pV/(\gamma - 1)$, this program will be carried through far enough to formulate a tentative stability condition. However, a new feature arises because of the unavoidable substantial variation of the coefficients (after linearizing) in distances comparable with Δx; the effect of this is discussed at the end of this section and in the next. It is convenient to take u, V, p, and q as dependent variables (the amplification matrix is of fourth order). We suppose that we are dealing with a region in x, t-plane in which $\partial u/\partial x \leq 0$ throughout, as in the steady-shock example of Section 12.10. (In a region in which $\partial u/\partial x \geq 0$, the pseudo-viscosity term does not enter, and the stability condition is as in Section 12.5.)

Note that we are permitted to assume that the variation of u does not alter the sign of $\delta u^n_{j+\frac{1}{2}}$ in the last equation of (12.45); this is clearly true for a point at which the zero order value of $\delta u^n_{j+\frac{1}{2}}$ is negative, because the first variation is supposed by definition to be smaller than any non-vanishing zero-order quantity. If the zero order value is zero, then the variation of q^n vanishes in any case to first order. It is found that two of the eigenvalues of $G(\Delta t, k)$ are $\lambda = 1$ and $\lambda = 0$, independent of Δt and of k. The other two eigenvalues satisfy the quadratic equation:

$$(\lambda - 1)\left\{\lambda - 1 - \frac{\lambda + 1}{2}(\gamma - 1)z\Delta t\right\} + \left(\frac{4\Delta t}{(\Delta x)^2}\sin^2\beta\right)$$
$$\times \left[\left(2(\lambda - 1) + \frac{\lambda + 1}{2}z\Delta t\right)l^2 z\left(1 + \frac{\gamma - 1}{2}z\Delta t\right) + \lambda\mu\Delta t\right] = 0,$$

where we have used the abbreviations

$$\mu = [\gamma p + (\gamma - 1)q]/V\rho_0^2,$$
$$z = -\frac{1}{V\rho_0}\frac{\partial u}{\partial x} = -\frac{1}{V}\frac{\partial V}{\partial t} > 0$$

(these are zero-order quantities which appear as coefficients in the equations of variation—they are to be treated as locally constant, and therefore their subscripts and superscripts have been omitted), and where, as before, $\beta = k\Delta x/2$, and $l = a\Delta x$ is the coefficient appearing in equation (12.41) for q. It is clear from this quadratic equation for λ that stability as Δt, $\Delta x \to 0$ for fixed l will require in any case that $\Delta t = O[(\Delta x)^2]$. If we take therefore $\Delta t/(\Delta x)^2 = $ constant and let $\Delta t, \Delta x \to 0$, the roots of the quadratic equation are

$$\lambda = 1 \text{ and } 1 - 8l^2 z \frac{\Delta t}{(\Delta x)^2} \sin^2 \beta$$

in this limit. Since z is positive, stability requires that

(12.48) $$8l^2 z \frac{\Delta t}{(\Delta x)^2} \leq 2.$$

In the first edition of this book, an argument was given which claimed to show that, with the inclusion of first-order terms, the roots of the equation are

$$\lambda = 1 + O(\Delta t) \text{ and } 1 - 8l^2 z \frac{\Delta t}{(\Delta x)^2} \sin^2 \beta + O(\Delta t).$$

Hence, (12.48) was taken as guaranteeing the von Neumann condition. This argument is now believed to be questionable, owing partly to an error in the equation in the first edition discovered by George White of Los Alamos, who has studied this stability question in detail, and partly to the omission from the equation of other first-order terms. In any case, the present stability discussion must be regarded as providing only a rough guide to stability, for reasons given in the next paragraph, and must be supplemented by empirical studies, such as those of Section 12.13. Therefore, we stop with the zero-order terms and take (12.48) as a tentative stability condition.

Quite apart from the question of rigor, the interpretation of (12.48) is a little vague for the following reason: the stability concept concerns the behavior of the solution as $\Delta t, \Delta x$ approach zero and begins to make sense only when Δt and Δx are much smaller than all times and distances characterizing the physical situation, and in particular when Δx is much smaller than the shock thickness, i.e. $\Delta x \ll l$. However, Δx is never made that small, because we intentionally always choose l to be comparable with Δx. For this reason (12.48) lacks not only rigor, but also precision; but it can serve as a rough guide in the choice of a mesh.

It might be thought that a better approach to the stability question would be to set $l = a\Delta x$ and then keep $\Delta t/\Delta x$ and a fixed as $\Delta t, \Delta x \to 0$. This has the disadvantage that the zero-order quantities p, q, V, u are

SEC. 12.12] STABILITY OF FINITE-DIFFERENCE EQUATIONS

represented by steeper and steeper curves in the shock region so that they cannot be thought of as locally almost constant there by any stretch of imagination, but it gives a valid result for the regions between shocks. It yields a condition which can be written as

$$(12.49) \qquad \sqrt{\gamma p V}\,\frac{\Delta t}{\rho_0 V \Delta x} \leq 1$$

which is recognized as the Courant-Friedrichs-Lewy condition, because $c = \sqrt{\gamma p V}$ is the adiabatic sound speed and $\Delta X = \rho_0 V \Delta x$ is the distance between two points differing by Δx in the Lagrange variable. *We thus see that the presence of the pseudo-viscosity terms does not alter the stability of the usual difference equations for smooth fluid flow.*

When (12.48) is to be used in practice, one must have an estimate for the zero-order quantity

$$z = -\frac{1}{V\rho_0}\frac{\partial u}{\partial x} = -\frac{1}{V}\frac{\partial V}{\partial t},$$

and we use the analytic solution (12.44) for the plane steady-state shock to provide such an estimate. The value of z thus obtained varies through the shock region, and clearly we must use the maximum of z. The maximum is easily found, and putting $l = a\Delta x$, the stability condition (12.48) takes the form

$$(12.50) \qquad \frac{s\Delta t}{\Delta x} \leq \frac{1}{2a}\sqrt{\frac{2\eta}{\gamma+1}}\,\frac{1}{\eta-1},$$

where s is the shock speed (more accurately, s is the time derivative of the shock's Lagrangean coordinate), and $\eta = V_2/V_1$ is the volume compression ratio of the shock.

In order to compare the two conditions, (12.49) and (12.50), which apply away from and within a shock, respectively, let p_f and V_f denote the pressure and specific volume immediately behind the shock[15] and call

$$L_f = \sqrt{\gamma p_f V_f}\,\frac{\Delta t}{\rho_0 V_f \Delta x};$$

---

[15] Obviously, if condition (12.49) is satisfied behind the shock it will be satisfied in front of it also, for the sound speed is less in front and the specific volume is greater there.

L_f is often called the "Courant number." After a short rearrangement of (12.50) using the Hugoniot conditions, the two requirements can be combined as follows:

$$L_f \leq 1,$$

(12.51)
$$L_f \leq \frac{\sqrt{\eta[\eta - (\gamma - 1)/(\gamma + 1)]}}{2a(\eta - 1)}.$$

For sufficiently weak shocks ($\eta - 1$ small), the first inequality of (12.51) restricts Δt, and for strong shocks the second inequality dominates. If an estimate of the shock strength is not available, one can simply assume the worst; the most severe restriction on L_f is for an infinitely strong shock, for which $\eta = (\gamma + 1)/(\gamma - 1)$, then we find

(12.52)
$$L_f \leq \sqrt{\gamma}/2a.$$

Accordingly, for a problem with very strong shocks, the pseudo-viscosity method reduces the permissible Δt by about a factor $\sqrt{\gamma}/2a$, which is around $1/3$ in practical problems. Otherwise, the reduction of the permissible Δt is less and given by (12.51). See addendum at end of chapter.

As noted above, these results lack not only rigor but also precision. Numerical experiments, described in the next section, suggest that the true restriction on L_f is slightly milder than the second inequality (12.51) for strong shocks, and slightly more severe for weak shocks.

12.13. Numerical Tests of the Pseudo-Viscosity Method

Numerical tests of the method have been performed in various ways by various people. In one series of tests, made with the Univac, equations (12.45) were used with $\alpha = 1$ and with $f(\mathscr{E}, V) = (\gamma - 1)\mathscr{E}/V$. The equations may be thought of as describing the motion of an ideal gas in a pipe. The gas was initially at rest at pressure p_0. A pressure Πp_0, where $\Pi > 1$, was applied at one end of the pipe, as by a piston exerting a constant force. The machine calculated the motion of the gas in a segment of the pipe starting from the piston end and represented by a net of 30 points. The pressure as a function of x for these 30 points at some instant might be as in Figure 12.2, where the presence of the shock is

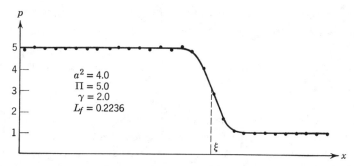

Fig. 12.2. Pressure profile in a steady shock as given by the difference equations.

clearly discernible. As time went on this shock progressed to the right, and the machine automatically kept track of its position, $x = \xi(t)$, defined as the point at which the pressure is halfway between p_0 and Πp_0; this point was found by inverse linear interpolation among the values of p at the net points. When the shock reached the 20th net point, the machine automatically added 10 net points at ambient pressure p_0 ahead and removed 10 net points adjacent to the piston. Thus the motion of the shock could be followed indefinitely, although only 30 points were being followed at any one time by the calculation.

There was provision in the code for typing into the machine values of Π, γ, a^2, and Δt at the beginning of a run. The machine then automatically followed the shock as described above. After an initial period to allow the shock to settle down to a steady shape and to get to a distance of at least 10 net points away from the piston, the machine started keeping track of $\xi(t)$ and other quantities for the next 100 cycles, i.e., for a time 100 Δt. The machine computed the shock speed from a least-squares fit of a straight line to $\xi(t)$ versus t, computed the specific volume V and the material velocity u behind the shock by averaging them for the left-most 5 points (also for the left-most 10 points), and computed the thickness of the shock for an estimate of the maximum slope of the pressure versus x. It then printed a short table on the supervisory control typewriter giving the values of these quantities (determined as described above), their probable errors (determined from the standard deviations during the 100 cycles), and their theoretical or correct values (determined from the Rankine-Hugoniot equations). The machine also wrote more extensive tables on magnetic tape for subsequent printing on a magnetic-tape-controlled typewriter (these tables gave the complete profiles of pressure, volume, material velocity, energy and entropy for the last 5 cycles) and then the machine waited for the parameters of the next problem to be typed in.

TABLE 12.I

Computer Print-Out for Shock Problem

type date	17 nov. 53		
.01 gamma	002000000000		
.01 Pi	005000000000		
.01 a sq.	004000000000		
L sub 0	005000000000		
quantity	theory	expt.	prob. error
shock speed	0.1414213	0.1414191	0.0000147
V (5 pts)	0.5000000	0.5025913	0.0000925
V (10 pts)		0.5013867	0.0000490
u (5 pts)	1.4142135	1.4143566	0.0001610
u (10 pts)		1.4141951	0.0001304
width (net intervals)		3.3651549	0.0023169
L_f	0.2236067		
$L_f$max	0.3535533		
next prob.			
.01 gamma, etc.			

Figure 12.2 shows the pressure profile during a typical cycle of a problem in which the pressure ratio Π was 5.0, γ was 2.0 and a was 2.0. The table printed by the machine during this problem is reproduced as Table 12.I. (The table as actually printed by the machine has been paraphrased very slightly to conform with the notation used here.) The date and the four numbers below it were typed by the operator in response to the requests at the left (this is a mechanism for getting numbers into the machine), and the rest of the table was typed automatically by the

SEC. 12.13] NUMERICAL TESTS OF PSEUDO-VISCOSITY METHOD

machine. The shock speed is expressed in net intervals per cycle—i.e., the printed quantity is

$$\frac{\Delta t}{\Delta x} \frac{d\xi(t)}{dt}.$$

The shock width is expressed in terms of net intervals, the specific volume V and the material velocity u behind the shock are expressed in certain units having to do with the specific volume and sound speed ahead of the shock. "L_f" denotes the value of the Courant number behind the shock, as in the preceding section, and "L_f max" denotes the value which L_f would have to have for the calculation to be on the verge of instability according to the criterion (12.51). It is seen that in this problem the

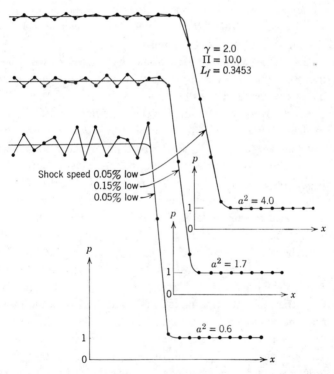

Fig. 12.3. Similar to Figure 12.2 but for a stronger shock and for three different values of the coefficient of pseudo-viscosity.

theoretical and experimental values agree quite well (the agreement on shock speed is fortuitously good in this particular problem).

Several dozen calculations were made with this code, with various values of the parameters. In all cases, if the constant a in the pseudo-viscosity was not less than about $3/4$, there was excellent agreement between "theory" and "experiment." Figure 12.3 shows pressure profiles for a typical cycle in each of three problems which differ as to the value of a used. For these problems, Π was 10.0, γ was 2.0 and Δt was such that the Courant number behind the shock was about $1/3$. It is seen that when a large amount of pseudo-viscosity is used the shock is thick and the pressure profile is smooth behind it; whereas when a small amount of pseudo-viscosity is used the shock is sharper but there are oscillations of pressure behind it. In choosing a value of a, one must compromise between the desiderata of a sharp pressure rise at the shock and a smooth profile behind it. In these three problems the shock speed agrees with the Hugoniot value to about 1 part per mil.

Figure 12.4 shows what happens when the viscosity term is omitted (by setting its coefficient to zero). The pressure oscillations behind the shock wave are severe and the shock speed is in error by 10%.

It should perhaps be pointed out that these oscillations are not instabilities. A calculation such as shown in Figure 12.4 can be continued indefinitely and the amplitude of the oscillation will remain about as shown. The oscillations can in fact be given a physical interpretation. The finite-difference equations describe roughly the motion of a certain model of the fluid. In this model the mass is concentrated into many thin parallel plates and the space between the plates is filled with a weightless compressible fluid which serves as elastic element. When the shock comes along the plates are set into motion much as are the molecules of a gas (a one-dimensional gas in this problem). It was pointed out by R. Peierls (unpublished work, Los Alamos) that these oscillations are analogous to the thermal agitation of the molecules and that they represent the internal energy which must appear in the shocked fluid according to the Hugoniot equations. The purpose of the pseudo-viscosity is to convert this energy of oscillation into true internal energy.

The Univac code described above has also been used to study the stability of the difference equations. For given values of Π, γ, and a, one

SEC. 12.13] NUMERICAL TESTS OF PSEUDO-VISCOSITY METHOD

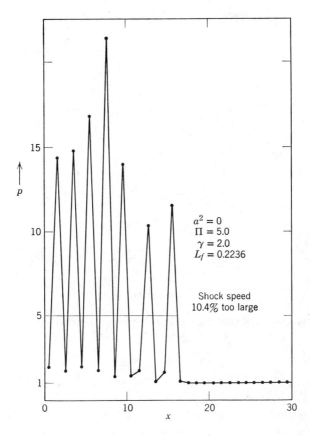

Fig. 12.4. Similar to Figure 12.2 but with the pseudo-viscosity term omitted entirely.

can repeat the problem with different values of Δt. If, for a given value of Δt, the problem runs smoothly and normally for the 100 cycles and gives agreement between "theory" and "experiment" to 1% or better, the equations are regarded as stable with that value of Δt. In all other cases the machine comes to an overflow stop because of some quantity getting out of scaling range long before the 100 cycles are completed. In these cases the equations are regarded as unstable. Figure 12.5 shows the result of investigations of this sort for cases with $\gamma = 2$ and $a = 2$. Three different shock strengths were studied, corresponding to compression

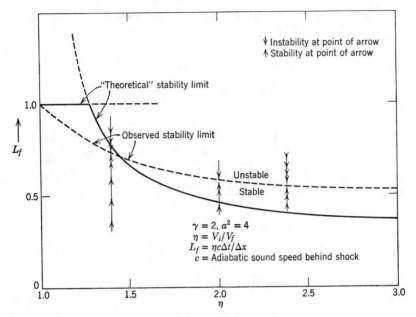

Fig. 12.5. Results of a numerical study of the stability of the difference equations with the pseudo-viscosity term included.

ratios $\eta = 1.4$, 2.0, and 2.383.[16] In each case an upward arrow point indicates a value of L_f for which the equations are stable and a downward arrow indicates a value for which they are unstable. The dashed line has been faired in between the arrow points to give an approximate observed stability limit. The solid line is the "theoretical" stability limit given by (12.51). We see that the equations are slightly more stable for strong shocks, and slightly less stable for weak shocks, than was expected.

12.14. The Lax-Wendroff Treatment of Shocks

The effect of the artificial viscosity is to cause short-wavelength components of a solution to be attenuated as the calculation progresses. Any difference scheme in which those components are attenuated may be

---

[16] For $\gamma = 2$ the largest possible compression ratio is 3.

SEC. 12.14]   THE LAX-WENDROFF TREATMENT OF SHOCKS

thought of as containing the effect of viscosity, even though no particular set of terms is necessarily regarded as representing the viscosity. Usually, one can rewrite the equations of such a scheme so that dissipative terms appear as a correction to an otherwise non-attenuating scheme.

The Lax-Wendroff scheme (12.22) for the equation

$$(12.53) \qquad \frac{\partial \mathbf{U}}{\partial t} + A \frac{\partial \mathbf{U}}{\partial x} = 0,$$

where $\mathbf{U}$ is a vector and A is a constant matrix, may be written as

$$(12.54) \qquad \frac{\mathbf{U}_j^{n+1} - \mathbf{U}_j^n}{\Delta t} + A \frac{\mathbf{U}_{j+1}^n - \mathbf{U}_{j-1}^n}{2\Delta x} = \frac{\Delta t}{2} A^2 \frac{\mathbf{U}_{j+1}^n - 2\mathbf{U}_j^n + \mathbf{U}_{j-1}^n}{(\Delta x)^2},$$

which might tempt one to regard the right member as an approximation to a dissipative term $\frac{1}{2}\Delta t \, A^2 \, \partial^2 \mathbf{U}/\partial x^2$, which therefore should be included in the differential equation. To do so would be incorrect, because the alleged dissipative term is of the same order as the truncation error of the left member; under these circumstances, one can shift terms back and forth between the two sides and obtain a variety of alleged corrections to the differential equation, some of which are not even dissipative. To make the process unique, we suppose that $\mathbf{U} = \mathbf{U}(x, t)$ is an exact solution of the equation

$$(12.55) \qquad \frac{\partial \mathbf{U}}{\partial t} + A \frac{\partial \mathbf{U}}{\partial x} = Q\mathbf{U},$$

and we try to determine the operator Q so that $\mathbf{U}$ satisfies

$$(12.56) \quad \mathbf{U}_j^{n+1} - \mathbf{U}_j^n + \frac{1}{2}\frac{A\Delta t}{\Delta x}(\mathbf{U}_{j+1}^n - \mathbf{U}_{j-1}^n)$$

$$- \frac{1}{2}\left(\frac{A\Delta t}{\Delta x}\right)^2 (\mathbf{U}_{j+1}^n - 2\mathbf{U}_j^n + \mathbf{U}_{j-1}^n) = O[(\Delta t)^4],$$

rather than merely $O[(\Delta t)^3]$, as it would be without the term $Q\mathbf{U}$ in (12.55). Assuming Q to be a differential operator with constant coefficients such that $Q\mathbf{U} = O[(\Delta t)^2]$, we expand the terms of (12.56) in Taylor's series about the point $j\Delta x$, $n\Delta t$ and equate to zero the sum of the terms of order $(\Delta t)^3$; we also make repeated use of the equation (12.55) itself, and we find that

(12.57) $$Q\mathbf{U} = -\tfrac{1}{6}A[(\Delta x)^2 - A^2(\Delta t)^2]\frac{\partial^3 \mathbf{U}}{\partial x^3}.$$

Unfortunately, the differential equation (12.55), with $Q\mathbf{U}$ given by (12.57), is not dissipative, although it is dispersive: different Fourier components of the solution travel with different speeds, but without change of amplitude.

In order to obtain a dissipative term, it is necessary to replace $O[(\Delta t)^4]$ in (12.56) by $O[(\Delta t)^5]$; then Q is replaced by $Q + Q'$, where the additional correction term $Q'\mathbf{U}$ is given by

(12.58) $$Q'\mathbf{U} = -\tfrac{1}{8}A^2\Delta t[(\Delta x)^2 - (A\Delta t)^2]\frac{\partial^4 \mathbf{U}}{\partial x^4},$$

which introduces dissipation into (12.55)—it causes the amplitude of the Fourier component $\exp(ikx)$ to decrease as $\exp(-\text{const.}\ k^4 t)$, assuming, as is necessary for stability, that $(\Delta x)^2 - (\lambda \Delta t)^2$ is positive for every eigenvalue λ of A.

A similar analysis of the von Neumann-Richtmyer difference equations (12.45) shows that in this case the lowest order correction (it is of order $(\Delta t)^2$) is both dispersive and dissipative.

It occasionally has been argued that the Lax-Wendroff equations must be in some sense less accurate than the centered or "leap-frog" equations because of the damping of those Fourier components $\exp(ikx)$ having $k\Delta x \sim 1$, which does not occur for the "leap-frog" equations or for the differential equations. This argument fails to take account of the phases of the Fourier coefficients, which are falsified, for $k\Delta x \sim 1$, by both the Lax-Wendroff and the leap-frog schemes, in fact by an amount of order $(k\Delta x)^3$—the same amount for both schemes—which is one order of magnitude larger than the falsification of the amplitudes (the damping). It seems to us that to retain the short-wave Fourier components with unchanged amplitude is unrealistic under these circumstances.

It should perhaps be pointed out that the dissipation does not represent any loss of energy or other conserved quantities; the kinetic energy that is removed from high-frequency oscillations is left in the system as internal energy of the fluid.

It seems preferable to study the effect of dissipation on the level of the difference equations rather than that of the differential equations.

SEC. 12.14] THE LAX-WENDROFF TREATMENT OF SHOCKS

Some progress has been made by Lax and Wendroff (1960) toward a theoretical understanding of the approximations to solutions with shocks obtained from dissipative difference equations of the Lax-Wendroff type (their results will be summarized at the end of this section), but the main evidence is still empirical.

As an example, we have computed the propagation of a plane steady-state shock in an ideal gas by the Lax-Wendroff equations, in a series of tests similar to that described in Section 12.13 for the older method. The two-step scheme (12.24a) and (12.24b) was used, with **U** and **F(U)** given by (12.21), and with the equation of state taken as $f(\mathscr{E}, V) = p = (\gamma - 1)\mathscr{E}/V$.

The initial-value problem studied was that of a piston moved into a tube of previously motionless gas by applying to the piston a constant force to give a pressure 5 times that of the gas ahead. The velocity profile near the shock front is shown for two cases in Figures 12.6 and 12.7, both for the instant at which the shock reached the 80th net-point down the tube from the piston (there were 100 net-points in all). The position of the shock was taken to be that abscissa for which the velocity was equal to

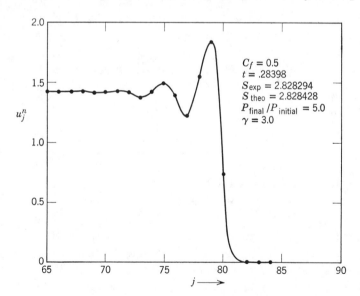

Fig. 12.6. Shock calculation by the Lax-Wendroff scheme.

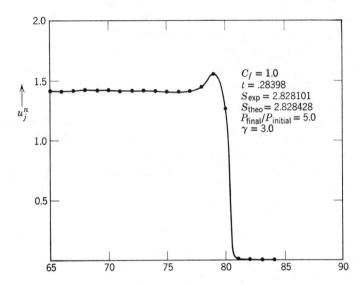

Fig. 12.7. Same as Figure 12.6, but for a larger value of Δt.

one half of its final value, as determined by linear interpolation. The shock velocity was determined from the time required for this position to advance from the 20th to the 80th net-point. It was found that this experimental shock velocity agreed with the theoretical one, computed from the first Hugoniot equation (12.37), to a small fraction of a percent.

The shock front is remarkably sharp, but behind it are oscillations, which are quite small when the Courant number behind the shock, given by

$$C_f = \frac{V_0}{V} \frac{c \Delta t}{\Delta x},$$

where V and c refer to conditions far behind, has its maximum value $C_f = 1.0$. The oscillations are considerably larger when $C_f = 0.5$.

Lax and Wendroff have shown (1960) how to reduce these oscillations by introducing additional dissipative or viscosity terms in the difference equations. More recently, it has been found (P. D. Lax, S. Burstein, A. Lapidus, private communication) that these additional terms are necessary to prevent non-linear instabilities in some problems. The reason is that if zero is an eigenvalue of the matrix A, then the correspond-

SEC. 12.14] THE LAX-WENDROFF TREATMENT OF SHOCKS

ing eigenvalue g of the amplification matrix G given by (12.25) equals 1 for all α (i.e., for all k) rather than < 1 for $k > 0$. In other words, the system is not dissipative in the sense of Kreiss (see Section 5.4), so that nonlinearities can destroy the stability. The eigenvalues of A are u and $u \pm c$, so that there is danger of instability near a stagnation point or a sonic point. The remedy of Lax and Wendroff (1960) was to add to the right member of (12.22) an artificial viscosity term of the form

$$\frac{\Delta t}{2\Delta x}[Q_{j+\frac{1}{2}}(\mathbf{U}_{j+1} - \mathbf{U}_j) - Q_{j-\frac{1}{2}}(\mathbf{U}_j - \mathbf{U}_{j-1})];$$

here, $Q_{j+\frac{1}{2}} = Q_{j+\frac{1}{2}}(\mathbf{U}_j, \mathbf{U}_{j+1})$ is a matrix which is so chosen as to give the desired dissipative effects when $\mathbf{U}_j$ and $\mathbf{U}_{j+1}$ are widely different, but so as to be negligible when they are nearly equal. The choice made by Lax and Wendroff is based on the following considerations: (1) if A were constant, then a change of dependent variables in the difference equations (12.54) which diagonalizes A would separate (12.54) into p (= 3 for fluid dynamics) independent or uncoupled scalar equations, and it seems advisable to insure that the dissipation would be effective for each of them; (2) for a single scalar equation, a reasonable dissipation and a satisfactory shock profile are obtained by putting $Q_{j+\frac{1}{2}}(U_j, U_{j+1}) = \varkappa|A(U_{j+1}) - A(U_j)|$, where $\varkappa$ is a dimensionless constant of the order of 1, and where $A = dF/dU$, F and U now being scalars. To meet these requirements, $Q_{j+\frac{1}{2}}$ should commute with A, which we interpret to mean $A_{j+\frac{1}{2}} = A(\frac{1}{2}U_j + \frac{1}{2}U_{j+1})$ in the variable coefficient case, and the ith eigenvalue of $Q_{j+\frac{1}{2}}$ should be $\varkappa|a_{j+1}^i - a_j^i|$, where a_j^i and a_{j+1}^i are the ith eigenvalues of $A(U_j)$ and $A(U_{j+1})$. These considerations lead to the choice

$$Q_{j+\frac{1}{2}} = g(A_{j+\frac{1}{2}}),$$

where $g(\cdot)$ is a polynomial of degree $p - 1$ (= 2 for fluid dynamics) which assumes the value $\varkappa|a_{j+1}^i - a_j^i|$ when its argument is equal to $a_{j+\frac{1}{2}}^i$, the ith eigenvalue of $A_{j+\frac{1}{2}}$; that is, it is a Lagrange interpolation polynomial. The coefficients of $g(\cdot)$ are functions of the components of $\mathbf{U}_j$ and $\mathbf{U}_{j+1}$.

If A were constant, $g(\cdot)$ would be identically zero, because then $|a_{j+1}^i - a_j^i|$ would vanish for each i; in general, the dominant term of the dissipation (the one containing the second difference of $\mathbf{U}$) is obtained by replacing

Q by a constant, which gives $(\Delta t/2\Delta x)Q\delta^2 U$. This adds to the amplification matrix (12.25) of the Lax-Wendroff equation a term $-(\Delta t/\Delta x)Q(1 - \cos \alpha)$; since the eigenvalues of Q are non-negative, and generally positive in a region of rapid variation of the coefficients such as a shock where nonlinear effects are most likely, the real parts (hence, also the moduli) of the eigenvalues g of G are generally reduced, even at points where an eigenvalue of A vanishes. If a and q are corresponding eigenvalues of A and Q, then the eigenvalue of G is

$$g = 1 - i \frac{a\Delta t}{\Delta x} \sin \alpha - \left[\left(\frac{a\Delta t}{\Delta x}\right)^2 + \left(\frac{q\Delta t}{\Delta x}\right)\right](1 - \cos \alpha).$$

In the smooth part of the flow, $q \ll a$, but in a shock, the quantity q, which is $\varkappa|a_{j+1} - a_j|$, may be $\sim a$. We assume, however, that even in a shock q does not exceed $\varkappa a^0$, where a^0 is the maximum (in modulus) eigenvalue of A. The ellipse given in the complex plane by g as α varies lies in the unit circle if the bracket above does not exceed 1. The stability condition therefore is

$$\left(\frac{a^0 \Delta t}{\Delta x}\right)^2 + \varkappa \frac{a^0 \Delta t}{\Delta x} < 1,$$

which gives, for Eulerian fluid dynamics,

$$\frac{a^0 \Delta t}{\Delta x} = (|v| + c)\frac{\Delta t}{\Delta x} < \left(1 + \frac{\varkappa^2}{4}\right)^{1/2} - \frac{\varkappa}{2}.$$

This is somewhat more restrictive than the Courant-Friedrichs-Lewy condition—for example, by a factor 0.78 if $\varkappa = 0.5$.

Lapidus has shown (private communication) that a slight modification of the Lax-Wendroff viscosity which has the same dominant term $(\Delta t/2\Delta x)Q\delta^2 U$ can be incorporated into the two-step procedure of Section 12.7 for a third-order system like fluid dynamics where $g(A) = g_0 + g_1 A + g_2 A^2$. Equations (12.24a, b) are replaced by

$$\mathbf{U}_{j+\frac{1}{2}}^{n+\frac{1}{2}} = \tfrac{1}{2}(\mathbf{U}_{j+1}^n + \mathbf{U}_j^n) - \frac{1}{2}\left(\frac{\Delta t}{\Delta x} + g_2^+\right)(\mathbf{F}_{j+1}^n - \mathbf{F}_j^n) - \tfrac{1}{2}g_1^+(\mathbf{U}_{j+1}^n - \mathbf{U}_j^n),$$

$$\mathbf{U}_j^{n+1} = \mathbf{U}_j^n - \frac{\Delta t}{\Delta x}(\mathbf{F}_{j+\frac{1}{2}}^{n+\frac{1}{2}} - \mathbf{F}_{j-\frac{1}{2}}^{n+\frac{1}{2}})$$

$$- \frac{1}{2}\frac{\Delta t}{\Delta x}[g_0^+(\mathbf{U}_{j+1}^n - \mathbf{U}_j^n) - g_0^-(\mathbf{U}_j^n - \mathbf{U}_{j-1}^n)],$$

SEC. 12.14]   THE LAX-WENDROFF TREATMENT OF SHOCKS

where g_0^+, g_1^+, g_2^+ are the coefficients of the polynomial that gives $Q_{j+\frac{1}{2}}$ in terms of $A_{j+\frac{1}{2}}$, and g_0^-, g_1^-, g_2^- are those of the polynomial that gives $Q_{j-\frac{1}{2}}$ in terms of $A_{j-\frac{1}{2}}$.

When the Lagrangean formulation (12.21) of the fluid-dynamical equations is used, the formulas become somewhat simplified. In this case, the eigenvalues of A are 0, $\pm c'$, where $c' = V_0 c/V$, c being the adiabatic sound speed. We write $c' = c'(\mathbf{U})$, since c depends on V and E. Then $g(\cdot)$ reduces to the square term only, and

$$Q_{j+\frac{1}{2}} = \varkappa \frac{|c'(\mathbf{U}_{j+1}^n) - c'(\mathbf{U}_j^n)|}{[c'(\mathbf{U}_{j+\frac{1}{2}}^n)]^2} (A_{j+\frac{1}{2}})^2.$$

By way of partial theoretical justification of the expectation that these procedures will give a correct description of flows with shocks, we quote two results of Lax and Wendroff (1960). The first is expressed in terms of the concept of a weak solution of a system of conservation laws, which will now be defined. First, if $\mathbf{U}(x, t)$ is any genuine solution of (12.16) with initial condition $\mathbf{U}(x, 0) = \mathbf{U}^0(x)$ (given), and if $\mathbf{W}(x, t)$ is any smooth, vector-valued function with the same number of components as $\mathbf{U}$ (namely, p) which vanishes outside a finite region of the x, t-plane (such a function is called a *test function*), then, on taking the scalar product of $\mathbf{W}(x, t)$ with equation (12.16), integrating over all x and all $t > 0$, and then integrating by parts, we find

$$\iint \left(\frac{\partial \mathbf{W}}{\partial t} \cdot \mathbf{U} + \frac{\partial \mathbf{W}}{\partial x} \cdot \mathbf{F}\right) dx\, dt + \int \mathbf{W}(x, 0) \cdot \mathbf{U}^0(x) dx = 0.$$

Any $\mathbf{U}(x, t)$, whether a genuine solution or not, that satisfies this equation for all test functions $\mathbf{W}(x, t)$ is called a *weak solution* of (12.16) with initial values $\mathbf{U}^0(x)$. It follows from this definition (see Lax, 1957) that at a jump discontinuity a weak solution satisfies the jump condition

$$S[\mathbf{U}] = [\mathbf{F}(\mathbf{U})],$$

where $[f]$ denotes the difference $f(x + 0) - f(x - 0)$ of any function $f(\cdot)$ across the jump, and S is the velocity with which the discontinuity moves. Since the components of $\mathbf{U}$ and of $\mathbf{F}(\mathbf{U})$ are the densities and flux densities of mass, momentum, and energy, this jump condition is equivalent to the Rankine-Hugoniot condition if the jump is a shock and to the condition of continuity of pressure and fluid velocity if the jump is a contact discontinuity. (These are the only possible types of simple jumps.) The results of Lax and Wendroff, referred to above, are

1. *If the solution* $\mathbf{U}_j^n = \mathbf{U}(j\Delta x, n\Delta t)$ *of the Lax-Wendroff equations* (12.22), *with or without the viscosity term, converges boundedly almost everywhere to a function* $\mathbf{V}(x, t)$, *then* $\mathbf{V}(x, t)$ *is a weak solution of the system of conservation laws* (12.16).

2. *In a certain approximation, any steady-state solution of the difference equations contains simply a transition, over two mesh intervals, between constant states that can be joined by a positive shock or by a contact discontinuity.*

12.15. The Method of S. K. Godunov

In 1959, Godunov described an ingenious method for one-dimensional problems with shocks. It is based on the Lagrangean equations

$$(12.59) \qquad \frac{\partial}{\partial t}\mathbf{U} + \frac{\partial}{\partial x}\mathbf{F}(\mathbf{U}) = 0,$$

where

$$(12.60) \qquad \mathbf{U} = \begin{bmatrix} V \\ u \\ E \end{bmatrix}, \qquad \mathbf{F}(\mathbf{U}) = V_0 \begin{bmatrix} -u \\ p \\ pu \end{bmatrix},$$

as in Section 12.7. It has in common with the two-step Lax-Wendroff method that, when the $\mathbf{U}_j^n$ are known for all j, approximate provisional values of $\mathbf{F}_{j+\frac{1}{2}}^{n+\frac{1}{2}}$ are first computed, and the $\mathbf{U}_j^{n+1}$ are then computed from the centered formula

$$(12.61) \qquad \mathbf{U}_j^{n+1} - \mathbf{U}_j^n = \frac{\Delta t}{\Delta x}(\mathbf{F}_{j+\frac{1}{2}}^{n+\frac{1}{2}} - \mathbf{F}_{j-\frac{1}{2}}^{n+\frac{1}{2}}).$$

According to the second equation (12.60), the provisional values of u and p only are needed. They are obtained from a physical model in which the distributions of u, p, V, E at $t = n\Delta t$ are approximated by step distributions such that these quantities have constant values in each interval $((j - \frac{1}{2})\Delta x, (j + \frac{1}{2})\Delta x)$. For a time of the order of $\Delta x/2c$, the

SEC. 12.15] THE METHOD OF S. K. GODUNOV

exact behavior of this model is known: each discontinuity immediately resolves into either two shock waves, two rarefaction waves, or one shock and one rarefaction. (In limiting cases, either wave or both may be of zero strength.) One of these waves moves toward increasing x (x is the Lagrangean coordinate) and the other toward decreasing x. Between the waves, in an interval of x that contains the initial discontinuity at, say $(j + \frac{1}{2})\Delta x$, u and p are constant in both space and time (until the waves from neighboring discontinuities have had time to run together), while V and E have jumps constituting a contact discontinuity at $(j + \frac{1}{2})\Delta x$. The constant values of u and p between the waves are taken for $u_{j+1/2}^{n+1/2}$ and $p_{j+1/2}^{n+1/2}$.

Since these values do not depend on Δt at all, it is clear that they cannot have the accuracy of the equation (12.24a) of the two-step Lax-Wendroff scheme: the error here is $O(\Delta x)$, because the magnitude of each assumed discontinuity is proportional to Δx, in the region where the true flow is smooth. The error can also be expressed as $O(\Delta t)$, because $\Delta t/\Delta x$ is assumed constant during refinements of the net; for comparison, the error of (12.24a) is $O[(\Delta t)^2]$. However, the Godunov procedure has the virtue of treating any real discontinuity the flow may contain more realistically than the viscosity methods do.

The formulas for the resolution of a discontinuity in fluid dynamics, with constant states on either side, are well known; they will now be derived in convenient form, assuming the equation of state to be that of a perfect gas. We denote the unknowns $p_{j+1/2}^{n+1/2}$ and $u_{j+1/2}^{n+1/2}$ temporarily by p^* and u^*. If the wave that moves to the right from the point $x = (j + \frac{1}{2})\Delta x$ is a shock, then the jump in fluid velocity is related to the jump in pressure by the momentum equation (12.35), which here takes the form:

(12.62) $$M^+(u^* - u_{j+1}^n) = p^* - p_{j+1}^n,$$

where M^+ is the mass of fluid swept over by the shock in unit time per unit area, namely, $\dot{\xi}/V_0$, where $x = \xi(t)$ is the shock position, and is given by the equation

(12.63) $$M^+ = \left(\frac{(\gamma + 1)p^* + (\gamma - 1)p_{j+1}^n}{2V_{j+1}^n}\right)^{1/2}$$

obtainable from (12.33) and (12.37) together with the assumed equation of state.

340 FLUID DYNAMICS IN ONE SPACE VARIABLE [CHAP. 12

If the wave moving to the right is not a shock but a rarefaction, then the central region where p and u have the constant values p^*, u^* is separated, in the x, t-plane, from the region where they have the constant values p_{j+1}^n, u_{j+1}^n by a fan, i.e., a region between two straight lines radiating from the point $P_0{}' = ((j + \frac{1}{2})x, n\Delta t)$; between these lines, u, V, p, ρ depend on x and t only through the similarity variable $[x - (j + \frac{1}{2})\Delta x]/(t - n\Delta t)$, which will be denoted by η. The second equation of the system (12.59) then gives

$$\eta \frac{du}{d\eta} = V_0 \frac{dp}{d\eta},$$

from which we obtain an equation of the same form as (12.62), namely,

(12.64) $$M^+(u^* - u_{j+1}^n) = p^* - p_{j+1}^n,$$

but where now M^+ is defined by the equation

(12.65) $$M^+ = \frac{1}{V_0} \frac{\int \eta \, du}{\int du} = \frac{1}{V_0} \bar{\eta},$$

in which the integration is extended from u^* to u_{j+1}^n; M^+ is the mass swept over in unit time per unit area by that characteristic within the rarefaction fan that has the slope $\bar{\eta}$.

In order to obtain an explicit expression for M^+, each of the first two equations (12.59) is written in the similarity variable η; each is solved for $du/d\eta$; the results are multiplied together and the square root is taken:

$$\frac{du}{d\eta} = \sqrt{-\frac{dV}{d\eta} \frac{dp}{d\eta}}.$$

Next, V is eliminated by means of the isentropic law

$$\frac{V}{V_{j+1}^n} = \left(\frac{p}{p_{j+1}^n}\right)^{-1/\gamma};$$

this leads to the equation

$$\frac{du}{d\eta} = \frac{2}{\gamma - 1} \frac{dc}{d\eta},$$

c being the sound speed $\sqrt{\gamma p V}$; hence, $2c/(\gamma - 1) - u$ is constant in the rarefaction wave. Use of the isentropic law once more leads to the desired expression

$$(12.66) \quad M^+ = \frac{\Delta p}{\Delta u} = \frac{\gamma - 1}{2\sqrt{\gamma p_{j+1}^n V_{j+1}^n}} \frac{p_{j+1}^n - p^*}{1 - (p^*/p_{j+1}^n)^{(\gamma-1)/2\gamma}}.$$

On combining formula (12.62), which is used for $p^* > p_{j+1}^n$, with formula (12.66), which is used for $p^* \leq p_{j+1}^n$, we have that

$$(12.67) \quad M^+ = \sqrt{p_{j+1}^n/V_{j+1}^n}\ \varphi(p^*/p_{j+1}^n),$$

where

$$(12.68) \quad \varphi(w) = \begin{cases} \sqrt{\dfrac{\gamma+1}{2} w + \dfrac{\gamma-1}{2}}, & \text{for } w > 1 \\[2mm] \dfrac{\gamma-1}{2\sqrt{\gamma}} \dfrac{1-w}{1 - w^{(\gamma-1)/2\gamma}}, & \text{for } w \leq 1. \end{cases}$$

For the wave moving to the left, we similarly have

$$(12.69) \quad -M^-(u^* - u_j^n) = p^* - p_j^n;$$

where

$$(12.70) \quad M^- = \sqrt{p_j^n/V_j^n}\ \varphi(p^*/p_j^n).$$

Equations (12.62) and (12.69), with M^+ and M^- expressed in terms of the function $\varphi(\cdot)$, determine p^* and u^*. Because of the complicated way in which p^* enters in $\varphi(\cdot)$, the equations are solved by an iterative procedure: when approximate values of M^+ and M^- are known (obtained from the preceding iteration or, in the case of the first iteration, from the preceding time step), an approximate value of p^* is obtained from the equation

$$(12.71) \quad \frac{p^* - p_j^n}{M^-} + \frac{p^* - p_{j+1}^n}{M^+} + u_{j+1}^n - u_j^n = 0,$$

which results from elimination of u^* from (12.62) and (12.69); this value of p^* is then substituted into (12.67) and (12.70) to give improved values of M^+ and M^-, and so on. After a suitable number of iterations, u^* is then obtained from (12.62) or (12.69) or from the equation

$$(12.72) \quad M^+(u^* - u_{j+1}^n) + M^-(u^* - u_j^n) = p_j^n - p_{j+1}^n,$$

obtained by eliminating p^* from (12.62) and (12.69). If the iteration is carried to convergence, all three of these equations give the same value of u^*; Godunov recommends use of (12.72), presumably because it gives better results when the iteration is not carried to complete convergence.

According to Godunov, this iterative procedure converges, unless one or both waves is a very strong rarefaction. In this case (which must be of rare occurrence), if $p^{*(i)}$ is the result of the ith iteration, one substitutes

$$\frac{\alpha p^{*(i-1)} + p^{*(i)}}{\alpha + 1}$$

for p^* into the function $\varphi(\cdot)$ to obtain the new values of M^+ and M^-, where α is a suitable positive constant. Godunov recommends for α the following expression, obtained from the preceding iteration:

$$\alpha = \text{Max}\left(\frac{\gamma - 1}{3\gamma} \frac{1 - z}{z^{(\gamma+1)/2\gamma} - 1} - 1, 0\right),$$

where z is an abbreviation for $p^{*(i-1)}/(p_j^n + p_{j+1}^n)$. (It is not clear to us why $p^{*(i)}$ should not appear instead of $p^{*(i-1)}$, here.)

If an equation of state other than the ideal-gas law is used, the function $\varphi(\cdot)$ is given by a formula different from (12.68), and some other formula is needed for the convergence parameter α—otherwise the procedure is the same.

Note that according to (12.68) $\varphi(w)$ is continuous at $w = 1$ and that $\varphi(1) = \sqrt{\gamma}$; this value corresponds to propagation with speed $c = \sqrt{\gamma p V}$ (i.e., speed cV_0/V relative to the Lagrangean coordinates) for both weak shock waves and weak rarefaction waves.

Figure 12.8 shows the profile of a plane, steady, strong shock, as computed by this method. The curve was copied from Godunov's paper, from which it appears that the profile is the same, within the accuracy of the graphical presentation, at successive time steps of the calculation. It is stated in the paper that the profile moves to the right with the correct theoretical shock speed—with what accuracy is unfortunately not stated. The method appears to have been extensively used in the Soviet Union.

In order to compare Godunov's method with the other methods that have been discussed in this chapter, for the smooth part of the flow, we

SEC. 12.15] THE METHOD OF S. K. GODUNOV 343

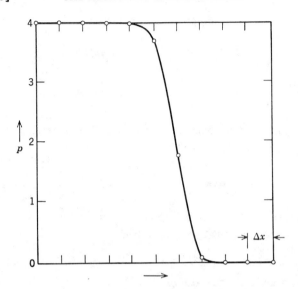

Fig. 12.8. Pressure profile of a strong shock (Godunov, 1959).

linearize equations (12.71) and (12.72) by setting $M^+ = M^- = c/V$ and $V = V_0$, which gives the equations

(12.73)
$$p^* = p_{j+\frac{1}{2}}^{n+\frac{1}{2}} = \tfrac{1}{2}(p_j^n + p_{j+1}^n) - \frac{c}{2V_0}(u_{j+1}^n - u_j^n),$$

$$u^* = u_{j+\frac{1}{2}}^{n+\frac{1}{2}} = \tfrac{1}{2}(u_j^n + u_{j+1}^n) - \frac{V_0}{2c}(p_{j+1}^n - p_j^n).$$

The corresponding linearized version of (12.59) reduces, for isentropic flow, to

(12.74)
$$\frac{\partial u}{\partial t} = -V_0 \frac{\partial p}{\partial x},$$

$$\frac{\partial p}{\partial t} = -\frac{c^2}{V_0}\frac{\partial u}{\partial x};$$

hence, (12.61) reduces to

(12.75)
$$u_j^{n+1} - u_j^n = -\frac{\Delta t}{\Delta x}V_0(p_{j+\frac{1}{2}}^{n+\frac{1}{2}} - p_{j-\frac{1}{2}}^{n+\frac{1}{2}}),$$

$$p_j^{n+1} - p_j^n = -\frac{\Delta t}{\Delta x}\frac{c^2}{V_0}(u_{j+\frac{1}{2}}^{n+\frac{1}{2}} - u_{j-\frac{1}{2}}^{n+\frac{1}{2}});$$

substitution of (12.73) here gives the final equations:

$$u_j^{n+1} - u_j^n = -\frac{\Delta t}{\Delta x} V_0 \tfrac{1}{2}(p_{j+1}^n - p_{j-1}^n) + \frac{c\Delta t}{2\Delta x}(u_{j+1}^n - 2u_j^n + u_{j-1}^n),$$
(12.76)
$$p_j^{n+1} - p_j^n = -\frac{\Delta t}{\Delta x}\frac{c^2}{V_0}\tfrac{1}{2}(u_{j+1}^n - u_{j-1}^n) + \frac{c\Delta t}{2\Delta x}(p_{j+1}^n - 2p_j^n + p_{j-1}^n).$$

These equations differ from the Lax-Wendroff equations only in having $\tfrac{1}{2}c\Delta t/\Delta x$ rather than $\tfrac{1}{2}(c\Delta t/\Delta x)^2$ as coefficient of the last term. In fact, they are equivalent to the Courant-Isaacson-Rees method (see Section 14.3) in the linearized case; in that method, the equations are written in characteristic form, and in each equation a backward or forward spatial difference quotient, respectively, is used when the velocity of the characteristic is positive or negative. If we write

$$\sigma_\pm = p \pm (c/V_0)u,$$

then equations (12.76) take the form

$$\sigma_{+\,j}^{\,n+1} - \sigma_{+\,j}^{\,n} = -\frac{c\Delta t}{\Delta x}(\sigma_{+\,j}^{\,n} - \sigma_{+\,j-1}^{\,n}),$$
(12.77)
$$\sigma_{-\,j}^{\,n+1} - \sigma_{-\,j}^{\,n} = +\frac{c\Delta t}{\Delta x}(\sigma_{-\,j+1}^{\,n} - \sigma_{-\,j}^{\,n}).$$

The Riemann invariants $\sigma_\pm$ are constant, according to (12.74), along forward and backward characteristics; equations (12.77) approximate the values of $\sigma_\pm$ at net-point $(n+1, j)$ by following the characteristics backward to time $n\Delta t$ and then interpolating linearly with respect to x between two neighboring net-points to find $\sigma_\pm$ at the foot of the characteristic. The Lax-Wendroff equations for the linearized problem can be interpreted in the same way, except that the interpolation is quadratic and uses all three net points $(j-1, j, j+1)$ at time $n\Delta t$.

The stability condition of Godunov's method is

$$\frac{c\Delta t}{\Delta x}\frac{V_0}{V} < 1,$$

as for the other Lagrangean methods.

Godunov points out that these equations, at least in the linearized form, have the property that if the dependent variables $\sigma_\pm$ are monotonic functions of x at time t, the calculated values at time $t + \Delta t$ are also monotonic. Therefore, at least for weak shocks, one does not expect the oscillations behind a shock which characterize viscosity methods. Lax and Wendroff (1964) gave an argument to show that for difference equations of a higher order of accuracy (such as the Lax-Wendroff equations) these oscillations cannot be avoided, although they can be minimized. Figure 12.8 indicates that Godunov's method also gives a sharp, clean pressure profile, without oscillations, for a strong shock.

12.16. Magneto-Fluid Dynamics

We conclude this chapter with a brief account of some problems in magneto-fluid dynamics, that is, the flow of an electrically conducting fluid in the presence of an electromagnetic field. The most important field of application is in plasma physics.[17] A plasma may be regarded, for our purposes, as an ionized gas of low density, consisting of negatively charged electrons and positively charged ions. In the simplest model, we consider the plasma as a single fluid of density ρ, pressure p, internal energy per gram $\mathscr{E}$, resistivity η and velocity $\mathbf{u}$, carrying a current $\mathbf{J}$ in a magnetic field $\mathbf{B}$ and electric field $\mathbf{E}$; we shall neglect electrostatic forces. The equations in Eulerian coordinates then consist firstly of the conservation laws of mass, momentum and energy—compare equations (12.2) to (12.4):

(12.78) $$\left(\frac{\partial}{\partial t} + \mathbf{u} \cdot \mathbf{\nabla}\right)\rho = -\rho \mathbf{\nabla} \cdot \mathbf{u},$$

(12.79) $$\rho\left(\frac{\partial}{\partial t} + \mathbf{u} \cdot \mathbf{\nabla}\right)\mathbf{u} = -\mathbf{\nabla}p + \mathbf{J} \times \mathbf{B},$$

(12.80) $$\rho\left(\frac{\partial}{\partial t} + \mathbf{u} \cdot \mathbf{\nabla}\right)\mathscr{E} = -p\mathbf{\nabla} \cdot \mathbf{u} + (\mathbf{E} + \mathbf{u} \times \mathbf{B}) \cdot \mathbf{J},$$

where we have assumed a scalar pressure, satisfying an equation of state of the form

(12.81) $$\mathscr{E} = F(p, \rho),$$

---

[17] For an introduction to this subject, see either Spitzer (1956) or W. B. Thompson (1962).

although in practice it is often necessary to use a pressure tensor because of the anisotropy introduced by the magnetic field. In addition, we have Maxwell's equations, in rationalized mks units and with the displacement current neglected,

(12.82) $$\nabla \cdot \mathbf{B} = 0,$$

(12.83) $$\mu_0 \mathbf{J} = \nabla \times \mathbf{B},$$

(12.84) $$\frac{\partial \mathbf{B}}{\partial t} + \nabla \times \mathbf{E} = 0$$

and the simple form of Ohm's law,

(12.85) $$\mathbf{E} + \mathbf{u} \times \mathbf{B} = \eta \mathbf{J},$$

where μ_0 is the permeability of free space. This last equation is used to eliminate $\mathbf{E}$ from equations (12.80) and (12.84); and then equation (12.83) could be used to eliminate $\mathbf{J}$, although it is often convenient to retain $\mathbf{J}$ as a subsidiary variable. If these substitutions are made, the right-hand side of (12.79) becomes

$$-\nabla\left(p + \frac{1}{2\mu_0} B^2\right) + \frac{1}{\mu_0} (\mathbf{B} \cdot \nabla)\mathbf{B},$$

where B is the magnitude of $\mathbf{B}$, while (12.84) becomes

(12.86) $$\left(\frac{\partial}{\partial t} + \mathbf{u} \cdot \nabla\right)\mathbf{B} = -\mathbf{B}(\nabla \cdot \mathbf{u}) + (\mathbf{B} \cdot \nabla)\mathbf{u} + (\eta/\mu_0)\nabla^2\mathbf{B}.$$

Some simplification results from assuming slab symmetry, for equation (12.82) implies that the x component of $\mathbf{B}$, B_x, is constant. But to compare the above equations with those of fluid dynamics, it is desirable to first consider the result of making two further simplifications—that $B_x = 0$ and that the resistivity $\eta = 0$. Then, $\mathbf{B} \cdot \nabla \equiv 0$, and equation (12.86) becomes identical to the continuity equation (12.78); the magnetic field is carried along with the fluid so that $\mathbf{B}/\rho$ is a function only of the Lagrangean space variable. Indeed, the whole system of equations reduce to those of fluid dynamics as given in (12.5), with the pressure replaced by

$p^* = p + (1/2\mu_0)B^2$ and the internal energy by $\mathscr{E}^* = \mathscr{E} + (1/2\mu_0)B^2/\rho$. In the stability conditions for explicit difference schemes, the local sound speed is then replaced by the magneto-sonic speed $(c^2 + c_A^2)^{1/2}$, where the Alfvén speed $c_A = B/(\mu_0\rho)^{1/2}$.

One of the main differences from fluid dynamics, so far as the solution of practical problems is concerned, is that regions of normal plasma density and field strength are often surrounded by regions of high fields and very low density, where the Alfvén speed is therefore very high. In these cases it is usually necessary to use an implicit difference scheme. Such schemes, with proper centering of all quantities, are not difficult to set up, but they involve considerably more computation in their solution. Because the equations are non-linear, an iterative procedure is involved, and at each stage a tri-diagonal system of equations has to be solved, as for the implicit system (10.6) for the wave equation. The reader is referred to Hain et al. (1960) for details.

Discarding the assumption of infinite conductivity increases the number of equations but makes them little more difficult to treat, except for the fact that a new and important physical phenomenon appears. This is the diffusion of the magnetic field through the fluid, described by the last term in equation (12.86). As a result, the convective terms in the equations have to be treated very accurately and, in one-dimensional problems, Lagrangean coordinates are therefore a great advantage. In experiments aimed at developing explicit two-dimensional schemes, where Lagrangean coordinates are more difficult to use, the "angled derivative" scheme has proved useful—see Roberts and Weiss (1966). In this scheme, proper centering is obtained by using mesh points $(n + 1, j)$ and (n, j) for time differences and $(n, j + 1)$, $(n + 1, j - 1)$ to give the correct component of the space difference. This is still an explicit system if the equations are solved in increasing order of j. As applied to isentropic, adiabatic fluid flow with no magnetic fields,

$$\frac{\partial \rho}{\partial t} + \frac{\partial}{\partial X}(\rho u) = 0$$

$$\frac{\partial u}{\partial t} + u\frac{\partial u}{\partial X} + (c^2/\rho)\frac{\partial \rho}{\partial X} = 0,$$

it gives the scheme

$$\left(1 + \frac{\Delta t}{2\Delta X} u_j^{n+\frac{1}{2}}\right) \frac{\rho_j^{n+1} - \rho_j^n}{\Delta t} + u_j^{n+\frac{1}{2}} \cdot \frac{\rho_{j+1}^n - \rho_{j-1}^{n+1}}{2\Delta X}$$

$$+ \frac{\rho_{j+1}^{n+1} + \rho_j^n}{2} \cdot \frac{u_{j+1}^{n+\frac{1}{2}} - u_{j-1}^{n+\frac{1}{2}}}{2\Delta X} = 0,$$

(12.87) $\left(1 + \frac{\Delta t}{2\Delta X} \cdot \frac{u_{j+1}^{n-\frac{1}{2}} + u_{j-1}^{n+\frac{1}{2}}}{2}\right) \frac{u_j^{n+\frac{1}{2}} - u_j^{n-\frac{1}{2}}}{\Delta t} + \frac{(u_{j+1}^{n-\frac{1}{2}})^2 - (u_{j-1}^{n+\frac{1}{2}})^2}{4\Delta X}$

$$+ (c^2/\rho)_j^n \frac{\rho_{j+1}^n - \rho_{j-1}^n}{2\Delta X} = 0;$$

here, the second equation is a linear, explicit equation for $u_j^{n+\frac{1}{2}}$, and the first then gives ρ_j^{n+1}. The amplification matrix of the linearized equations has the characteristic equation $(\lambda - \alpha)^2 + \beta^2 \lambda = 0$, where

$$\alpha = [1 + \tfrac{1}{2}ul(1 - e^{ik\Delta X})]/[1 + \tfrac{1}{2}ul(1 - e^{-ik\Delta X})],$$

$l = \Delta t/\Delta X$, and $\beta = cl \sin k\Delta X/[1 + \tfrac{1}{2}ul(1 - e^{-iK\Delta X})]$. Thus, the von Neumann condition is satisfied if $|\beta| < 2|\alpha|$, that is,

$$l^2(c^2 - u^2) \sin^2 k\Delta X < [2 + ul(1 - \cos k\Delta X)]^2.$$

From this it is not difficult to see that the stability condition is

(12.88) $$(c - u) \frac{\Delta t}{2\Delta X} < 1;$$

this agrees with the Courant-Friedrichs-Lewy condition since only the "backward" characteristic is treated explicitly and in our scheme the space mesh is effectively doubled.

A further problem, related to the need for accurate calculation of the diffusion of the magnetic field through the fluid, arises in the presence of magneto-fluid dynamic shocks. When the fluid is infinitely conducting, shocks are described by jump conditions, the de Hoffman-Teller shock relations, which are analogous to the Rankine-Hugoniot conditions treated above. But, in general, the magnetic field has a structure in the shock region whose form can be very important in physical problems (this is particularly true when the restriction $\mathbf{B} \cdot \mathbf{\nabla} \equiv 0$ is relaxed). Therefore, it has to be accurately described, and a variable mesh spacing, which is adjusted as the shock moves, is often used.

We will illustrate this problem of the shock structure by considering a slightly different physical model. As stated at the beginning of this section, all the equations so far presented have been based on regarding a plasma as a single fluid. One of the results of taking into account its two-fluid character is a pair of equations of motion—one for each fluid. When added together, these result in equation (12.79) but, when combined differently, they yield an equation for the current $\mathbf{J}$ which replaces the rather arbitrarily assumed Ohm's Law (12.85). Suppose we incorporate in system (12.78) to (12.84) the equation of this kind which is appropriate when the plasma collision frequency is very low. If we retain, for simplicity here, the assumption that $B_x = 0$ and use the Lagrangean coordinate x introduced in Section 12.4, the resulting set of equations for slab symmetry—compare (12.8) and (12.9); see Morton (1964) for a derivation—are:

(12.89)
$$\frac{\partial V}{\partial t} - \frac{\partial u}{\partial x} = 0,$$

$$\frac{\partial u}{\partial t} + \frac{\partial}{\partial x}(p + \tfrac{1}{2}B^2) = 0,$$

$$\frac{\partial \mathscr{E}}{\partial t} + p\frac{\partial V}{\partial t} = 0, \quad \mathscr{E} = F_1(p, V),$$

$$\frac{\partial}{\partial t}\left[\frac{\partial^2 \mathbf{B}}{\partial x^2} - V\mathbf{B}\right] = 0.$$

Here, u is the velocity component in the x direction, only the y and z components of $\mathbf{B}$ are non-zero and some normalization has been carried out. With these equations, the field and its first derivative are continuous even through shocks and the fluid variables are discontinuous only if the compression is sufficiently great.

The usual finite difference equations, as described in Section 12.5, can be used for the first three equations, and then the last can be replaced by

$$\frac{\delta^2 B_j^{n+1}}{(\Delta x)^2} - (VB)_j^{n+1} = \frac{\delta^2 B_j^n}{(\Delta x)^2} - (VB)_j^n.$$

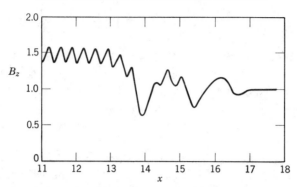

Fig. 12.9. A typical waveform obtained from the "two-fluid" equations when $B_x \neq 0$.

A study of stability then shows phenomena very similar to those encountered in the equation of coupled sound and heat flow in Section 10.4. The energy method leads quite naturally to the sufficient condition $(c^2 + c_A^2)^{1/2} (\Delta t/V\Delta x) < 1$, met already for the case of infinite conductivity. But the characteristics of the system (12.89) are determined only by the adiabatic sound speed c, and a Fourier analysis together with experimental checking shows that the practical stability criterion is

$$(12.90) \qquad \left[c^2 + \left(\frac{(\Delta x)^2 V}{4 + (\Delta x)^2 V}\right) c_A^2\right]^{1/2} \frac{\Delta t}{V\Delta x} < 1.$$

In shocked regions, the von Neumann-Richtmyer artificial viscosity may be used as in pure fluid dynamics; as in that case, the stability range is then substantially reduced. Rather more care in its use is necessary, however. For Figure 12.9 shows that real oscillations can occur in the solution, which are very similar to those shown in Figure 12.3 as due to insufficient artificial viscosity. We have already mentioned, however, that accurate solutions are required in the shocked region so that too much viscosity cannot be tolerated.

Addendum to Section 12.12. G. N. White of Los Alamos has communicated to us the following extensively used stability condition for the viscosity method:

$$L + 4a^2 |V_{j+1/2}^{n+1} - V_{j+1/2}^{n}|/V_{j+1/2}^{n} < 1,$$

where L is the left member of (12.49) and the second term is a difference approximation to half the left member of (12.48).

CHAPTER 13

Multi-Dimensional Fluid Dynamics

13.1. Introduction

For the initial-value problems of fluid dynamics in two or more space variables, the Lagrangean formulation with differencing leads to serious loss of accuracy, owing to the distortion of the Lagrangean point net as time goes on, unless a new net is defined from time to time to restore rectangularity; this leads to difficult and rather inaccurate interpolations. Nevertheless, some good results have been obtained, starting with the work of Kolsky (1955), who first worked out the formulation of the difference equations in detail—see also Blair *et al.* (1959), Goad (1960), and Schultz (1964b).

The Eulerian formulation encounters difficulties if there are non-rigid boundaries or interfaces between fluids of differing thermodynamic properties because there is no very simple way of telling which kind of fluid is to be found, at a given instant, at a given net-point. Considerable progress has been made recently (mostly in problems where this difficulty does not arise) by use of the Lax-Wendroff equations; see, for example, Burstein (1963, 1965), some of whose results will be mentioned briefly in Section 13.5.

Various methods have been developed using a mixed Eulerian-Lagrangean formulation—see, for example, Frank and Lazarus (1964) and Noh (1964). These methods are often advantageous when the character of the flow is radically different in two different directions.

For all these methods, many variants of the artificial viscosity treatment of shocks have been introduced—see Schultz (1964a) and Noh (1964) and the references cited by them. Two-dimensional versions of the Lax-Wendroff terms are described in Section 13.5.

Methods using characteristics and bi-characteristics have been developed by several people, for example, Bruhn and Haack (1958), Richardson (1964), Talbot (1963), and Ansorge (1963)

DeSanto and Keller (1961) have studied the stability of a laminar boundary layer by the use of difference methods for the Navier-Stokes equation in two space variables and time. Because of the strong advection in this problem, it appears to be highly desirable to use a scheme that is implicit relative to the direction parallel to the boundary. Methods for the Navier-Stokes equation were also devised by Fromm (1964).

The particle-in-cell and related methods, developed by Harlow and his associates (Harlow, 1964) at Los Alamos, have been rather successful in many problems. These methods are based not on differencing the partial differential equations, but on certain elementary physical models. In spite of the inherent low accuracy, even when a large expenditure of computer time is made, these methods have turned out to be surprisingly versatile.

In multi-dimensional problems it is often unclear whether the physical problems one wishes to solve are properly posed.[1] The linearized problem of Taylor instability (or of Helmholtz instability) for incompressible fluids is not properly posed—the solution does not exist unless the initial data (describing the initial configuration of the interface) have infinitely many derivatives, and, even when the solution exists, it does not depend continuously on the initial data. It is possible that these problems were properly posed before the linearization, and rather likely that they are properly posed when compressibility is taken into account, and almost certain that they are properly posed if effects like viscosity, surface tension, heat conduction are included. But the basic instability remains in the sense that small irregularities can become highly magnified in a rather short time. This is an instability of a physical nature, and no amount of refinement of the difference equations will remove it. The only way to avoid trouble is to ensure that the representation of initial shapes and flows is just as smooth as the physical model represented, and that no irregularities are subsequently introduced. For example, if a spherical interface is subject to conditions of Taylor instability, it is totally unsatisfactory to represent the interface by some approximate pseudosphere in a rectangular point net, unless the net is exceedingly fine, because the representation effectively introduces irregularities on the

---

[1] This is also true of steady-state problems—see, for example, Morawetz (1956).

SEC. 13.1] INTRODUCTION

scale of the mesh size, and this may be the very scale of irregularity which becomes enormously amplified in a short time by Taylor instability. E. Teller (private communication) has suggested that this difficulty can be removed by adding terms to the equations representing the effect of surface tension on the interface. The amount of surface tension would be chosen artificially, so as to stabilize irregularities on the scale of the mesh size without much affecting irregularities of larger scale, and would have to depend on the instantaneous acceleration of the interface. Such a scheme might take care of Taylor instability, but Helmholtz instability may occur on a contact discontinuity that has been spontaneously generated by the flow (as happens in Mach reflection) and for this case one would not know in advance where to put the artificial surface tension.

We seem to be far from having any universal methods for multi-dimensional fluid dynamics, and it seems likely that for some time there will be nearly as many methods as problems.

In this chapter, attention will be confined mainly to the development of methods in which shocks and other singularities are treated in more detail than in artificial-viscosity or particle-in-cell methods. This choice has been made for two reasons: first, economy usually dictates the use of a rather coarse computational point net in multi-dimensional problems, and procedures like artificial viscosity spread the singularities out to such an extent that detail is largely lost; second, the singularities that can occur in multi-dimensional flows are of great variety and complexity (a few of them are discussed in Section 13.6), and an accurate numerical description of them requires tailoring the procedure to the peculiarities of a given problem.

The development of methods of the type considered here is far from complete at the present time, and the discussion below is necessarily rather sketchy and provisional, in many places. Nevertheless, it is our belief that effective progress is likely to take place in this direction. The opinion has often been expressed that what is needed for multi-dimensional problems is computing machines of greatly increased speed and capacity, whereas it is our opinion that we are more often limited by inadequacies of the mathematical (including numerical) methods than by inadequacies of the computers. We need to know more about the nature of possible singularities, we need to know how to formulate the piecewise analytic initial-value problems in more-or-less well posed forms, we need to find analytic and numerical computing procedures which take these things into account, and we need to investigate the stability and accuracy of these procedures by a combination of empirical and analytical techniques.

13.2. The Multi-Dimensional Fluid-Dynamic Equations

The equations for the smooth parts of the flow were given in the Eulerian form as equations (12.1) to (12.4). The corresponding conservation-law form, analogous to (12.16) to (12.18), for the case of two Cartesian variables X and Y, is

$$(13.1) \qquad \frac{\partial}{\partial t} \mathbf{U} + \frac{\partial}{\partial X} \mathbf{F}(\mathbf{U}) + \frac{\partial}{\partial Y} \mathbf{G}(\mathbf{U}) = 0,$$

where

$$(13.2) \quad \mathbf{U} = \begin{bmatrix} \rho \\ m \\ n \\ e \end{bmatrix}, \quad \mathbf{F}(\mathbf{U}) = \begin{bmatrix} m \\ (m^2/\rho) + p \\ mn/\rho \\ (e+p)m/\rho \end{bmatrix}, \quad \mathbf{G}(\mathbf{U}) = \begin{bmatrix} n \\ mn/\rho \\ (n^2/\rho) + p \\ (e+p)n/\rho \end{bmatrix};$$

here, m and n are the X and Y components, respectively, of momentum per unit volume, e is the total energy per unit volume, and p and ρ are the pressure and the density. In terms of the fluid velocity components u and v, $m = \rho u$ and $n = \rho v$; if $\mathscr{E}$ is the internal energy per unit volume, then $e = \rho[\mathscr{E} + (u^2 + v^2)/2]$, and the equation of state becomes

$$(13.3) \qquad p = f(\mathscr{E}, V) = f\left(\frac{e}{\rho} - \frac{m^2 + n^2}{2\rho^2}, \frac{1}{\rho}\right).$$

Shock fronts, contact discontinuities, and the like are generally located on surfaces moving either with or through the fluid. We shall classify the main types of discontinuities. We suppose that P is a point of such a surface in the neighborhood of which the surface has a continuously turning tangent plane. We suppose that ρ, p, $\mathbf{u}$ have simple jumps, the values on the two sides at P being denoted by subscripts 1 and 2. Let $\boldsymbol{\lambda}$ be a unit vector normal to the surface at P, pointing into the region

SEC. 13.2]　MULTI-DIMENSIONAL FLUID-DYNAMIC EQUATIONS

corresponding to subscript 2. Let S denote the speed of the surface, relative to the Eulerian coordinate system, measured normally to the surface at P; that is, if a line is drawn through P in the direction λ, S is the speed of the point of the intersection of the surface with that line. Since $\lambda \cdot \mathbf{u}$ is the component of fluid velocity in that same direction, the mass of fluid that flows through unit area of the surface per unit time is $\rho(S - \lambda \cdot \mathbf{u})$; conservation of mass requires that

(13.4) $$M \stackrel{\text{def.}}{=} \rho_1(S - \lambda \cdot \mathbf{u}_1) = \rho_2(S - \lambda \cdot \mathbf{u}_2).$$

The change, per unit time, of the momentum of this mass is $M(\mathbf{u}_1 - \mathbf{u}_2)$; the force acting on it is simply $\lambda(p_1 - p_2)$, because unit area of the surface is involved. Therefore, $M(\mathbf{u}_1 - \mathbf{u}_2) = \lambda(p_1 - p_2)$, by Newton's second law; from this we derive two equations

(13.5) $$M(\lambda \cdot \mathbf{u}_1 - \lambda \cdot \mathbf{u}_2) = p_1 - p_2,$$

(13.6) $$M(\lambda \times \mathbf{u}_1 - \lambda \times \mathbf{u}_2) = 0.$$

From the last equation, we see that there are two possibilities: (1) a shock front with $M \neq 0$, in which case *the tangential component of* $\mathbf{u}$ *is continuous* across the front or (2) a contact discontinuity generally combined with a slip surface, with $M = 0$, across which $p_1 = p_2$, $S = \lambda \cdot \mathbf{u}_1 = \lambda \cdot \mathbf{u}_2$, so that the *pressure and the normal component of* $\mathbf{u}$ *are continuous*.

For $M \neq 0$, the energy equation is

(13.7) $$M\left(\mathscr{E}_1 + \frac{u_1^2}{2} - \mathscr{E}_2 - \frac{u_2^2}{2}\right) = p_1 \lambda \cdot \mathbf{u}_1 - p_2 \lambda \cdot \mathbf{u}_2,$$

where u_1 and u_2 stand for the magnitudes of $\mathbf{u}_1$ and $\mathbf{u}_2$ (this equation is satisfied automatically if $M = 0$). The Rankine-Hugoniot jump conditions for a shock can be derived from (13.4) to (13.7); in terms of the specific volume V, they are:

(13.8a) $$\frac{1}{V_1}(S - \lambda \cdot \mathbf{u}_1) = \frac{1}{V_2}(S - \lambda \cdot \mathbf{u}_2) = \sqrt{\frac{p_1 - p_2}{V_2 - V_1}},$$
$$\mathscr{E}_1 - \mathscr{E}_2 = \tfrac{1}{2}(p_1 + p_2)(V_2 - V_1),$$
$$\lambda \times (\mathbf{u}_1 - \mathbf{u}_2) = 0;$$

these equations differ from (12.37) not only because of the increased number of dimensions but also because they refer to an Eulerian coordinate system, so that S enters instead of $d\xi/dt$. By elimination of S, we find

$$(13.8b) \qquad \boldsymbol{\lambda}\cdot(\mathbf{u}_1 - \mathbf{u}_2) = \sqrt{(p_1 - p_2)(V_2 - V_1)},$$

which is the analogue of the second equation of (12.37). As in one space variable, negative shocks ($M < 0$) are ruled out by considerations of entropy or stability.

The numerical treatment of an interface is more complicated than in one space variable (see Section 12.6), because it is generally accompanied by slippage, i.e., by a jump of $\boldsymbol{\lambda} \times \mathbf{u}$.

At the head or foot of a rarefaction fan, no boundary conditions are required beyond the continuity of the flow variables $\mathbf{u}$, p, ρ. We shall not attempt to classify other possible singularities of multi-dimensional flows, but a few of them will be mentioned in Section 13.6.

13.3. Properly and Improperly Posed Problems

As stated in Chapter 3, a linear initial-value problem is *properly posed* if (1) the genuine solutions are dense in $\mathscr{B}$ for all t in some interval $0 \leq t \leq T$ and (2) the genuine solution operator is bounded for t in that interval; that is, there is a constant $K > 0$ such that every genuine solution $u(t)$ satisfies the inequality $\|u(t)\| < K\|u(0)\|$ for $0 \leq t \leq T$; that is, the solution depends continuously on the initial data. For non-linear problems, condition (1) is retained unchanged, but (2) is reformulated as follows: Let $u(t)$ be a genuine solution with initial element $u_0 = u(0)$; let $u(t) + \delta u(t)$ be obtained from an initial element $u_0 + \delta u_0$ by taking $\delta u(t)$ as a genuine solution of the linearized problem (linearized with respect to $u(t)$ as the zero-order solution) with initial element δu_0; if there is a $K > 0$ such that $\|\delta u(t)\| \leq K\|\delta u_0\|$ in some interval $0 \leq t \leq T$, for all such $\delta u(t)$, the problem is called *properly posed with respect to the initial element u_0*. In other words, the problem that results from linearizing about $u(t)$ must be properly posed.

As a first example of an improperly posed problem, we consider the phenomenon of Helmholtz instability [for a discussion of this and the related phenomenon of Taylor instability, see Birkhoff (1962)]. If an

SEC. 13.3] PROPERLY AND IMPROPERLY POSED PROBLEMS

ideal incompressible fluid has uniform motion with speed U in the $+X$ direction at all points above the X, Y-plane and uniform motion with the same speed U in the $-X$ direction at all points below this plane, then this motion is a stationary solution of the equations of fluid dynamics, but it is an unstable solution. One may consider a perturbation in which the flow is still irrotational both above and below the slip surface and asymptotically constant for $Z \to \pm \infty$, but in which the slip surface has a small-amplitude corrugation given by $Z = Z_0 \sin kX = Z_0 \sin(2\pi/\lambda)X$, where $Z_0 \ll \lambda$, at time $t = 0$. The solution of this initial-value problem shows that at a later time t the amplitude of the corrugation $Z_0(t)$ is of the form $ae^{\omega t} + be^{-\omega t}$, where a and b are constants, and $\omega = kU$. This solution is valid as long as $|Z_0(t)| \ll \lambda$. By taking λ small enough (k large enough), $|Z_0(t)|/|Z_0(0)|$ can be made arbitrarily large for any fixed $t > 0$. Note that for this purpose we do not have to violate the assumption $|Z_0(t)| \ll \lambda$ on which the approximate solution $ae^{\omega t} + be^{-\omega t}$ was based; we merely have to choose k so that $e^{\omega t} = e^{kUt}$ is sufficiently large and then choose a so that $ae^{\omega t} \ll 2\pi/k$ for the given t.

For this solution, the departure of the fluid velocity $\mathbf{u} = \mathbf{u}(X, Z, t)$ from the zero-order value $(\pm U, 0, 0)$ is also proportional to $Z_0(t)$ both above and below the slip surface. Consequently, for any choice of norm for the perturbation based on the departures $\delta \mathbf{u}$ and δp of the fluid velocity and of the pressure from their zero order values, the genuine solution operator of the linearized problem is necessarily unbounded.

The problem of Taylor instability is similar, except that there are two different fluids, initially at rest, of densities ρ and ρ' above and below the X, Y-plane; if $\rho > \rho'$, and if the entire system is in uniform acceleration a in the $+Z$ direction (or, equivalently, is in a gravitational field a in the $-Z$ direction), then small corrugations grow in a similar manner, but now the exponential growth rate is given by

$$\omega = \sqrt{[(\rho - \rho')/(\rho + \rho')]\, a|k|}.$$

Here, also, the solution operator of the linearized problem is unbounded for any $t > 0$.

If the fluid velocity and the density both have discontinuities at $Z = 0$ and, in addition, the interface is subject to a surface tension with coefficient

v (Birkhoff's phrase "interfacial tension" is here more appropriate), then the growth rate is given by

$$(13.9) \qquad \omega^2 = \frac{\rho\rho' k^2}{(\rho + \rho')^2}(U - U')^2 + \frac{\rho - \rho'}{\rho + \rho'} a|k| - \frac{v|k|^3}{\rho + \rho'}$$

(see Birkhoff, p. 61). If $v \neq 0$ (i.e., $v > 0$), the effect of the surface tension is to stabilize the short-wavelength corrugations in such a way that the problem is now properly posed. Note that the stationary solution is still unstable (relative to longer wavelengths). There is no objection to this and, in fact, one may well wish to calculate the growth of such instabilities numerically; what is objectionable in the problems with $v = 0$ is the *unbounded* instability as $\lambda \to 0$ ($k \to \infty$).

A consequence of the unboundedness of the solution operator of the problems with $v = 0$ is that if the initial shape of the interface is represented by a Fourier series or integral, say

$$Z = \sum_{(n)} Z_0(n) \sin nk_0 X$$

(with cosine terms, too, if you wish), then no solution exists at all, for any $t > 0$, unless the coefficients $Z_0(n)$ decrease rapidly (in fact, at least exponentially) with increasing n. In the linearized theory, the interface must therefore be an analytic surface for a solution to exist. Fine-grained roughness, even of very small amplitude, results in immediate fine-scale mixing of the two fluids and increasing diffuseness of the interface, as observed physically, for $v = 0$.

The effect of compressibility has still to be taken into account. It is clear on physical grounds that compressibility will be important only if the time $\lambda/c, = 2\pi/kc$, required for a sound wave to travel from one corrugation to the next, is at least comparable with the time $1/\omega$ for the amplitude to increase noticeably. From equation (13.9) it is then seen that compressibility cannot influence Taylor instability at short wavelengths, but is expected to have some influence, although probably not a dominant one, on Helmholtz instability (here the two times mentioned vary at the same rate as $k \to \infty$). We state without proof that, if the slip speed $|U - U'|$ exceeds the sound speed c by a certain factor, then the Helmholtz instability is stabilized, whereas, for smaller slip speeds, the problem remains improperly posed for $v = 0$.

SEC. 13.3] PROPERLY AND IMPROPERLY POSED PROBLEMS

If the two fluids are different, say air and water, surface tension surely plays a role, and it ought to be included in the formulation of a problem. However, slip surfaces can appear in a single fluid, as a result of the motion itself. An example is provided by the phenomenon of Mach reflection, which can be described with the aid of Figure 13.1—see Courant and

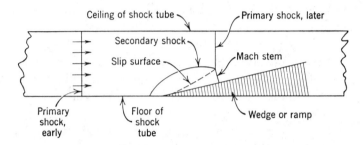

Fig. 13.1. Mach reflection.

Friedrichs (1948) and Duff (1962). A plane shock wave moving from left to right down a shock tube meets a wedge or ramp on the floor of the shock tube. A disturbance of the flow starts from the tip of the wedge when the shock wave touches it and spreads out subsequently; this disturbance has a curved secondary shock as its frontier. Under certain conditions of wedge angle and primary shock strength, the primary and secondary shocks meet at a certain point, called the *triple point*, somewhat above the wedge; from this point on down to the wedge, the two shocks are coalesced into a single (also curved) shock called the *Mach stem*. Downstream from the triple point is a slip surface, across which the pressure and the normal velocity component are continuous, while the density and the tangential velocity component have jumps. A fluid element immediately below the slip plane was compressed by a single shock, and a fluid element immediately above was compressed by two shocks in succession to the same final pressure; the compression by a single shock produces a larger entropy increase because the entropy increase produced by a shock varies roughly as the cube of the shock strength; hence, the temperature and the specific volume are higher just below the slip line than just above; hence, also, the speed of flow away from the shocks is higher just below, to achieve mass conservation.

The photographs reproduced in the paper of Duff (1962) show that this slip surface is indeed highly unstable relative to short-wavelength disturbances and that somewhat downstream it is completely washed out. In some earlier photographs, for example, those of Bleakney and Taub (1949), the slip surface appears quite smooth. It would seem that, under some conditions, if the gas ahead of the primary shock is sufficiently quiescent, then any small disturbances that may be present initially do not grow to an observable extent in the exceedingly short time interval during which the phenomenon takes place; however, there can be no doubt that the instability at short wavelengths is really present.

In a numerical calculation, one cannot count on the initial disturbances being small because the finite-difference procedures necessarily introduce perturbations of a wavelength comparable with the spacing of the points in the computational net. It was proved in Section 3.9 that a finite difference scheme for a problem with an unbounded solution operator is necessarily unstable. For a calculation to be successful for an improperly posed problem, the difference equations must have a high order of accuracy in the space variables, the initial data must be smooth, and the circumstances must be such that the entire calculation requires only a relatively small number of cycles (say a dozen or so at most) in the variable t. Such calculations have been made by Van Dyke in the detached shock problem. In general, however, there would seem to be no hope for finite-difference methods unless the physical formulation of a problem is somehow altered so as to make it properly posed, for example, by including terms that represent surface tension (on each slip surface or contact discontinuity present) of sufficient magnitude to stabilize all disturbances with wavelength comparable to the mesh spacing.

13.4. The Two-Step Lax-Wendroff or L-W Method

For smooth flows, the Lax-Wendroff method gives excellent results. As in the case of one space variable (see Section 12.14) it can sometimes be used when shocks are present if the spatial resolution required is not very great and shock profiles like those of Figures 12.6 and 12.7 are acceptable. In the generalization [Richtmyer (1962)] of the two-step

SEC. 13.4] THE TWO-STEP LAX-WENDROFF OR L-W METHOD

method (12.24a) and (12.24b) to two Cartesian spatial variables X and Y, it is convenient to change the notation slightly from that of Chapter 12: fractional indices will be avoided by thinking of the unit cell of the calculational net as one having dimensions $2\Delta X$, $2\Delta Y$, and $2\Delta t$ in the X, Y, and t directions. Provisional values at $t = (n + 1)\Delta t$ are first obtained from the equation

$$(13.10) \quad \mathbf{U}_{jl}^{n+1} = \tfrac{1}{4}(\mathbf{U}_{j+1\,l}^{n} + \mathbf{U}_{j-1\,l}^{n} + \mathbf{U}_{j\,l+1}^{n} + \mathbf{U}_{j\,l-1}^{n})$$
$$- \frac{\Delta t}{2\Delta X}(\mathbf{F}_{j+1\,l}^{n} - \mathbf{F}_{j-1\,l}^{n}) - \frac{\Delta t}{2\Delta Y}(\mathbf{G}_{j\,l+1}^{n} - \mathbf{G}_{j\,l-1}^{n}),$$

where $\mathbf{F}_{jl}^{n}$ is an abbreviation for $\mathbf{F}(\mathbf{U}_{jl}^{n})$ etc., and where $\mathbf{U}$, $\mathbf{F}(\mathbf{U})$, and $\mathbf{G}(\mathbf{U})$ are given by (13.2); the final values at $t = (n + 2)\Delta t$ are then obtained from the equation

$$(13.11) \quad \mathbf{U}_{jl}^{n+2} = \mathbf{U}_{jl}^{n} - \frac{\Delta t}{\Delta X}(\mathbf{F}_{j+1\,l}^{n+1} - \mathbf{F}_{j-1\,l}^{n+1}) - \frac{\Delta t}{\Delta Y}(\mathbf{G}_{j\,l+1}^{n+1} - \mathbf{G}_{j\,l-1}^{n+1}).$$

Equations (13.10) and (13.11) give no coupling between the set of net-points (in space time) having even values of $n + j + l$ and the set having odd values; half the net-points can therefore be omitted, if desired.

Substitution of (13.10) into (13.11) for the case $\mathbf{F}(\mathbf{U}) = A\mathbf{U}$ and $\mathbf{G}(\mathbf{U}) = B\mathbf{U}$, where A and B are constant matrices, gives an equation similar to (12.23) for $\mathbf{U}^{n+2}$ in terms of $\mathbf{U}^{n}$. This equation involves a nine-point cluster of space points, centered about the point j, k as shown in Figure 13.2; $\partial/\partial X$ is evaluated four times, using the point pairs labeled 1, 2, 3, and 4, and the results are averaged ($\partial/\partial Y$ is treated analogously); $\partial^2/\partial X^2$ is evaluated using the three points labeled A ($\partial^2/\partial Y^2$ is treated analogously); finally, $\partial^2/\partial X \partial Y$ is evaluated using the four points labeled B.

As in one space variable, these equations have error $O(\Delta^3)$, and the finite-difference analogues of the integrated conservation laws hold exactly. We shall show that the von Neumann stability condition and the hypothesis of the theorem in Section 4.11 are satisfied, for the case $\Delta Y = \Delta X$, if

$$(13.12) \quad (|\mathbf{u}| + c)\frac{\Delta t}{\Delta X} < \frac{1}{\sqrt{2}},$$

where c is the local adiabatic sound speed and $\mathbf{u}$ is the fluid velocity.

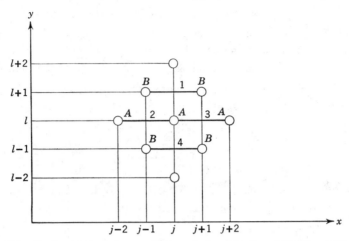

Fig. 13.2. Cluster of net points occurring in a single application of (13.10) and (13.11).

To linearize the system (13.10) and (13.11), $\mathbf{F}(\mathbf{U})$ and $\mathbf{G}(\mathbf{U})$ are replaced by $A\mathbf{U}$ and $B\mathbf{U}$, where the matrices $A = A(\mathbf{U})$ and $B = B(\mathbf{U})$ are the Jacobians of $\mathbf{F}(\mathbf{U})$ and $\mathbf{G}(\mathbf{U})$; i.e., $A_{ij} = \partial F_i/\partial U_j$ etc. Let G denote the amplification matrix of the combined system; that is, if $\mathbf{U}_{j\,l}^n$ is replaced in (13.10) and (13.11) by $\mathbf{U}_0^n \exp[ik_x j\Delta X + ik_y l\Delta X]$, where, for each n, $\mathbf{U}_0^n$ is a constant vector, then G is defined by setting $\mathbf{U}_0^{n+2} = G\mathbf{U}_0^n$. We find that

$$(13.13) \quad G = I - i(\cos\alpha + \cos\beta)\frac{\Delta t}{\Delta X}(A\sin\alpha + B\sin\beta)$$

$$- 2\left[\frac{\Delta t}{\Delta X}(A\sin\alpha + B\sin\beta)\right]^2,$$

where $\alpha = k_x \Delta X$ and $\beta = k_y \Delta X$.

The matrices A and B are rather complicated, and it is convenient to discuss instead the simpler matrices A' and B' obtained from the original Eulerian form (12.2), (12.3), and (12.4) of the equations, which can be written for two variables as

$$(13.14) \qquad \frac{\partial \mathbf{U}'}{\partial t} + A'\frac{\partial \mathbf{U}'}{\partial X} + B'\frac{\partial \mathbf{U}'}{\partial Y} = 0,$$

where

$$U' = \begin{bmatrix} \rho \\ u \\ v \\ p \end{bmatrix},$$

(13.15)
$$A' = A'(U') = \begin{bmatrix} u & \rho & 0 & 0 \\ 0 & u & 0 & 1/\rho \\ 0 & 0 & u & 0 \\ 0 & \rho c^2 & 0 & u \end{bmatrix},$$

$$B' = B'(U') = \begin{bmatrix} v & 0 & \rho & 0 \\ 0 & v & 0 & 0 \\ 0 & 0 & v & 1/\rho \\ 0 & 0 & \rho c^2 & v \end{bmatrix},$$

after introduction of the sound speed c according to (12.7). The linearized version of (13.2) can be obtained from that of (13.14) by a similarity transformation of the form

(13.16) $\quad d\mathbf{U} = P d\mathbf{U}', \quad A = PA'P^{-1}, \quad B = PB'P^{-1}.$

Since the eigenvalues of the amplification matrix G are invariant under a similarity transformation, the von Neumann condition for (13.2) is the same as that for (13.14), and we may replace A and B by A' and B' in equation (13.13) for the amplification matrix.

We call $C = A' \sin \alpha + B' \sin \beta$, and we denote by θ the angle between the X-axis and a direction whose direction cosines are proportional to $\sin \alpha$ and $\sin \beta$, i.e.,

$$\cos \theta = \frac{\sin \alpha}{\sqrt{\sin^2 \alpha + \sin^2 \beta}}.$$

Then,

$$C = \sqrt{\sin^2 \alpha + \sin^2 \beta} \begin{bmatrix} u' & \rho \cos \theta & \rho \sin \theta & 0 \\ 0 & u' & 0 & \rho^{-1} \cos \theta \\ 0 & 0 & u' & \rho^{-1} \sin \theta \\ 0 & \rho c^2 \cos \theta & \rho c^2 \sin \theta & u' \end{bmatrix},$$

where $u' = u\cos\theta + v\sin\theta$. The eigenvalues of C are

$$\lambda = \sqrt{\sin^2\alpha + \sin^2\beta} \begin{Bmatrix} u' \\ u' \\ u' + c \\ u' - c \end{Bmatrix},$$

and if we call $\mu = \lambda \Delta t/\Delta X$, the eigenvalues of G are $g = 1 - i\mu(\cos\alpha + \cos\beta) - 2\mu^2$. Hence,

$$|g|^2 = (1 - 2\mu^2)^2 + \mu^2(\cos\alpha + \cos\beta)^2$$
$$\leq (1 - 2\mu^2)^2 + 2\mu^2(\cos^2\alpha + \cos^2\beta)$$
$$= 1 - 2\mu^2[\sin^2\alpha + \sin^2\beta - 2\mu^2]$$
$$= 1 - 2\mu^2(\sin^2\alpha + \sin^2\beta)\left[1 - 2\left(\frac{\Delta t}{\Delta X}\right)^2 \begin{Bmatrix} u' \\ u' \\ u' + c \\ u' - c \end{Bmatrix}^2\right].$$

The largest possible value of the quantities in the curly bracket is $(|\mathbf{u}| + c)$; hence, $|g|^2 \leq 1$ if (13.12) is satisfied, as was to be proved.

According to (13.13), G is a (quadratic) polynomial in the matrix $A\sin\alpha + B\sin\beta$, which is equal to PCP^{-1} by (13.16); therefore, G has the same eigenvectors as PCP^{-1}. The matrix P is easily found from (13.16), and

$$\det P = \rho^3 \frac{\partial}{\partial p} \mathscr{E}(p, \rho),$$

where $\mathscr{E}(p, \rho)$ is the specific internal energy; also, P happens to be triangular, so that $\det (P^{-1}) = (\det P)^{-1}$. For normal materials, $\partial \mathscr{E}/\partial p$ is bounded away from zero for the range of conditions normally encountered—it is equal to $1/(\gamma - 1)$ for a perfect gas. Therefore, the determinant of the normalized eigenvectors of G is bounded away from zero if and only if the same is true of C. The normalized eigenvectors of C are easily found from the above form of C, and their determinant is

$$\Delta = \frac{2\rho}{c(1 + \rho^2 c^2 + \rho^2/c^2)}.$$

(The peculiar appearance of this expression arises from the fact that $\mathbf{U}$ and $\mathbf{U}'$ are vectors only in the mathematical sense and their components have differing dimensions.) We see that for fixed p, ρ, u, v, the determinant of the normalized eigenvectors of $G = G(\mathbf{k}, \Delta t)$ does not depend on $\mathbf{k}$ or Δt; hence, the theorem of Section 4.11 applies, and the von Neumann condition is sufficient as well as necessary for stability of the linearized equations. Non-linear effects will be discussed in the next section.

Kasahara (1965) has considered certain difference schemes very similar to the Lax-Wendroff scheme but differing therefrom in that certain quantities are centered at time levels $n + \frac{1}{2}$ and $n - \frac{1}{2}$, which seems rather natural from the point of view of accuracy and which simplifies some of the equations. (This is in the same spirit as the system (12.10) for the one-dimensional Lagrangean system, wherein, from the point of view of centering of the time differences, the velocities u_j^{n+1} and u_j^n really refer to the time levels $n + \frac{1}{2}$ and $n - \frac{1}{2}$, respectively.) Kasahara found the rather surprising result that certain schemes of this type are unconditionally unstable in certain parts of the flow. This shows how careful one has to be when modifying a standard difference scheme.

13.5. The Viscosity Term for the L-W Method

Non-linear instability is even more of a hazard in two dimensions than in one. According to (13.13), the amplification matrix G has an eigenvalue equal to 1 if the matrix $A \sin \alpha + B \sin \beta$ has an eigenvalue equal to 0, which is the case if $u' = 0$ or $u' = \pm c$, where u' is the component of the fluid velocity in the direction with direction cosines proportional to $\sin \alpha$ and $\sin \beta$; there is always a value of the ratio $R = \sin \alpha/\sin \beta$ for which $u' = 0$, and in a supersonic flow region there are two values of R for which $u' = \pm c$. Corresponding to each point X, Y, t we have either one or three curves in the α, β-plane, hence in the k_x, k_y-plane, along which G has the eigenvalue 1; such a curve is a locus $\sin \alpha/\sin \beta = $ const. The equations are nowhere fully dissipative in the sense of Kreiss.

For some problems, however, according to the numerical experiments of Burstein (1963), the equations behave in a stable way and give accurate results, provided the usual stability criterion is satisfied. In other problems, according to Burstein (1965) and Lapidus (private communication), instabilities show up in regions of rapid change (presumably owing to the non-linearities or to the rapid variation of the coefficients) and quickly swamp the entire solution, unless a viscosity term similar to the one discussed in Section 12.14 is used. Figure 13.3 shows the main

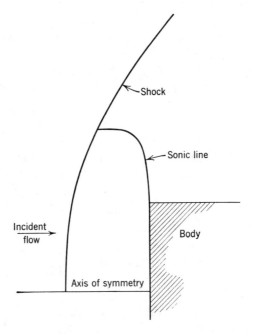

Fig. 13.3. Detached-shock calculation (Burstein).

features of a flow between a detached shock and a rectangular body, as calculated by Burstein using the Lax-Wendroff method. In this problem there are two regions of very rapid change—the shock layer (which has a finite thickness in the calculation) and the neighborhood of the corner of the body, where the flow changes very rapidly from subsonic to supersonic—and the calculation is unstable in both of them unless a viscosity term is included in the equations.

The main objective of such a calculation is to find the steady-state flow by solving an initial-value problem whose asymptotic solution for large t is

SEC. 13.5] THE VISCOSITY TERM FOR THE L-W METHOD

expected to be the desired flow, but the stability problem is presumably the same in calculations of true non-steady flows. Burstein found that when the viscosity term described below was used, the known features of the asymptotic flow were given quite accurately: in the calculation of Figure 13.3 (free stream Mach number = 4.3, perfect gas with $\gamma = 1.4$, viscosity coefficient $\varkappa = 0.5$), the flow field was constant to about four decimals after about 2,000 cycles, the stagnation density was then correct to slightly better than 1%, the stagnation pressure was correct to a very small fraction of 1%, etc.

Burstein took as the viscosity term simply the sum of two one-dimensional viscosity terms; that is, if the Lax-Wendroff one-dimensional viscosity is written as $(\Delta t/2\Delta X)\delta(Q\delta \mathbf{U})$, where δ is, as usual, the centered difference operator, Burstein takes

$$(13.17) \qquad \frac{\Delta t}{2\Delta X}[\delta_X(Q_X\delta_X\mathbf{U}) + \delta_Y(Q_Y\delta_Y\mathbf{U})],$$

where the matrices Q_X and Q_Y are defined as for the one-dimensional case but by taking only the X and the Y variations, respectively, into account. This procedure has been satisfactory in all cases tried so far.

It is difficult to compute the eigenvalues of G when the Burstein viscosity is used, for although Q_X commutes with A and Q_Y with B, their contribution to G, whose dominant term is proportional to $Q_X \sin^2 \alpha + Q_Y \sin^2 \beta$, does not generally commute with $A \sin \alpha + B \sin \beta$. Some preliminary experiments of Lapidus (private communication) indicate that a simpler viscosity term, which avoids this difficulty and is also simpler to use, is also satisfactory in practice. Lapidus retains only the constant term g_0 of the quadratic polynomial $g(A)$ which appears in the expression for Q given in Section 12.14; he also simplifies the term by including only differences of the fluid speed u, v. His formula is therefore (13.17) with Q_X and Q_Y given by

$$(Q_X)_{j+1\,l} = \varkappa'|u_{j+2\,l} - u_{j\,l}|I,$$
$$(Q_Y)_{j\,l+1} = \varkappa'|v_{j\,l+2} - v_{j\,l}|I.$$

Now, the contribution of the viscosity terms to the amplification matrix is simply a negative multiple of the unit matrix; therefore, if zero is an eigenvalue of $A \sin \alpha + B \sin \beta$, the corresponding eigenvalue of G is less than 1.

The above expressions have been written in a form suitable for the two-step procedure, in which we have agreed to take $2\Delta X$, $2\Delta Y$, and $2\Delta t$ as the fundamental increments. Then (13.17) is replaced by

$$(13.18) \qquad \frac{\Delta t}{2\Delta X}[\Delta_{0X}(Q_X\Delta_{0X}\mathbf{U}) + \Delta_{0Y}(Q_Y\Delta_{0Y}\mathbf{U})],$$

where, for any $f(X, Y)$, $2\Delta_{0X}f(X, Y) = f(X + \Delta X, Y) - f(X - \Delta X, Y)$, and similarly for Δ_{0Y}. In the two-step procedure, the expression (13.18), with $\mathbf{U}$ interpreted as $\mathbf{U}^n$, is simply added to the right member of (13.11).

13.6. Piecewise Analytic Initial-Value Problems

In Section 13.3 it was shown that the linearized initial-value problem of Helmholtz instability (also that of Taylor instability) has no solution for any $t > 0$ unless the interface is initially analytic, and that, when a solution exists, it does not depend continuously on the initial data (initial shape of the interface) unless surface tension is present. In general, for multi-dimensional problems, we are very far from knowing whether a solution exists, except in special cases, and whether it depends continuously on the initial data if it exists. We shall present here a rough conjecture relative to existence. Concerning continuous dependence, we remark merely that if one knows enough about the solution to know in particular that Taylor or Helmholtz instability is likely to be encountered, one may modify the formulation by including surface tension or some other stabilizing mechanism before attempting a numerical calculation.

The conjecture is roughly that if the initial data are piecewise-analytic in space, then the initial-value problem has a unique solution, which is piecewise-analytic in space-time, at least for some interval $0 \leq t \leq T$. Clearly, the class of initial-value problems must be defined, and "piecewise-analytic" must be defined. Since the truth of the conjecture surely depends on these definitions, and since we do not know now what is required, we shall be intentionally somewhat vague.

The class of problems probably includes the idealized problems of compressible fluid dynamics with an analytic equation of state. The formulation is understood to consist of the partial differential equations for the smooth part of the flow, the jump conditions described in Section

SEC. 13.6]   PIECEWISE ANALYTIC INITIAL-VALUE PROBLEMS   369

13.2 for shocks and contact discontinuities, standard boundary conditions on external surfaces, and the prohibition of negative shocks. Body forces like gravity, if present, are supposed to be described by a specified analytic, curl-free vector field.

The initial data are called *piecewise analytic* if space can be divided into a number of cells or regions in each of which the flow functions are real analytic functions, the interfaces between cells are analytic surfaces, and the curves on which these surfaces meet are analytic curves.

Similarly, the solution is called *piecewise analytic* if space-time is similarly divided into cells in which the flow functions are analytic. This does not imply that the subdivision of space into cells remains for $t > 0$ what it was at $t = 0$; new interfaces can appear, as when the incident shock arrives at the tip of the wedge in the Mach reflection problem (see Figure 13.1) or when a shock front appears spontaneously in the interior of a flow.

We have been vague intentionally about one point, which may turn out to be crucial. We have required, for example, that if $\mathbf{X}_0$ is an interior point of one of the regions, and if $f(\mathbf{X})$ is one of the flow functions, then $f(\mathbf{X})$ must be real analytic in $\mathbf{X}$ at $\mathbf{X}_0$, that is, that there is some sphere $S: \|\mathbf{X} - \mathbf{X}_0\| < a$ in which $f(\mathbf{X})$ has a convergent power series in the components of $\mathbf{X} - \mathbf{X}_0$; we have not said whether $f(\mathbf{X})$ must be analytic at a point $\mathbf{X}_0$ of an interface bounding this region (in this case, the power series would represent $f(\mathbf{X})$ only in that part of the sphere S that lies in the region; in the rest of S it would represent an extrapolation of $f(\mathbf{X})$ into one or more adjoining regions); similarly, we have not said whether the surfaces have to be analytic up to and including or merely up to the curves that bound them; finally, we have not said whether the curves have to be analytic up to and including or merely up to their end points. Intuition suggests that if $\mathbf{X}_0$ is an interior point of one of the analytic surfaces bounding the region, i.e., not on an edge or corner, we should require the flow quantities to be analytic at $\mathbf{X}_0$. If $\mathbf{X}_0$ is on an edge, we cannot make such a requirement, because there are known cases in which the analyticity breaks down at an edge. For example, at the triple point in the Mach reflection problem, i.e., the point where the three shocks meet (this point of the diagram represents an edge perpendicular to the drawing in the real three-dimensional flow), it is known that the flow is not analytic: the functions $X(s)$ and $Y(s)$ representing the coordinates along one of the curved shocks (or the slip line) cannot be expanded in power series of the arc length s at the triple point, and the flow functions $f(X, Y)$ cannot be expanded in powers of X and Y about this point. Expansions in fractional powers, logarithms, etc. have been attempted, but so far without success; the nature of the singularity is not known. Similarly, at the tip of the wedge, under some circumstances, there is subsonic corner flow. The theory of subsonic, compressible non-isentropic corner flow has not been worked out in detail, but it is likely that the singularity can be described by an expansion in fractional powers of the space variables. In these examples, the primary flow variables remain bounded as the singularity is approached, and perhaps this should be a general requirement. The example of a centered rarefaction fan shows that the primary flow quantities do not always have unique limiting values as the singularity is approached.

For problems in one space variable and time, some progress has been made toward establishing a conjecture of this type, for example by Goldner (1949), by Rozhdestvenskii (1960), and by Logemann (1965). Logemann considered a second-order hyperbolic system similar to the (third-order) system of fluid dynamics, together with the Rankine-Hugoniot and entropy conditions; he considered a problem in which the dependent variables $u(x, t)$, $v(x, t)$ are smooth in x, for $t = 0$, except at $x = x_0$, where they have simple jumps. The corresponding problem with $u(x, 0)$ and $v(x, 0)$ constant except at $x = x_0$ is well known; the solution contains either two shocks, two centered rarefaction fans, or one of each, originating at $x = x_0$ and moving into the fluid in opposite directions; between these waves, u and v are constant; Logemann has proved that a similar solution exists and is unique, generally, under suitable circumstances; the front of a shock and the head and tail of a rarefaction move with variable speeds, in the general case, and u and v are generally no longer constant, although smooth, between the waves. It seems likely that progress on the multi-dimensional problems will be much slower.

Similar conjectures can be made for other classes of problems, for example, for problems of the Navier-Stokes equation, problems of magneto-fluid-dynamics, and problems of fluid dynamics with heat flow or chemical reactions.

As a special case, consider the spontaneous generation of a shock. If a

SEC. 13.6]   PIECEWISE ANALYTIC INITIAL-VALUE PROBLEMS   371

piston, moving with a speed that increases continuously from zero, is pushed into a tube containing a gas initially at rest and at a pressure $p > 0$, a shock forms, as indicated schematically in Figure 13.4. The

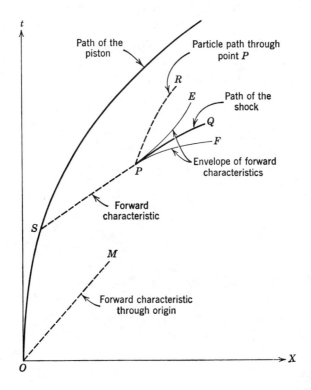

Fig. 13.4. Spontaneous generation of a shock.

shock is of zero strength at the point P. Under suitable assumptions about the motion of the piston, there are no other shocks. If the equation of state is a relation between the pressure p and the density ρ, independent of the entropy, and the path of the piston $X(t)$ is analytic for $t > 0$, then the path of the shock (the curve PQ) is the only singularity of the flow, except for a high-order singularity on that forward characteristic OM that starts from the instant at which the piston was set in motion. In this case the entire problem can be treated by the method of characteristics, as is done, for example, in Courant and Friedrichs (1948). Each primary flow quantity $f(X, t)$ is analytic up to and on the curve PQ, on both sides of it, except at the point P itself, where there is a singularity. The surface obtained by plotting $f(X, t)$ in the space (X, t, f) is folded in such a way that f is formally triple-valued in the region between the envelopes PE and PF; along the shock path PQ, in between, a jump is made from the top sheet of this surface (which gives the correct values of f to the left of PQ) to the bottom sheet (which gives f to the right). For a more general equation of state, there is an additional singularity on the particle path PR, where the third derivative of the entropy with respect to X becomes infinite.

As a second special case, consider Helmholtz instability at a blunt edge like that shown at P on the initial interface in Figure 13.5.

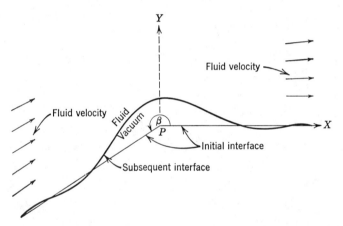

Fig. 13.5. Conjectured possible configuration of interface in Helmholtz instability.

As pointed out in Section 13.3, if one attempts to find the solution by linearization (which includes the assumption that the interface is nearly flat) and Fourier analysis, the attempt fails completely unless the interface is initially an analytic surface. We conjecture that an analytic (or piecewise analytic) solution exists nevertheless, for t in some interval $(0, t_0)$, for the idealized problem in which the initial interface is analytic except at the edge. (Roughness would presumably still lead to fine-scale mixing, as described toward the end of Section 13.3.) We further

SEC. 13.6] PIECEWISE ANALYTIC INITIAL-VALUE PROBLEMS

idealize the problem by neglecting compressibility, viscosity, and surface tension and by supposing that at $t = 0$ the velocity field is that of the well-known solution for incompressible flow around the corner of a rigid body; in terms of the coordinates X, Y shown in Figure 13.5, the initial velocity potential φ is the real part of $Z^{\pi/\beta}$, where $Z = X + iY$, and where β is the angle of the wedge, as shown. For $t > 0$, the pressure on the interface is assumed everywhere zero, so that it becomes a free surface. The dynamical equations of incompressible flow are assumed to hold in the fluid above the interface. There is also a dynamical boundary condition on the interface, which says that the normal component of fluid velocity must equal the normal component of the velocity of the interface itself. Finally, at large distances within the fluid the velocity potential is required to agree asymptotically with its initial value, namely, the real part of $Z^{\pi/\beta}$.

The problem as thus formulated contains no length scale; if a solution exists and is unique, it must be self-similar, i.e., the velocity potential $\varphi(X, Y, t)$ must be of the form $g(t)\Phi(\xi, \eta)$, where $\xi = X/f(t)$ and $\eta = Y/f(t)$, and similarly the vector position $Z = X + iY = Z(s, t)$ on the interface at time t of the particle that was initially at arc length s from the corner must be of the form $Z(s, t) = f(t)\zeta[s/f(t)]$. Instantaneous photographs of the interface taken at two instants t_1 and t_2 would be identical except for the degree of enlargement. If these expressions are substituted into the partial differential equations and boundary conditions, it is found that the problem indeed reduces to a boundary-value problem involving the functions $\Phi(\xi, \eta)$ and $\zeta(\sigma)$, provided that $f(t) = t^\alpha$, where $\alpha = (2 - \pi/\beta)^{-1}$ and $g(t) = t^\gamma$, where $\gamma = \pi\alpha/\beta$.

It is conjectured that the interface appears somewhat as shown in Figure 13.5. Since $0 < \alpha < 1$ for $\beta > \pi$, the curve representing the interface starts spreading out from the edge at $t = 0$ with initially infinite speed. It seems likely that the curve has oscillations about the initial straight lines, with rapidly decreasing amplitude (and possibly increasing wavelength), as one proceeds away from the origin along the curve in either direction. However, the question of existence and uniqueness of the solution of the reduced problem is at present open.

Other configurations than the one shown are also conceivable, for example, in which the free surface has a cusp and possibly a vortex sheet penetrating into the fluid from the cusp. In any case, however, the configuration would presumably be subject to the similarity principle.

13.7. A Program for the Development of Methods

The validity of a conjecture of the kind described in the foregoing section would be of value not only in establishing the existence and uniqueness of solutions, but also in suggesting a program for the development of numerical methods; such a program will now be outlined briefly.

In the smooth parts of the flow (regions of analyticity), one may use a rectangular point-net in space-time and approximate the differential equations by difference equations, which could in principle be of an arbitrarily high order of accuracy, since the functions are analytic, but will usually be the two-step Lax-Wendroff method, or something similar, in practice. Discontinuities and singularities of all kinds, as far as their natures are known in advance, would be handled by special methods, like shock fitting, in order to tie together the finite-difference calculations in the various regions and to satisfy jump conditions to an accuracy consistent with the accuracy of the calculation of the smooth flow.

If $N(= 2, 3,$ or $4)$ is the dimension of space-time, singularities associated with a hypersurface of dimension $N - 1$ can be: (1) a shock front, (2) a contact discontinuity (which may also be a slip surface), or (3) a surface on which the primary flow quantities are continuous but their derivatives of some order are not. We shall take the attitude that it is not worthwhile to devise special methods for singularities of type (3), at least at the present time (although to do so would be easier than shock-fitting); instead, we suppose that the rectangular-grid calculation is carried across such a hypersurface, as though no singularity were present. In Section 13.9, a simplified case of shock fitting is described; a more general case of shock fitting and methods for contact discontinuities are also mentioned very briefly.

Examples of singularities associated with hypersurfaces of dimension $N - 2$ are: for $N = 2$ (one space variable), (1) the collision of two parallel plane or concentric spherical shocks, (2) the instantaneous coincidence of a shock with a contact discontinuity or an interface between two media, (3) the collision of two free surfaces, (4) the collapse of a converging spherical shock onto the center of symmetry. For $N = 3$, examples are: (5) the triple point where three shocks meet in Mach reflection, (6) the intersection of two non-co-planar shocks. In examples (1) to (4), the nature of the singularity is known, and power series or other expansions can be used to devise special numerical fitting procedures; we shall not describe any of these procedures, because they are usually quite special to a particular problem. In examples (5) and (6), the nature of the singularity is not known.

13.8. Characteristics in Two-Dimensional Flow

In the discussion of one-dimensional shock fitting in Section 12.9, it was noted that information flows in a certain sense along the characteristics or characteristic curves in the X, t-plane and that only one of the three characteristics through a given point was in such a direction as to be able to convey information up to the shock from the flow behind it. Those considerations will now be generalized to two-dimensional flow.

The one-dimensional equations in characteristic form were given as equations (12.30) of Chapter 12. In each one of these three equations, all dependent quantities are differentiated in the same direction in the X, t-plane. In two-dimensional flow, there is only one linear combination of the original partial differential equations in which all quantities appearing are differentiated in a single direction; this is the equation $D\mathscr{E} + pD(1/\rho) = 0$, where D denotes differentiation along the particle path; it says that entropy is constant along a particle path in the smooth part of the flow. Even this equation is a peculiarity of the fluid-dynamical equations; the most that one can hope for, for a general hyperbolic system in three independent variables, is to find linear combinations of the original equations such that the dependent quantities are differentiated in directions that are coplanar. Such equations are said to be in *characteristic form* for the two-dimensional flow, and the corresponding plane is a *characteristic plane*. It will be shown below that there are infinitely many characteristic planes through a given point (X_0, Y_0, t_0); namely, any plane tangent to the particle trajectory at (X_0, Y_0, t_0) is characteristic, and any plane tangent at (X_0, Y_0, t_0) to the sonic cone with apex at (X_0, Y_0, t_0) is characteristic.

To find the possible orientations of a characteristic plane, the four equations (12.2) to (12.4) (the second of these is really two equations, for it now has two components) are multiplied by coefficients α, β, γ, δ, respectively, and added, and the internal energy $\mathscr{E}$ is expressed in terms of p and ρ by the equation of state. The resulting equation is of the form

(13.19) $$\boldsymbol{\eta}_1 \cdot \nabla p + \boldsymbol{\eta}_2 \cdot \nabla \rho + \boldsymbol{\eta}_3 \cdot \nabla u + \boldsymbol{\eta}_4 \cdot \nabla v = 0,$$

where ∇ denotes here the vector operator whose components are $(\partial/\partial X, \partial/\partial Y, \partial/\partial t)$. The $\boldsymbol{\eta}_i$ are linear in α, β, γ, δ and depend also on p, ρ, u, v, hence on X, Y, t. The requirement that the $\boldsymbol{\eta}_i$ be coplanar is that there exist a non-zero vector $\boldsymbol{\zeta}$ such that

(13.20) $$\boldsymbol{\zeta} \cdot \boldsymbol{\eta}_i = 0 \quad (i = 1, 2, 3, 4);$$

$\boldsymbol{\zeta}$ is then the normal to the characteristic plane. For fixed X, Y, t, these are four linear equations in the four unknowns α, β, γ, δ; they have a non-trivial solution only if $\Delta = 0$, where $\Delta = \Delta(\boldsymbol{\zeta})$ is the determinant of the system and is a polynomial of at most the fourth degree in the three components of $\boldsymbol{\zeta}$. Since the length of $\boldsymbol{\zeta}$ is arbitrary, the equation $\Delta = 0$ determines a one-parameter family of possible orientations of the characteristic plane through the given point X, Y, t.

The equations in characteristic form obtained by the procedure described in the preceding paragraph are:

(13.21) $$\frac{Dp}{Dt} - c^2 \frac{D\rho}{Dt} = 0,$$

(13.22) $$\boldsymbol{\mu} \cdot \left(\frac{D\mathbf{u}}{Dt} + \frac{1}{\rho}\nabla p\right) = 0,$$

(13.23) $$\rho c \left(\lambda \frac{D}{Dt} + c\nabla\right) \cdot \mathbf{u} + \left(\frac{D}{Dt} + c\boldsymbol{\lambda} \cdot \nabla\right)p = 0;$$

all the vectors appearing in these equations are two-dimensional (in the X, Y-plane) and, in particular, $\mathbf{u} = (u, v)$, and $\nabla = (\partial/\partial X, \partial/\partial Y)$; furthermore, c is the adiabatic sound speed (a function of p and ρ), and $\boldsymbol{\lambda}$ and $\boldsymbol{\mu}$ are arbitrary unit vectors in the X, Y-plane; finally, D/Dt denotes the operator

(13.24) $$\frac{D}{Dt} = \frac{\partial}{\partial t} + \mathbf{u} \cdot \nabla = \frac{\partial}{\partial t} + u \frac{\partial}{\partial X} + v \frac{\partial}{\partial Y},$$

which effects differentiation along the paths (world lines) of the particles. In equation (13.22), the directions of differentiation are restricted to a plane tangent to the particle path and parallel to $\boldsymbol{\mu}$; in (13.23), they are restricted to a plane tangent to the sonic cone and such that the intersection of this plane with the X, Y-plane is perpendicular to $\boldsymbol{\lambda}$ (see Figure 13.6).

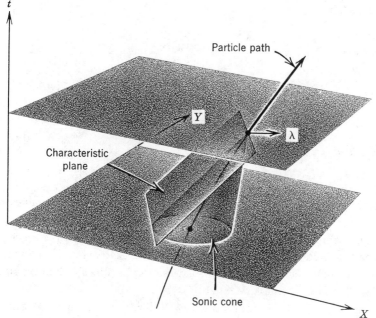

Fig. 13.6. Space-time diagram showing a particle path and a sonic cone.

These equations and the corresponding characteristic planes will appear in the next section.

13.9. Shock Fitting in Two Dimensions

To take the simplest case first, we suppose that a shock is moving into previously undisturbed material, so that ahead of it the pressure, the density, and the two fluid velocity components have known constant values p_2, ρ_2, u_2, and v_2 (in the notation of Section 13.2), whereas behind the shock they are functions of X, Y, and t; here, as always, X and Y are Cartesian Eulerian coordinates. At any instant $t = n\Delta t$, the shock front lies on a curve in the X, Y-plane; in the numerical work, this curve is represented by a sequence of points with coordinates $(X, Y) = (\xi_l^n, \eta_l^n) = \boldsymbol{\xi}_l^n$ along it, $l = 1, 2, \ldots$, as indicated in Figure 13.7, where the rectangular

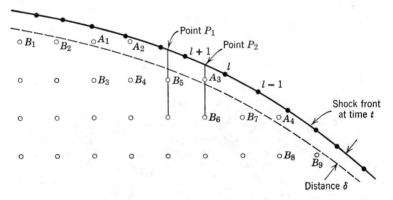

Fig. 13.7. Arrangement of net points and shock points.

net points are also shown and are indicated by circles. The values of p, ρ, u, v at point $\boldsymbol{\xi}_l^n$ and time $n\Delta t$ are denoted by p_l^n, ρ_l^n, u_l^n, v_l^n (these are the values immediately behind the shock—those in front are p_2, ρ_2, u_2, v_2), whereas the values at the net points are denoted by p_{jk}^n etc., as in Section 13.4. The unit vector in the X, Y-plane normal to the shock and pointing into the fluid ahead is denoted by $\boldsymbol{\lambda}_l^n$.

We suppose that for time $t = n\Delta t$ all quantities are known at the shock points and at the net points behind the shock and that we wish to find these quantities at time $t = (n + 1)\Delta t$. The new position $(\xi_l^{n+1}, \eta_l^{n+1}) = \boldsymbol{\xi}_l^{n+1}$ of each shock point must also be found; this position is somewhat arbitrary so long as it lies on the shock front at time $t = (n + 1)\Delta t$,

SEC. 13.9] SHOCK FITTING IN TWO DIMENSIONS

somewhere between the points ξ_{l-1}^{n+1} and ξ_{l+1}^{n+1}, and is fairly near the old position ξ_l^n, i.e., the only component of the velocity $\dot{\xi}_l$ that has physical meaning is the component in the direction λ_l. For simplicity, we therefore set $\dot{\xi}_l = S_l \lambda_l$, where S_l is the shock speed appearing in equations (13.8a); in difference form, this can be approximated as

$$(13.25) \qquad \frac{1}{\Delta t}(\xi_l^{n+1} - \xi_l^n) = \tfrac{1}{2}(S_l^{n+1}\lambda_l^{n+1} + S_l^n \lambda_l^n).$$

We note at this point that the entire shock-fitting procedure is highly implicit. The two components of (13.25) provide two equations in five unknowns; together with other equations given below they constitute a considerable system of simultaneous non-linear equations which must be solved in practice by iterative approximation.

From the Rankine-Hugoniot jump conditions (13.8a) and (13.8b) we find

$$(13.26) \qquad S_l^{n+1} - \lambda_l^{n+1} \cdot \mathbf{u}_2 = \frac{1}{\rho_2}\sqrt{\frac{p_l^{n+1} - p_2}{1/\rho_2 - 1/\rho_l^{n+1}}},$$

$$(13.27) \qquad \lambda_l^{n+1} \cdot (\mathbf{u}_l^{n+1} - \mathbf{u}_2) = \sqrt{(p_l^{n+1} - p_2)(1/\rho_2 - 1/\rho_l^{n+1})},$$

$$(13.28) \qquad F(p_l^{n+1}, \rho_l^{n+1}) - F(p_2, \rho_2) = \tfrac{1}{2}(p_l^{n+1} + p_2)(1/\rho_2 - 1/\rho_l^{n+1})$$

$$(13.29) \qquad \lambda_l^{n+1} \times (\mathbf{u}_l^{n+1} - \mathbf{u}_2) = 0,$$

where $F(p, \rho)$ is equal to $\mathscr{E}$ by the equation of state (12.1). The only unknowns in these equations refer to time $t = (n + 1)\Delta t$. The unit vector λ_l^{n+1} can be expressed in terms of the positions of the shock point l and its two nearest neighbors by first computing a tangent vector λ^* to the curve at ξ_l^{n+1} by

$$(13.30) \qquad \lambda^* = \alpha(\xi_{l+1}^{n+1} - \xi_l^{n+1}) + \beta(\xi_l^{n+1} - \xi_{l-1}^{n+1}),$$

where α and β are coefficients (functions of the ξ's) so chosen as to make λ^* a unit vector and to weight the two neighbors of ξ_l^{n+1} inversely as their distances from it. (This is equivalent to fitting a certain quadratic curve to the three points and taking λ^* as its tangent at ξ_l^{n+1}.) Then, the vector λ, given by

$$(13.31) \qquad (\lambda_X, \lambda_Y) = (-\lambda_Y^*, \lambda_X^*),$$

is the desired unit normal λ_l^{n+1}; if the indices l increase in the direction along the curve which has the shocked region on its left, λ_l^{n+1} is the normal outward from this region.

Counting the vectorial equations (13.25), (13.30), and (13.31) twice and the intermediate equations once, we now have 10 equations for the 11 unknowns, namely, S, p, ρ, and the components of $\mathbf{u}$, $\boldsymbol{\xi}$, $\boldsymbol{\lambda}$, $\boldsymbol{\lambda}^*$, at shock point l and time $(n+1)\Delta t$. (Note in this connection that (13.29) counts as one equation, since only its Z component conveys information—it might have been written as $\boldsymbol{\lambda}^* \cdot (\mathbf{u}_l^{n+1} - \mathbf{u}_2) = 0$.) As in one-dimensional shock-fitting, the additional equation needed is obtained by requiring that one of the differential equations be satisfied, in a suitable difference approximation, not only at the net points but also at each shock point. One must decide (1) what differential equation to use and (2) what difference approximation to it to use. In addition, one must decide (3) how to obtain the new values p_{jk}^{n+1} etc. at net points so close to the shock that the standard difference equation (e.g., Lax-Wendroff) cannot be used. (In practice, the p_{jk}^{n+1} at these and also the regular net-points are computed before the shock-fitting is begun.) It is clear that these questions can be answered in a great variety of ways; the answers given below are mainly for illustration, because the relative advantages of the many possible procedures have not yet been fully explored and because it seems likely that no one procedure will be suitable for all problems.

As for one-dimensional shock-fitting, if the differential equation chosen for augmenting the jump conditions, as described above, is one of the equations in characteristic form (13.21), (13.22), or (13.23), then a simple physical argument indicates a unique choice for this equation. In one-dimensional flow, information flows, in a sense, in the characteristic directions in the X, t-plane; in this same sense, information flows in the two-dimensional problems along the characteristic planes. Since we need information from the flow behind the shock, we choose a characteristic plane through the point $\boldsymbol{\xi}_l^{n+1}$, $(n+1)\Delta t$ of space-time so that its past portion lies as completely as possible in the region behind the shock, and this choice is unique: we must choose a plane, tangent to the sonic cone, whose intersection with the plane $t = (n+1)\Delta t$ is tangent to the shock front, and which, if the t-axis is thought of as vertical, is tipped forward as much as possible toward the shock. In other words, we must choose equation (13.23) with $\boldsymbol{\lambda}$ taken as the normal to the shock, namely, $\boldsymbol{\lambda} = \boldsymbol{\lambda}^{n+1}$.

For simplicity, we shall describe only a first-order difference approximation to (13.23), corresponding to the replacement of (13.25) by the simpler equation $\boldsymbol{\xi}_l^{n+1} - \boldsymbol{\xi}_l^n = S_l^n \boldsymbol{\lambda}_l^n \Delta t$. The divergence of the velocity can be written as $\nabla \cdot \mathbf{u} = (\boldsymbol{\lambda} \cdot \nabla)(\boldsymbol{\lambda} \cdot \mathbf{u}) + (\boldsymbol{\lambda}^* \cdot \nabla)(\boldsymbol{\lambda}^* \cdot \mathbf{u})$, where $\boldsymbol{\lambda}^*$ is the unit vector (in the X, Y-plane) orthogonal to $\boldsymbol{\lambda}$. Then (13.23) can be written as

$$\frac{\delta}{\delta t} p + \rho c \frac{\delta}{\delta t} (\boldsymbol{\lambda} \cdot \mathbf{u}) + (\boldsymbol{\lambda}^* \cdot \nabla)(\boldsymbol{\lambda}^* \cdot \mathbf{u}) = 0,$$

where

$$\frac{\delta}{\delta t} \stackrel{\text{def.}}{=} \frac{\partial}{\partial t} + (\mathbf{u} + c\boldsymbol{\lambda}) \cdot \nabla$$

is the operator of differentiation with respect to time along the line of tangency of the characteristic plane and the sonic cone.

In Figure 13.8 the curves C^n and C^{n+1} represent two successive positions

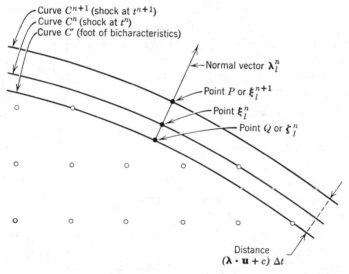

Fig. 13.8. Diagram showing quantities that appear in equation (13.32).

of the shock, and the curve C' is the locus of points at a distance $(\mathbf{u} \cdot \boldsymbol{\lambda} + c) \Delta t$ behind the later position of the shock. If P and Q denote the points $\boldsymbol{\xi}_l^{n+1}, t^{n+1}$, and $\boldsymbol{\zeta}_l^n, t^n$, of space-time, where $\boldsymbol{\zeta}_l^n$ is the intersection of $\boldsymbol{\lambda}_l^n$

(extended back) with curve C', then $\delta f/\delta t$ can be approximated as $[f(P) - f(Q)]/\Delta t$ for any function f, while $\boldsymbol{\lambda}^* \cdot \nabla$ is differentiation with respect to arc-length along curve C' at point Q and is denoted by d/ds. Therefore, (13.23) is approximated as

(13.32) $\quad p(P) - p(Q) + \rho c(Q)\boldsymbol{\lambda}_l^n \cdot (\mathbf{u}(P) - \mathbf{u}(Q)) + ((\Delta/\Delta s)\boldsymbol{\lambda}^* \cdot \mathbf{u})_Q = 0$,

where $\Delta/\Delta s$ is a difference quotient approximation to d/ds along C'. For this equation, the values of p, ρc, and $\boldsymbol{\lambda}^* \cdot \mathbf{u}$ are needed at the points ζ_l^n of C'; they are obtained by a suitable interpolation at time t^n using values at the net-points and at the shock points ξ_l^n.

Stability has been found to require exclusion from the procedure of any net-points that lie too close to the shock; one therefore draws a curve (shown dashed in Figure 13.7) at a fixed distance δ behind the shock, where $\delta \sim \Delta x$, and one calls net-points lying between this curve and the shock *net-points of type A*; the values of p_{jk}^n etc. at these points are not used in the calculation; however, when all other quantities are known at $t = (n+1)\Delta t$, interpolated values of p_{jk}^{n+1} etc. are supplied to each net-point of type A, to be used later if, owing to the advance of the shock, the point jk is no longer of type A on the next cycle.

First-order methods, which appear to be more stable than second-order ones, may seem inconsistent with the second-order accuracy of the Lax-Wendroff equations, but G. W. Strang has shown (private communication) that the convergence to the true solution as Δx, etc. $\to 0$ is still of second order, for problems of this type, even if boundaries are treated with only first-order accuracy. This is roughly because the ratio of the number of boundary points to the number of net-points is a small quantity of first order in the limit. Such a procedure is in any case clearly better than the artificial-viscosity methods which treat the boundary (shock) with only zero-order accuracy.

Any net-point one or more of whose neighbors is missing (i.e., is on the other side of the dashed line) is said to be *of type B* and requires special treatment. To calculate p_{jk}^{n+1} etc. for example at B_5 in Figure 13.7 one first obtains values at the point labeled P_1 on the shock by interpolation with respect to arc length and then uses P_1 in lieu of the missing neighbor (the one with indices $jk+1$), after a slight modification of the Lax-Wendroff formulas to take the non-uniform spacing into account.

If the material ahead of the shock is not in a constant state, the flow quantities immediately ahead, denoted above by p_2, ρ_2, u_2, v_2, will depend on l and n and must presumably be obtained somehow from the net-point calculation ahead. The details remain to be worked out and tested.

Fitting procedures for a slip surface, which may be a contact discontinuity of entropy or a material interface, have been attempted, but so far without success. In this case, there are nine unknowns: p, ρ, u, v on each side and the normal speed S of the surface itself. There are three jump conditions: p is the same on both sides, and the normal component of **u** is the same on both sides and is equal to S. Presumably, therefore, the jump conditions must be supplemented by three differential equations on each side. Suitable surface-tension terms must be included in the equations to stabilize Taylor and Helmholtz instabilities of wavelengths comparable with Δx.

Many additional details of a two-dimensional calculation vary from problem to problem so as to make general or universal procedures unfeasible; special procedures must be used near the confluence of three or more shock fronts or the intersection of a shock front with a rigid wall, and so on. A question of a somewhat topological nature that arises is the following: if several curves, representing shocks etc., divide the X, Y-plane at an instant t into several regions, one must have a subroutine for telling in which of the regions a given net-point jk lies; this can be done in various ways. Finally, for a diverging or converging shock, it is desirable to have procedures for occasionally increasing or decreasing the number of points ξ_l^n on the curve, so as to keep their spacing comparable with Δx. All these things are primarily a matter of bookkeeping, whereas the basic fitting procedures for the internal and external boundaries still need further development, analysis, and testing.

13.10. The Problem of the Atmospheric Front

The meteorological problem to be discussed here is roughly that of the motion of a polar cap of cold air on the earth, shown schematically in Figure 13.9. The cold air is separated from the warmer air above by a fairly sharp interface or *front*, across which the temperature difference is

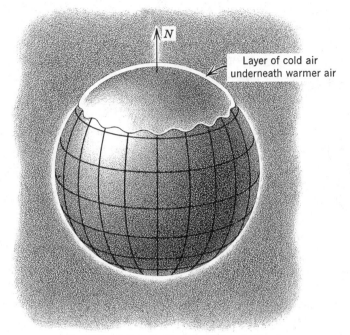

Fig. 13.9. Schematic diagram of the polar front.

of the order of 5 to 10°C. The intersection of the front with the ground is along a generally wavy line in middle latitudes called a "cold front" or a "warm front" on a weather map depending on whether it is moving in such a way that stationary people become cooler or warmer. The front slopes upward to the north with an inclination of a (generally small) fraction of a degree, so that at the pole the layer of cold air has a thickness of several miles.

In the lowest approximation, the polar cap is in equilibrium, with Coriolis forces (caused by the earth's rotation) balanced by gravity forces. In this approximation the warm air above is to be thought of as having a steady zonal circulation, from west to east, relative to the cold air below, so that at the front there is a discontinuity of the east-west component of wind velocity as well as of temperature. The Coriolis force per unit volume of air is $\rho \mathbf{v} \times \boldsymbol{\omega}$, where $\mathbf{v}$ is the wind velocity relative to the rotating earth and $\boldsymbol{\omega}$ is the earth's angular velocity; hence, the Coriolis force is different for the warm and cold air masses. On the other hand, the cold air is denser than the warm air, so that if there were no Coriolis forces, hydrostatic equilibrium would require the front to be everywhere horizontal. The equilibrium under both forces is discussed by Kasahara *et al.* (1964) and in the sources referred to by him. Roughly, in this approximation, the warm air rotates slightly more rapidly about the earth's axis than the cold air and is subject to larger centrifugal forces; this causes it to tend to accumulate at the lower latitudes and to force the colder air to higher latitudes.

Kasahara, Isaacson, and Stoker (1964, 1965) have developed a numerical method for computing the motion of a simplified model of the front, when not in equilibrium, as an initial-value problem. The full three-dimensional problem with compressibility would be intractable (even if heat sources, convection, and the like were ignored) but, fortunately, there are good simplified models. In the one used here (Stoker, 1953, 1957), the following approximations are made: (1) the calculation is made on a small enough portion of the earth's surface (say a rectangle about the size of the United States), with suitable additional boundary conditions, so that one can neglect the curvature of the surface and the variation of the Coriolis parameter $f = 2\omega \sin \varphi$, φ being the latitude (this is sometimes called the *tangent-plane approximation*); (2) vertical accelerations are neglected, so that the pressure at each point is given by the elementary hydrostatic formula; (3) compressibility is neglected, because wind speeds are much smaller than sound speed; (4) each layer has a constant density; (5) the warm air layer is assumed negligibly influenced by disturbances in the cold layer so that its velocity and the total height of the two layers take their equilibrium values throughout the calculation. This leads to the system of equations

$$(13.33) \qquad \frac{Du}{Dt} + g\left(1 - \frac{\rho'}{\rho}\right)\frac{\partial h}{\partial X} = fv,$$

$$(13.34) \qquad \frac{Dv}{Dt} + g\left(1 - \frac{\rho'}{\rho}\right)\frac{\partial h}{\partial Y} = f\left(\frac{\rho'}{\rho}\bar{u}' - u\right),$$

$$(13.35) \qquad \frac{Dh}{Dt} = 0,$$

386   MULTI-DIMENSIONAL FLUID DYNAMICS   [CHAP. 13

where D/Dt is the operator

$$\frac{D}{Dt} = \frac{\partial}{\partial t} + u\frac{\partial}{\partial X} + v\frac{\partial}{\partial Y},$$

X and Y are Cartesian coordinates pointing east and north, respectively, $u = u(X, Y, t)$ and $v = v(X, Y, t)$ are the X and Y components of the fluid (wind) velocity, $h = h(X, Y, t)$ is the vertical thickness of the cold air layer, and the other quantities are constants: ρ'/ρ is the density ratio of the warm to cold air, $\bar{u}'$ is the constant west-to-east speed of the warm air, and g is the acceleration of gravity.

In the mixed initial-boundary-value problem treated by Kasahara et al., the flow is assumed periodic in longitude with a period such that all quantities are the same at corresponding points of the east and west boundaries of the rectangle shown in Figures 13.10 and 13.11; at the north

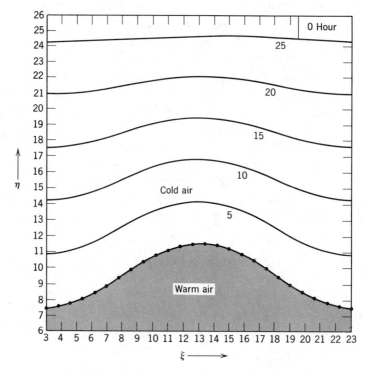

Fig. 13.10. Height contour pattern of cold air at $t = 0$.

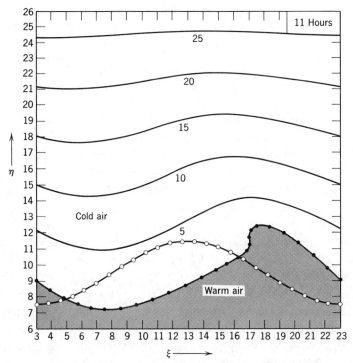

Fig. 13.11. Height contour pattern of cold air at $t = 11$ hours.

boundary a rigid wall is assumed; no boundary condition is needed on the south side of the rectangle because the front does not reach down that far.

Let C denote the curve on which the front intersects the earth's surface (the X, Y-plane) at time t, i.e., the locus $h(X, Y, t) = 0$. From the point of view of the differential equations, no boundary condition is needed on C except that the differential equations (13.33) to (13.35) should hold on and above C; in the numerical calculation, it is convenient to place points on C, as in the figures, with coordinates $(\xi_l, \eta_l) = \boldsymbol{\xi}_l = \boldsymbol{\xi}_l(t)$, and follow the motion of these points and the values of $(u_l, v_l) = \mathbf{u}_l = \mathbf{u}_l(t)$ at them, as in shock-fitting. The effective boundary conditions are obtained from (13.33) and (13.34) and are

(13.36) $$\frac{d}{dt}\boldsymbol{\xi}_l = \mathbf{v}_l$$

(13.37) $$\frac{d}{dt}\mathbf{v}_l = -g\left(1 - \frac{\rho'}{\rho}\right)\nabla h + \gamma\mathbf{v}_l + \mathbf{K},$$

where

$$\gamma = \begin{bmatrix} 0 & f \\ -f & 0 \end{bmatrix}, \quad \mathbf{K} = f\frac{\rho'}{\rho}\bar{u}'\begin{bmatrix} 0 \\ 1 \end{bmatrix},$$

and where ∇h is to be taken at ξ_l.

The method of Kasahara *et al.* uses the Lax-Wendroff equations (with minor variations) at each regular net-point, that is, at each point, not on the northern boundary, all eight of whose neighbors lie above the curve C. Points near C are divided into type A and type B points and are treated very much as in shock-fitting. However, Kasahara found it possible and worthwhile to make all the calculations to second-order accuracy. The curve C is then regarded as a succession of arcs, each given by a cubic which is either of the form $\eta = f(\xi)$ or $\xi = g(\eta)$, depending on the slope (a quadratic would presumably have sufficed), and all interpolating and differencing are performed with comparable accuracy. Since the direction of ∇h at a point ξ_l on C is necessarily that of the normal λ, one simply follows a line in this direction inward from ξ_l until it intersects a net line (parallel to X or Y) at a point, say $\mathbf{Z}$. By interpolation in the net, one finds h and $\lambda \cdot \nabla h$ at $\mathbf{Z}$; these values and the value $h = 0$ at ξ_l permit quadratic fitting of h along the line from ξ_l to $\mathbf{Z}$ and thereby finding ∇h at ξ_l. The finite difference forms of (13.36) and (13.37) are then taken as

$$\xi_l^{n+1} - \xi_l^n = \tfrac{1}{2}(\mathbf{v}_l^{n+1} + \mathbf{v}_l^n)\Delta t$$

$$\mathbf{v}_l^{n+1} - \mathbf{v}_l^n = \tfrac{1}{2}(\boldsymbol{\psi}_l^{n+1} + \boldsymbol{\psi}_l^n)\Delta t,$$

where $\boldsymbol{\psi}$ denotes the right member of (13.37). As in shock-fitting, the entire procedure is highly implicit, and an iterative process, described in detail by Kasahara *et al.* (1964), is used for solving the equations.

In one of the calculations by this method, the following initial data were used: $h(X, Y, 0)$ is as shown in Figure 13.10, and $\mathbf{u}(X, Y, 0) \equiv 0$; the configuration of the cold air mass 8 hours later, as calculated, is shown in Figure 13.11. By energy tests and the like, comparison with observations, etc., it was concluded that the method is stable and quite accurate.

REFERENCES

Ansorge, R. (1963): *Die Adams-Verfahren als Charakteristikenverfahren höherer Ordnung zur Lösung von hyperbolischen Systemen halblinearer Differentialgleichungen.* Numer. Math., vol. 5, p. 443

Aronson, D. G. (1963): *On the stability of certain finite difference approximations to parabolic systems of differential equations.* Numer. Math., vol. 5, pp. 118 and 290

Ashenhurst, R. L., and Metropolis, N. C. (1959): *Unnormalized floating-point arithmetic.* J. Assoc. Comput. Mach., vol. 6, p. 415

Bagrinovskii, K. A., and Godunov, S. K. (1957): *Difference schemes for multi-dimensional problems.* Dokl. Akad. Nauk USSR, vol. 115, p. 431

Banach, S. (1932): *Théorie des opérations linéaires.* Hafner Publishing Company, New York

Becker, R. (1922): *Stosswelle und Detonation.* Z. Physik, vol. 8, p. 321

Belotserkovskii, O. M., and Chushkin, P. I. (1962): *The numerical method of integral relations.* Zh. Vychisl. Mat. i Mat. Fiz., Akad. Nauk USSR, vol. 2, p. 731

Bethe, H. A., Fuchs, K., von Neumann, J., Peierls, R., and Penny, W. G. (1944): *Shock hydrodynamics and blast waves.* Report AECD-2860, Los Alamos Sci. Lab.

Birkhoff, G. (1962): *Helmholtz and Taylor instability.* Proceedings of Symposia in Appl. Math., Amer. Math Soc., vol. 13, p. 55

Birkhoff, G., and Varga, R. S. (1959): *Implicit alternating direction methods.* Trans. Amer. Math. Soc., vol. 92, p. 13

Blair, A., Metropolis, N. C., von Neumann, J., Taub, A. H., and Tsingou, M. (1959): *A study of a numerical solution to a two-dimensional hydrodynamical problem.* Math. Comp. (formerly MTAC), vol. 13, p. 145

Bleakney, W., and Taub, A. H. (1949): *Interaction of shock waves.* Rev. Mod. Phys., vol. 21, p. 584

Brillouin, M. (1931): *Sur quelques problèmes non résolus de la physique mathématique classique. Propagation de la fusion.* Ann. Inst. H. Poincaré, vol. 1, p. 285

Bruce, G. H., Peaceman, D. W., Rachford, H. H. Jr., and Rice, J. D. (1953): *Calculation of unsteady-state gas flow through porous media.* Trans. Amer. Inst. of Mining and Met. Engnrs., vol. 198, p. 79

Bruhn, G., and Haack, W. (1958): *Ein Charakteristikenverfahren für dreidimensionale instationäre Gasströmungen.* ZAMP, vol. 9b, p. 173

Buchanan, M. L. (1963a): *A necessary and sufficient condition for stability of difference schemes for second order initial value problems.* J. Soc. Indust. Appl. Math., vol. 11, p. 474

Buchanan, M. L. (1963b): *A necessary and sufficient condition for stability of difference schemes for initial value problems.* J. Soc. Indust. Appl. Math., vol. 11, p. 919

BURSTEIN, S. Z. (1963): *Numerical calculation of multidimensional shocked flows.* Report NYO 10433, Courant Inst. Math. Sci., New York University

BURSTEIN, S. Z. (1965): *Finite difference calculations for hydrodynamic flows containing discontinuities.* Report NYO 1480-33, Courant Inst. of Math Sci.,. New York University

BUTLER, D. S. (1960): *The numerical solution of hyperbolic systems of partial differential equations in three independent variables.* Proc. Roy. Soc., vol. A255, p. 232

CALKIN, J. W. (1942): (On the spontaneous formation of shocks during a fluid motion). Unpublished report

CARLSON, B. G. (1953): (The S_n method). Unpublished Los Alamos Report and subsequent reports

CARSLAW, H. S. (1945): *Mathematical theory of the conduction of heat in solids.* Dover Publications, New York

CHANDRASEKHAR, S. (1944a): *On the radiative equilibrium of a stellar atmosphere, I.* Astroph. J., vol. 99, p. 180

CHANDRASEKHAR, S. (1944b); *On the radiative equilibrium of a stellar atmosphere, II.* Astroph. J., vol. 100, p. 82

COLLATZ, L. (1951): *Numerische Behandlung von Differentialgleichungen.* Springer-Verlag, Berlin

COURANT, R., and FRIEDRICHS, K. O. (1948): *Supersonic flow and shock waves.* Interscience Publishers, New York

COURANT, R., FRIEDRICHS, K. O., and LEWY, H. (1928): *Über die partiellen Differenzengleichungen der mathematischen Physik.* Math. Ann., vol. 100, p. 32

COURANT, R., and HILBERT, D. (1953): *Methods of mathematical physics,* vol. 1. Interscience Publishers, New York

COURANT, R., and HILBERT, D. (1962): *Methods of mathematical physics,* vol. 2, Interscience Publishers, New York

COURANT, R., ISAACSON, E., and REES, M. (1952): *On the solution of nonlinear hyperbolic differential equations by finite differences..* Communications Pure and Appl. Math., vol. 5, p. 243

COURANT, R., and LAX, P. D. (1949): *On nonlinear partial differential equations with two independent variables.* Communications Pure and Appl. Math., vol. 2, p. 255

CRANDALL, S. H. (1954): *Numerical treatment of a fourth-order parabolic partial differential equation.* J. Assoc. Computing Machinery, vol. 1, p. 111

CRANK, J., and NICHOLSON, P. (1947): *A practical method for numerical integration of solutions of partial differential equations of heat-conduction type.* Proc. Cambridge Philos. Soc., vol. 43, p. 50

DAHLQUIST, G. G. (1963): *Stability questions for some numerical methods for ordinary differential equations.* Proceedings of Symposia in Appl. Math.; *Experimental arithmetic, high speed computing and mathematics,* Amer. Math. Soc., p. 147

DE SANTO, D. F., and KELLER, H. B. (1961): *Numerical studies of transition from laminar to turbulent flow over a flat plate.* Coll. of Eng. Res. Div. Report, New York University

DOUGLAS, J. (1955): *On the numerical integration of* $\frac{\partial^2 u}{\partial x^2} + \frac{\partial^2 u}{\partial y^2} = \frac{\partial u}{\partial t}$ *by implicit methods.* J. Soc. Industrial and Appl. Math., vol. 3, p. 42

REFERENCES

DOUGLAS, J. (1956): *On the relation between stability and convergence in the numerical solution of linear parabolic and hyperbolic equations.* J. Soc. Industrial and Appl. Math., vol. 4, p. 20

DOUGLAS, J. (1957): *A note on the alternating direction implicit method for the numerical solution of the heat equation.* Proc. Amer. Math. Soc., vol. 8, p. 409

DOUGLAS, J. (1962): *Alternating direction methods for three space variables.* Numer. Math., vol. 4, p. 41

DOUGLAS, J., and GUNN, J. (1964): *A general formulation of alternating direction methods, I.* Numer. Math., vol. 6, p. 428

DOUGLAS, J., and RACHFORD, H. H. (1956): *On the numerical solution of the heat conduction problems in two and three variables.* Trans. Amer. Math. Soc., vol. 82, p. 421

DUFF, R. E. (1962): *Slip line instability.* Proceedings of Symposia in Appl. Math, Amer. Math. Soc., vol. 13 (*Hydrodynamic Instability*), p. 77

DU FORT, E. C., and FRANKEL, S. P. (1953): *Stability conditions in the numerical treatment of parabolic differential equations.* Math. Tables and other Aids to Computation, vol. 7, p. 135

D'YAKONOV, E. G. (1962): *On several difference schemes for the solution of boundary problems.* Zh. Vychisl. Mat. i Mat. Fiz., Moscow, vol. 2, p. 57

D'YAKONOV, E. G. (1964): *Difference schemes of second order accuracy with a splitting operator for parabolic equations without mixed partial derivatives.* Zh. Vychisl. Mat. i Mat. Fiz., Moscow, vol. 4, p. 935

EDDY, R. P. (1949): *Stability in the numerical solution of initial value problems in partial differential equations.* Naval Ordnance Laboratory Memo. 10232

EHRLICH, R., and HURWITZ, H. JR. (1954): *Multigroup methods for neutron problems.* Nucleonics, vol. 12, p. 23

ELLIOTT, L. A. (1962): *Shock fronts in two-dimensional flow.* Proc. Roy. Soc., vol. A267, p. 558

EVANS, G. W., BROUSSEAU, R., and KIERSTEAD, R. (1954): *Instability considerations for various difference equations derived from the diffusion equation.* Stanford Research Institute, Stanford, California

FLANDERS, D. A., and SHORTLEY, G. (1950): *Numerical determination of fundamental modes.* J. Appl. Phys., vol. 21, p. 1326

FORSYTHE, G. E., and WASOW, W. R. (1960): *Finite difference methods for partial differential equations.* John Wiley, New York

FRANK, R. M., and LAZARUS, R. B. (1964): *Mixed Eulerian Lagrangian method.* Methods in Comput. Phys., Academic Press, New York, vol. 3, p. 47

FRANKEL, S. P. (1950): *Convergence rates of iterative treatments of partial differential equations.* Math. Tables and other Aids to Computation, vol. 4, p. 65

FRIEDRICHS, K. O. (1954): *Symmetric hyperbolic linear differential equations.* Communications Pure and Appl. Math., vol. 7, p. 345

FRIEDRICHS, K. O. (1958): *Symmetric positive linear differential equations.* Comm. Pure Appl. Math., vol. 11, p. 333

FROMM, J. E. (1964): *The time-dependent flow of an incompressible fluid.* Methods in Comput. Phys., Academic Press, New York, vol. 3, p. 346

GÅRDING, L. (1953): *Dirichlet problem for linear elliptic partial differential equations.* Math. Scand., vol. 1, p. 53

GARY, J. (1962a): *Numerical computation of hydrodynamic flows which contain a shock.* Report NYO 9603, Courant Inst. Math. Sci., New York University

GARY, J. (1962b): *On certain finite difference schemes for the equations of hydrodynamics.* Report NYO 9188, Courant Inst. Math. Sci., New York University

GERSHGORIN, S. (1931): *Über die Abgrenzung der Eigenwerte einer Matrix.* Izvestiya Akad, Nauk, USSR, Ser. Math., vol. 7, p. 749

GOAD, W. B. (1960): *WAT: a numerical method for two-dimensional unsteady fluid flow.* Report LAMS 2365, Los Alamos Sci. Lab.

GODUNOV, S. K. (1959): *A finite difference method for the numerical computation and discontinuous solutions of the equations of fluid dynamics.* Mat. Sb., vol. 47, p. 271

GODUNOV, S. K. (1962): *The method of orthogonal sweeps for the solution of difference equations.* Zh. Vychisl. Mat. i Mat. Fiz., Moscow, vol. 2, p. 972

GODUNOV, S. K., and RYABENKII, V. S. (1963a): *Special criteria of stability of boundary-value problems for non-self-adjoint difference equations.* Uspekhi Mat. Nauk, vol. 18, p. 3

GODUNOV, S. K., and RYABENKII, V. S. (1963b): *Canonical forms of systems of linear difference equations with constant coefficients.* Zh. Vychisl. Mat. i Mat. Fiz., Moscow, vol. 3, p. 211

GODUNOV, S. K., and RYABENKII, V. S. (1964): *Introduction to the theory of difference schemes.* Interscience Publishers, New York, also Fizmatigiz, Moscow (1962)

GÖLDNER, S. R. (1949): *Existence and uniqueness theorems for two systems of non-linear hyperbolic differential equations for functions of two independent variables.* Ph.D. thesis, New York University

HADAMARD, J. (1923): *Lectures on Cauchy's problem in linear partial differential equations.* Yale University Press

HAIN, K., HAIN, G., ROBERTS, K. V., ROBERTS, S. J., and KÖPPENDÖRFER, W. (1960): *Fully ionized pinch collapse.* Z. Naturforsch, vol. 15a, p. 1039

HARLOW, F. H. (1964): *The particle-in-cell computing method for fluid dynamics.* Methods in Comput. Phys., Acad. Press, vol. 3, p. 319

HERSCH, R. (1963): *Mixed problems in several variables.* J. Math. Mech., vol. 12, p. 317

HILDEBRAND, F. B. (1952): *On the convergence of numerical solutions of the heat-flow equation.* J. Math. Physics, vol. 31, p. 35

HOPF, E. (1950): *The partial differential equation* $u_t + uu_x = \mu u_{xx}$. Comm. Pure Appl. Math., vol. 3, p. 201

HOSKIN, N. E. (1964): *Solution by characteristics of the equations of one-dimensional unsteady flow.* Methods in Comput. Phys., Academic Press, New York, vol. 3, p. 265

JOHANNSON, O., and KREISS, H. O. (1963): *Über das Verfahren der zentralen Differenzen zur Lösung des Cauchy-Problems für partielle Differentialgleichungen.* Nordisk Tidskr. Informations-Behandling, vol. 3, p. 97

JOHN, F. (1952): *On the integration of parabolic equations by difference methods.* Communications Pure and Appl. Math., vol. 5, p. 155

JOHN, F. (1953): *A note on "improper" problems in partial differential equations.* Communications Pure and Appl. Math., vol. 8, p. 591

REFERENCES

JUNCOSA, M. L., and YOUNG, D. M. (1953): *On the order of convergence of solutions of a difference equation to a solution of the diffusion equation.* J. Soc. Industrial and Appl. Math, vol. 1, p. 111

KANTOROWITCH, L. W. (1948): *Functional analysis and applied mathematics.* Uspekhi Mat. Nauk, USSR, vol. 3, p. 89

KASAHARA, A., ISAACSON, E., and STOKER, J. J. (1964): *Numerical studies of frontal motion in the atmosphere, I.* Report NYO 1480-6, Courant Inst. Math. Sci., New York University

KASAHARA, A. (1965): *On certain finite-difference methods for fluid dynamics.* Monthly Weather Review, Jan. 1965, p. 27

KATO, T. (1960): *Estimation of iterated matrices, with application to the von Neumann condition.* Numer. Math., vol. 2, p. 22

KELLER, H. B. (1956): *Difference equations for hyperbolic systems.* Seminar, New York University

KELLER, H. B. (1958): *Approximate solutions of transport problems—part I: steady-state elastic scattering in plane and spherical geometry.* J. Soc. Indust. Appl. Math., vol. 6, p. 452

KELLER, H. B. (1960a): *Approximate solutions of transport problems—part II: convergence and applications of the discrete ordinate method.* J. Soc. Indust. Appl. Math., vol. 8, p. 43

KELLER, H. B. (1960b): *On the pointwise convergence of the discrete ordinate method.* J. Soc. Indust. Appl. Math., vol. 8, p. 560

KELLER, H. B. (1960c): *Convergence of the discrete ordinate method for the anisotropic scattering transport equation.* Symposium on the numerical treatment of ordinary differential equations, integral and integro-differential equations, Rome; Birkhäuser, Basel, p. 292

KELLER, H. B., and WENDROFF, B. (1957): *On the formulation and analysis of numerical methods for time dependent transport equations.* Communications Pure and Appl. Math., vol. 10, p. 567

KOLSKY, H. (1955): Report LA 1867, Los Alamos Sci. Lab.

KONOVALOV, A. N. (1964): *Applications of the method of splitting to the numerical solution of dynamical problems of the theory of elasticity.* Zh. Vychisl. Mat. i Mat. Fiz., vol. 4, p. 760

KREIN, M. G. (1958): *Integral equations on a half-line with kernel depending on the difference of the arguments.* Usp. Mat. Nauk, Moscow, vol 13, p. 3

KREISS, H. O. (1959): *Über die Differenzapproximation hoher Genauigkeit bei Anfangswertproblemen für partielle Differentialgleichungen.* Numer. Math., vol. 1, p. 186

KREISS, H. O. (1960): *Über die Lösung von Anfangswertaufgaben für partielle Differentialgleichungen mit Hilfe von Differenzengleichungen.* Trans. Roy. Inst. Tech., Stockholm, No. 166

KREISS, H. O. (1962): *Über die Stabilitätsdefinition für Differenzengleichungen die partielle Differentialgleichungen approximieren.* Nordisk Tidskr. Informations-Behandling, vol. 2, p. 153

KREISS, H. O. (1963): *Über implizite Differenzenmethoden für partielle Differentialgleichchungen.* Numer. Math., vol. 5, p. 24

KREISS, H. O. (1964) : *On difference approximations of the dissipative type for hyperbolic differential equations.* Comm. Pure Appl. Math., vol. 17, p. 335

KREISS, H. O. (1965): *On difference approximations.* Symposium on numerical solution of partial differential equations, Univ. of Maryland, May, 1965

LAASONEN, P. (1949) : *Über eine Methode zur Lösung der Wärmeleitungsgleichung.* Acta Math., vol. 81, p. 309

LADYZHENSKAYA, O. A. (1952) : *Solution of Cauchy's problem for hyperbolic systems by the method of finite differences.* Leningrad Gos. Univ. Uch. Zap., vol. 144 (Ser. Mat.), p. 192

LADYZHENSKAYA, O. A. (1953) : *The mixed problem for a hyberbolic equation.* Izdat Tekhn.-Teor. Lit., Moscow

LADYZHENSKAYA, O. A. (1957a) : *The method of finite differences in the theory of partial differential equations.* Uspekhi Mat. Nauk, vol. 12, p. 123

LADYZHENSKAYA, O. A. (1957b) : *On the construction of discontinuous solutions of quasi-linear hyperbolic equations when the "coefficient of viscosity" tends to zero.* Trudy Moskov. Mat. Obshch., vol. 6, p. 465

LAX, A. (1956): *On Cauchy's problem for partial differential equations with multiple characteristics.* Comm. Pure Appl. Math., vol. 9, p. 135

LAX, P. D. (1953): *Nonlinear hyperbolic equations.* Comm. Pure Appl. Math., vol. 6, p. 231

LAX, P. D. (1954) : *Weak solutions of nonlinear hyperbolic equations and their numerical computation.* Comm. Pure Appl. Math., vol. 7, p. 159

LAX, P. D. (1957) : *Hyperbolic systems of conservation laws, II.* Comm. Pure Appl. Math., vol. 10, p. 537

LAX, P. D. (1960) : *The scope of the energy method.* Bull. Amer. Math. Soc., vol. 66, p. 32

LAX, P. D. (1961) : *On the stability of difference approximations to solutions of hyperbolic equations with variable coefficients.* Comm. Pure Appl. Math., vol. 14, p. 497

LAX, P. D., and KELLER, J. B. (1951) : *The initial and mixed boundary value problem for hyperbolic systems.* Report LAMS-1205, Los Alamos Sci. Lab.

LAX, P. D., and NIRENBERG, L. (1966). (in press)

LAX, P. D., and RICHTMYER, R. D. (1956) : *Survey of the stability of linear finite difference equations.* Comm. Pure Appl. Math., vol. 9, p. 267

LAX, P. D., and WENDROFF, B. (1960) : *Systems of conservation laws.* Comm. Pure Appl. Math., vol. 13, p. 217

LAX, P. D., and WENDROFF, B. (1962) : *On the stability of difference schemes.* Comm. Pure Appl. Math., vol. 15, p. 363

LAX, P. D., and WENDROFF, B. (1964) : *Difference schemes for hyperbolic equations with high order of accuracy.* Comm. Pure Appl. Math., vol. 17, p. 381

LEES, M. (1960a) : *A priori estimates for the solution of difference approximations to parabolic partial differential equations.* Duke Math. Jour., vol. 27, p. 297

LEES, M. (1960b) : *Energy inequalities for the solution of differential equations.* Trans. Amer. Math. Soc., vol. 94, p. 58

LEES, M. (1961) : *Alternating direction and semi-explicit difference methods for parabolic partial differential equations.* Numer. Math., vol. 3, p. 398

REFERENCES

LEHNER, J., and WING, G. M. (1955): *On the spectrum of an unsymmetric operator arising in the transport theory of neutrons.* Comm. Pure Appl. Math., vol. 8, p. 217

LEITH, C. E. (1964): *Lagrangian advection in an atmospheric model.* Report UCRL 7822, Lawrence Radiation Lab., University of California, Livermore

LEUTERT, W. (1951): *On the convergence of approximate solutions of the heat equation to the exact solution.* Proc. Amer. Math. Soc., vol. 2, p. 433

LEUTERT, W. (1952): *On the convergence of unstable approximate solutions of the heat equation to the exact solution.* J. Math. Physics, vol. 30, p. 245

LEWY, H. (1927): *Über das Anfangswertproblem bei einer hyperbolischen nichtlinearen partiellen Differentialgleichung zweiter Ordnung mit zwei unabhängigen Veränderlichen.* Math. Ann., vol. 98, p. 179

LEWY, H. (1948): *On the convergence of solutions of difference equations.* Essays presented to R. Courant on his 60th birthday. Interscience Publishers, New York, p. 211

LIEBMANN, L. (1918): *Die Angenäherte Ermittelung Harmonischer Funktionen und konformer Abbildungen.* Sitz. Bayer, Akad. Wiss. Math-Phys. Klasse, p. 385

LOGEMANN, G. (1965): *Existence and uniqueness of rarefaction waves.* Ph.D. thesis, New York University

LUDFORD, G. S. S., POLACHEK, R. J., and SEEGER, R. (1953): *On unsteady flow of viscous fluids.* J. Appl. Phys., vol. 24, p. 490

MACDUFFEE, C. C. (1946): *The theory of matrices.* Chelsea Publishing, Bronx, New York

MARCHUK, G. I., and YANENKO, N. N. (1965): *Applications of the method of splitting (fractional steps) to the solution of problems of mathematical physics.* IFIP Congress, New York

MARTIN, M. H. (1955): *An existence and uniqueness theorem for unsteady one-dimensional anisentropic flow.* Comm. Pure Appl. Math., vol. 8, p. 367

MILLER, J. J. H., and STRANG, W. G. (1965): *Matrix theorems for partial differential and difference equations.* Stanford University Tech. Report CS28, Stanford, California

MILNE, W. E. (1953): *Numerical solution of differential equations.* John Wiley, New York

MITCHELL, A. R., and PEARCE, R. P. (1962): *High accuracy difference formulae for the numerical solution of the heat conduction equation.* The Computer Journal, vol. 5, p. 142

MORAWETZ, C. S. (1956): *On the non-existence of continuous transonic flows past profiles.* Comm. Pure Appl. Math., vol. 9, p. 45

MORIMOTO, H. (1962): *Stability in the wave equation coupled with heat flow.* Numer. Math., vol. 4, p. 136

MORTON, K. W. (1964a): *On a matrix theorem due to H. O. Kreiss.* Comm. Pure Appl. Math., vol. 17, p. 375

MORTON, K. W. (1964b): *Finite amplitude compression waves in a collision-free plasma.* Physics of Fluids, vol. 7, p. 1800

MORTON, K. W., and SCHECHTER, S. (1965): *On the stability of finite difference matrices.* J. Soc. Indust. Appl. Math. Numer. Anal., Ser. B, vol. 2, p. 119

NOH, W. F. (1964): *CEL: a time-dependent two-space-dimensional, coupled Eulerian-Lagrangian code.* Methods in Comput. Phys., Academic Press, New York, vol. 3, p. 117

VON NEUMANN, J. (1943): *Mathematische Grundlagen der Quantenmechanik.* Dover Publications, New York

VON NEUMANN, J., and RICHTMYER, R. D. (1950): *A method for the numerical calculations of hydrodynamical shocks.* J. Appl. Phys., vol. 21, p. 232

O'BRIEN, G. G., HYMAN, M. A., and KAPLAN, S. (1951): *A study of the numerical solution of partial differential equations.* J. Math. Physics, vol. 29, p. 223

OLEINIK, O. A. (1956): *The problem of Cauchy for non-linear differential equations of the first order with discontinuous initial conditions.* Trudy Moskov. Mat. Obshch., vol. 5, p. 433

OLEINIK, O. A. (1957): *Discontinuous solutions of non-linear equations.* Uspekhi Mat. Nauk, vol. 12, p. 3

OLEINIK, O. A., and VVEDENSKAYA, N. D. (1957): *The solution of Cauchy problems and boundary value problems for non-linear equations in the class of discontinuous functions.* Doklad. Akad. Nauk, vol. 113, p. 503

PARLETT, B. N. (1966): *Accuracy and dissipation in difference schemes.* Comm. Pure Appl. Math., vol. 19, p. 111

PARTER, S. V. (1962): *Stability, convergence, and pseudo-stability of finite difference equations for an over-determined problem.* Numer. Math., vol. 4, p. 277

PEACEMAN, D. W., and RACHFORD, H. H. JR. (1955): *The numerical solution of parabolic and elliptic differential equations.* J. Soc. Industrial Appl. Math., vol. 3, p. 28

PEARCE, R. P., and MITCHELL, A. R. (1962): *On finite difference methods of solution of the transport equation.* Math. Comp., vol. 16, p. 155

PEARCY, C. (1966): *A short proof of the Halmos inequality.* (in press)

PHILLIPS, N. A. (1959): *An example of non-linear computational instability.* The Atmosphere and the Sea in Motion, Bert Bolin, ed., Rockefeller Institute Press, New York, p. 501

PHILLIPS, R. S. (1957): *Dissipative hyperbolic systems.* Trans. Amer. Math. Soc., vol. 86, p. 109

PHILLIPS, R. S. (1959): *Dissipative operators and hyperbolic systems of partial differential equations.* Trans. Amer. Math. Soc., vol. 90, p. 193

QUARLES, D. (1964): Ph.D. Thesis, New York University

RICHARDSON, D. J. (1964): *The solution of two-dimensional hydrodynamic equations by the method of characteristics.* Methods in Comput. Phys., Academic Press, New York, vol. 3, p. 295

RICHTMYER, R. D. (1960): *Proposed fluid-dynamics research.* Unpublished memorandum, Courant Inst. Math. Sci., New York University

RICHTMYER, R. D. (1961): *Progress report on the Mach reflection calculation.* Report NYO 9764, Courant Inst. Math. Sci., New York University

RICHTMYER, R. D. (1962): *A survey of difference methods for non-steady fluid dynamics.* NCAR Technical note 63–2, Nat'l Center for Atmos. Res., Boulder, Colo.

RICHTMYER, R. D. (1964): *The stability criterion of Godunov and Ryabenkii for difference schemes.* Report NYO 1840–4, Courant Inst. Math. Sci., New York University

REFERENCES

RICHTMYER, R. D., and MORTON, K. W. (1964): *Stability studies for difference equations: (I) non-linear instability (II) coupled sound and heat flow.* Report NYO 1480-5, Courant Inst. Math. Sci., New York University

RIESZ, F., and Sz. NAGY, B. (1952): *Leçons d'analyse fonctionelle.* Budapest

ROBERTS, K. V., and WEISS, N. O. (1966): *Convective difference schemes.* Math. Comp., vol. 20, p. 272

ROSE, M. E. (1956): *On the integration of non-linear parabolic equations by implicit difference methods.* Quart. Appl. Math., vol. 14, p. 237

ROZHDESTVENSKII, B. L. (1960): *Discontinuous solutions of systems of quasilinear equations of hyperbolic type.* Uspekhi Mat. Nauk, vol. 15, p. 59

RYABENKII, V. S. (1964): *Necessary and sufficient conditions for the well-conditionedness of boundary problems for systems of ordinary difference equations.* Zh. Vychisl. Mat. i Mat. Fiz., vol. 4, p. 242.

RYABENKII, V. S., and FILLIPOV, A. F. (1960): *Über die Stabilität von Differenzengleichungen.* Deutscher Verlag der Wissenschaften, Berlin

SAMARSKII, A. A. (1964): *Economical difference schemes for systems of equations of parabolic type.* Zh. Vychisl. Mat. i Mat. Fiz., vol. 4, p. 927

SAUL'EV, V. K. (1957): *On a method of numerical integration of the equation of diffusion.* Doklady Akad. Nauk USSR, vol. 115, p. 1077

SAUL'EV, K. (1958): *On methods of increased accuracy in two-sided approximations to the solution of parabolic equations.* Doklady Akad. Nauk USSR, vol. 118, p. 1088

SAUER, R. (1963): *Differenzenverfahren für hyperbolische Anfangswertprobleme bei mehr als zwei unabhängigen Veränderlichen mit Hilfe von Nebencharaakteristiken.* Numer. Math., vol. 5, p. 55

SCHULTZ, W. D. (1964a): *Tensor artificial viscosity for numerical hydrodynamics.* J. Math. Phys., vol. 5, p. 133

SCHULTZ, W. D. (1964b): *Two-dimensional Lagrangian hydrodynamical difference equations.* Methods in Comput. Phys., Academic Press, New York, vol. 3, p. 1

SCHUR, I. (1909): *Über die charakteristischen Wurzeln einer linearen Substitution mi einer Anwendung auf die Theorie der Integralgleichung.* Math. Ann., vol. 66, p. 488

SEIDMAN, T. I. (1963): *On the stability of certain difference schemes.* Numer. Math., vol. 5, p. 201

SHORTLEY, G. (1952): *Use of Tschebyscheff-polynomial operators in the numerical solution of boundary-value problems.* J. Appl. Phys., vol. 24, p. 392

SHORTLEY, G., and WELLER, R. (1938): *The numerical solution of Laplace's equation.* J. Appl. Phys., vol. 9, p. 334

SOFRONOV, I. D. (1962): *On the difference solution of the heat-flow equation in curvilinear coordinates.* Zh. Vychisl. Mat. i Mat. Fiz., vol. 3, p. 786

SOUTHWELL, R. V. (1946): *Relaxation methods in theoretical physics.* Oxford University Press

SPITZER, L. JR. (1956): *Physics of fully ionized gases.* Interscience Publishers, New York

STARK, R. H. (1956): *Rates of convergence in numerical solution of the diffusion equation.* J. Assoc. Comput. Mach., vol. 3, p. 29

STETTER, H. J. (1961): *On the convergence of characteristic finite-difference methods of high accuracy for quasi-linear hyperbolic equations.* Numer. Math., vol. 3, p. 321

STOKER, J. J. (1953): *Dynamical theory for treating the motion of cold and warm fronts in the atmosphere.* Report IMM 200, Courant Inst. Math. Sci., New York University

STOKER, J. J. (1957): *Water waves.* Interscience Publishers, New York

STONE, M. H. (1932): *Linear transformations in Hilbert space.* Amer. Math. Soc. Colloq. Publications

STRANG, W. G. (1960): *Difference methods for mixed boundary-value problems.* Duke Math. J., vol. 27, p. 221

STRANG, W. G. (1962): *Polynomial approximation of Bernstein type.* Trans. Amer. Math. Soc., vol. 105, p. 525

STRANG, W. G. (1963): *Accurate partial difference methods I: linear Cauchy problems.* Arch. Rational Mech. Anal., vol. 12, p. 392

STRANG, W. G. (1964a): *Accurate partial difference methods II: non-linear problems.* Numer. Math., vol. 6, p. 37

STRANG, W. G. (1964b): *Wiener-Hopf difference equations.* J. Math. Mech., vol. 13, p. 85

STRANG, W. G. (1965): *Necessary and insufficient conditions for well-posed Cauchy problems.* J. Differential Equations, vol. 2, p. 107

TALBOT, G. P. (1963): *Application of the numerical method of characteristics in three independent variables to shock-thermal layer interaction problems.* Report, Royal Armament Research and Development, U.K.

THOMAS, L. H. (1944): *Note on Becker's theory of the shock front.* J. Chem. Phys., vol 12, p. 449

THOMAS, L. H. (1949): *Numerical solution of partial differential equations of parabolic type.* Seminar of Scientific Computation. International Business Machines Corp.

THOMAS, L. H. (1950): *Stability of partial difference equations.* Symposium on theoretical compressible flow. Report No. 1132, U.S. Naval Ordnance Laboratory

THOMAS, L. H. (1954): *Computation of one-dimensional compressible flows including shocks.* Communications Pure and Appl. Math., vol. 7, p. 195

THOMÉE, V. (1962): *A stable difference scheme for the mixed boundary problem for a hyperbolic first order system in two dimensions.* J. Soc. Indust. Appl. Math., vol. 10, p. 229

THOMÉE, V. (1965): *Stability of difference schemes in the maximum norm.* J. Differential Equations, vol. 1, p. 273

THOMÉE, V. (1966): *Stability of difference schemes in L^p.* (in press)

THOMPSON, R. J. (1964): *Difference approximation for inhomogeneous and quasi-linear equations.* J. Soc. Indust. Appl. Math., vol. 12, p. 189

THOMPSON, W. B. (1962): *An introduction to plasma physics.* Pergamon Press, Oxford

TODD, J. (1955): *A direct approach to the problem of stability in the numerical solution of partial differential equations.* Report No. 4260, National Bureau of Standards

TÖRNIG, W. (1963): *Über Differenzenverfahren in Rechteckgittern zur numerischen Lösung quasi-linearer hyperbolischer Differentialgleichungen.* Numer. Math., vol. 5, p. 353

REFERENCES

TRULIO, J. G. (1964): *The strip code and the jetting of gas between plates.* Methods in Comput. Phys., vol. 3, p. 69

TURNER, J., and WENDROFF, B. (1964): *An unconditionally stable implicit difference scheme for the hydrodynamical equations.* Report LA-3007, Los Alamos Sci. Lab.

URM, V. YA. (1961): *On the necessary and sufficient conditions for stability of systems of difference equations.* Dokl. Akad. Nauk, USSR, vol. 139, p. 40

VAN NORTON, R. (1962): *On the real spectrum of a mono-energetic neutron transport operator.* Comm. Pure Appl. Math., vol. 15, p. 99

VARGA, R. S. (1962): *Matrix iterative analysis.* Prentice Hall, Englewood Cliffs, New Jersey

WENDROFF, B. (1960a): *On the convergence of the discrete ordinate method.* J. Soc. Indust. Appl. Math., vol. 8, p. 508

WENDROFF, B. (1960b): *On central difference equations for hyperbolic systems.* J. Soc. Indust. Appl. Math., vol. 8, p. 549

WHITTAKER, E. T., and WATSON, G. N. (1927): *Modern analysis.* 4th ed. Cambridge University Press

WICK, G. C. (1943): *Über ebene Diffusionsprobleme.* Z. Physik, vol. 121, p. 702

WIDLUND, O. B. (1965): *On the stability of parabolic difference schemes.* Math. Comp., vol. 19, p. 1

WIDLUND, O. B. (1966): *Stability of parabolic difference schemes in the maximum norm.* Numer. Math., vol. 8, p. 186

YANENKO, N. N. (1959): *On a difference method of calculation of the multi-dimensional equation of heat flow.* Doklady Akad. Nauk, vol. 125, p. 1207

YANENKO, N. N. (1964): *On weak approximation of systems of differential equations.* Sibirsk. Mat. Zh., vol. 5, p. 1430

YOUNG, D. M. (1954): *Iterative methods for solving partial differential equations of elliptic type.* Doctoral Disseration, Harvard University (1950). Trans. Amer. Math. Soc., vol. 76, p. 92

ZYGMUND, A. (1952): *Trigonometrical series.* Chelsea Publishing, Bronx, New York

Index

Absolutely continuous function, 38
Absorbing medium, 220
Addition theorem for Legendre polynomials, 224
Adiabatic sound speed, 291
Age-diffusion equation, 194, 207, 210
Alfvén speed, 347
Algebraic stability condition, 60
Alternating-direction methods, 211 ff, 216
Alternative definitions of stability, 58, 95 ff
Amplification factor, 67, 71, 194, 211, 215, 266
Amplification matrix, 67, 187, 262, 263, 274, 275, 303
Angled derivative, 347
Artificial viscosity, 311 ff, 319, 350, 365
  see also *Pseudo-viscosity*
Associated Legendre polynomial, 225
Asymptotic behavior of Fourier coefficient, 22
Atmospheric front, 383
Auxiliary Banach space, 169
Axial symmetry, 221

Banach space, 30 ff
  choice of, 65
Backward time difference, 17
Boltzmann equation, 218
Bound of an operator, 34
Bound of a matrix, 69
Boundary conditions, 39, 41, 56, 131, 134, 137, 148, 154, 163, 186, 214, 220, 221, 223, 229, 237, 249, 272, 276, 280 ff, 288, 296
Boundary point, 134
Bounded operator, 34
Buchanan stability criterion, 80, 85, 89

Carlson's S_n method, 244 ff
Carlson's scheme, 240 ff
Cauchy convergence theorem, 31
Cauchy sequence, 31
Centering, 262
Central difference, 17
Change of norm, 57
Characteristic form, 305 ff
Characteristic plane, 375
Characteristics, 309
  method of, 248
  theory of, 309
  in two-dimensional flow, 375
Class of genuine solutions, 40, 44
Closed operator, 37, 49, 54
Closed set, 32
Complete space, 31
Conservation of energy, 261, 284
Conservation-law form, 300
Conservation laws, 337
Conservation of mass, momentum, and energy, 302
Consistency, 174, 177, 179
Consistency condition, 43, 45, 67 ff, 101, 171, 175, 187
Consistent approximation, 44
Consistent operator, 127
Contact discontinuity, 306, 355
Continuous dependence on initial data, 41
Convergence, 44, 45, 172, 179
Convergence-in-the-mean, 29
Coupled sound and heat flow, 143, 264 ff
Courant-Friedrichs-Lewy condition, 262, 323
Courant, Isaacson, and Rees, method of, 290
Courant number, 324, 334
Crank-Nicholson scheme, 142, 151, 189, 212

INDEX

Cross-section, 219
Current, 345
Cylindrical symmetry, 221

Decay factor, 10
Dense, 32
Detached shock, 366
Difference notation, 134
Differential operator, 33, 63
Diffusion, 185 ff
Diffusion constant, 186
Diffusion equation, 4, 48, 176, 208, 211
Direct integration method for the transport equation, 246 ff
Discrete-ordinate methods, 227, 242
Dispersive, 332
Dissipation, 94
Dissipation in fluid dynamics, 311 ff
Dissipative, 109, 113, 119, 130, 270, 304, 335
Dissipative difference schemes, 108
Dissipative terms, 331, 334
Domain, 33
Du Fort-Frankel equation, 176 ff, 190, 192
Dynamic elasticity, 271

Effectively explicit, 296
Elastic vibrations, 271
Electric field, 345
Embedding of mesh function, 135, 149
Energy inequality, 147
Energy method, 13, 60, 132 ff, 137 ff, 143, 164, 285
Equation of continuity, 259, 289
Equation of motion, 259
Equations of high accuracy, 168
Equation of state, 259, 289, 301, 345
Equivalence of the spherical harmonic equations and the Wick-Chandrasekhar equations, 235
Equivalent norms, 57, 136
Error growth, 133
Eulerian equations, 288, 289 ff, 300
Explicit difference system, 17
Extension of an operator, 33
Extension theorem, 34, 41

Exterior points, 134
Extra boundary condition, 141, 157

Filling in of mesh function, 30, 135
Finite-difference approximation, 42 ff, 45
Finite-difference equations, 7, 65 ff
Finite-difference form of conservation laws, 305
First variation of an operator, 126, 127
Five-point difference equation, 86
Five-term recurrence relation, 276
Fluid dynamics in one space variable, 288 ff
Forward and backward space differences, 238 ff
Forward and backward sweeps, 198 ff
Forward difference, 14
Forward time difference, 17
Fourier integrals, 61
Fourier series, 6 ff, 61
Fourier transform method, 60
Four-point difference scheme, 137
Fourteen-point formula, 210
Fractional-step methods, 216 ff
Friedrichs' condition, 120
Function space, 28

Gårding partition of unity, 106, 118
Gary's equations, 306
Gauss quadrature, 234
Generalized solution, 6, 53 ff
Generalized solution operator, 41
Generalized states of a system, 29
Genuine solution, 40, 53
Gerschgorin's theorem, 76, 114
Global stability, 97
Godunov-Ryabenkii stability criterion, 151 ff, 153, 156 ff
Godunov-Ryabenkii theorem, 160, 167
Godunov's method, 338 ff
Gram determinant, 85
Grid, 7
Growth factor, 10, 187

Hadamard criterion, 41
Heat capacity, 186
Heat flow, 4, 185 ff, 211

Heat flow in a star, 186
Heat flux density, 185
Helmholtz instability, 352
Heuristic stability argument, 205, 206
Hexagonal point net, 215
High-accuracy schemes, 208
High-order formulas, 24
Hilbert norm, 32
Hilbert space, 61
Hugoniot relations, 314, 316
Hyperbolicity, 109
Hyperbolic systems, 108, 119 ff, 125, 146, 183, 228, 232, 242, 295

Implicit difference equations, 16, 17, 108, 183, 194, 198, 203, 207, 215, 233, 239, 263, 265, 274, 275, 295, 347
Improperly posed problems, 59, 92, 261, 294, 356
Infinitesimal generator of a semi-group, 49
Inhomogeneous problems, 52 ff
Initial-boundary-value problems, 131 ff, 137, 164
Initial value problems, 3, 60 ff
  properly posed, 63 ff
Inner product, 61
Instability, 9
Instantaneous state of a physical system, 28
Integro-differential equation, 220
Interfaces in fluid dynamics, 298 ff, 306
Interior points, 134
Internal boundary conditions, 4, 307
Isothermal sound speed, 264

John's theorem, 107
Jump conditions, 337
Jump conditions at a shock, 306

Kato's condition, 88
Kreiss's matrix theorem, 73 ff, 90, 112, 133
Kreiss theorem on weak stability, 96

L_2 norm, 100
Laasonen equation, 189

Lagrangean coordinate, 293
Lagrangean equations, 288, 293 ff, 300, 338
Laminar boundary layer, 352
Lattice, 7, 62
Law of heat conduction, 185
Lax equivalence theorem, 45 ff, 56, 171
Lax-Keller equations, 291
Lax-Nirenberg theorem, 121, 130
Lax-Wendroff condition, 88, 288, 300 ff
Lax-Wendroff treatment of shocks, 330
Legendre polynomials, 224
Lehner and Wing, work of, 257
Lelevier's method, 292
Limit point, 32
Linear difference equations, 39
Linear hyperbolic systems, 183
Linear operators, 28 ff, 34
Linear transformation, 34
Lipschitz continuous, 83
Local amplification matrix, 109
Local normal modes, 153, 165
Local stability, 91, 97
Local stability condition, 131
Lower order terms, 107, 195 ff

Mach reflection, 359
Magnetic field, 345
Magneto-fluid dynamics, 345
Magneto-fluid dynamic shocks, 348
Magneto-sonic speed, 347
Marginal stability, 130
Maximum norm, 29, 31, 100, 102, 103
Maximum principle, 48
Maxwell's equations, 346
Mean free path, 221
Mesh size, 7
Minimally semi-bounded, 147
Mixed problems, 131 ff, 137, 146, 156, 164
Modified von Neumann condition, 270
Morimoto's theorem, 266
Multi-diagonal equation, 183
Multi-dimensional fluid dynamics, 351
Multi-level difference equations, 43, 108, 157, 168 ff, 179
Multiplying medium, 220

404 INDEX

Navier-Stokes equation, 352
Necessary condition for stability, 187
Negative shocks, 315
Nested eigenvalues, 81
Nested sequence of numbers, 80
Net, 7
Net functions, 135
Neutron diffusion, 218
Neutron groups, 221
Neutron velocity, 219
Nine-point formula, 210
Non-linear equations, 124
Non-linear hyperbolic systems, 183
Non-linear instability, 94, 128 ff, 334
Non-linear problems, 91 ff, 119, 201 ff
Non-linear terms, 137
Norm, 29, 31, 132, 273
 of a matrix, 84
Normal matrix, 70, 72
Normal mode, 151, 257
Normal mode analysis, 133
Normed space, 31
Nuclear reactors, 207
Null element of a Banach space, 32

Opacity coefficient, 186
Operator, 33
Order of accuracy, 67 ff, 174, 212
Order of magnitude notation, 20
Oscillations behind a shock, 328, 334, 345

Parabolic equations, 94, 100, 183, 206
Parseval relation, 62
Particle-in-cell method, 352
Periodicity conditions, 7, 260
Perturbation, 97
Phase errors, 332
Phase space, 218
Piecewise analytic initial-value problem, 368
Plasma, 345
Practical stability condition, 146, 184, 231, 232, 238, 266, 269 ff, 284 ff, 350
Principal error terms, 126

Principal part of a difference operator, 101, 110
Principle of uniform boundedness, 34, 46
Properly posed, 206
Properly posed initial-value problems, 39 ff, 45, 63 ff, 179, 261
Pseudo-viscosity method, 313 ff, 324 ff
Pure initial-value problems with constant coefficients, 60

Quasi-linear system, 125

Radiation flow, 186
Radiation transfer, 218
Range of an operator, 33
Rankine-Hugoniot equations, 308
Rarefaction wave, 306
Rate of convergence, 19, 22 ff, 133
Recurrence relation, 199
Reference configuration, 293
Refinement of net, 15
Regular hyperbolic, 119
Regular point, 157
Relaxation method, 207
Resolvent, 157
Resolvent condition, 75, 79
Reversed heat flow equation, 59, 65
Riemann invariants, 310, 344
Root-mean-square norms, 29
Rounding errors, 24
Running-wave, 201, 202, 205
Ryabenkii's lemma, 158

S_n method of Carlson, 244
Saul'ev equation, 191, 192, 193
Scattering medium, 220
Schur's theorem, 77
Semi-bounded operator, 147
Semi-group, 41
Several space variables, 206 ff
Shock fitting, 308 ff
 in two dimensions, 378
Shocks, 306 ff, 338
Significant-digit arithmetic, 25
Simultaneous equations, 19, 207, 275
Slab symmetry, 221

INDEX

Slip surface, 359
Solution of implicit equations, 198 ff
Solution of implicit equations of arbitrary order, 275 ff
Sound and heat flow, 143
Sound waves, 72, 85, 259 ff
Space variables, 28
Specific volume, 294
Spectral radius, 69
Spectrum of a family of operators, 153
Sphere in a Banach space, 33
Spherical harmonic, 84, 224 ff
Spherical harmonic approximations of order L, 226
Spherical symmetry, 221
Splitting methods, 216 ff
Spontaneous development of shocks, 311, 370
Stability, 9, 45 ff, 68 ff, 73, 101, 133, 194, 262
Stability condition, 12, 69, 103, 149, 208, 240, 297, 298, 336, 344, 348
Stability and perturbations, 58
Stability of the pseudo-viscosity method, 328
Stability, sufficient conditions for, 72, 83 ff
Stability threshold, 129
Stable, 171, 179
Stable family of matrices, 73
Standard reference configuration, 293
Stefan-Boltzmann constant, 186
Strang's theorem, 92
Strong stability, 99
Strong stability condition, 110
Subsonic corner flow, 370
Sufficient condition for stability, 133, 187, 263, 274
Summation by parts, 14, 136
Surface tension, 357
Symbol, 109, 121
 difference scheme, 120
Symbol of operator C, 103
Symmetric hyperbolic systems, 94
Systems of linear equations, 183

Two-level formula, 43
Two-step Lax-Wendroff method, 300, 302 ff, 338, 360
Taylor instability, 352
Temperature gradient, 185
Test function, 337
Thermal conductivity, 185, 186
Thin beam, 271
Thomée theorem, 100
Three-level equation, 168
Time-dependent coefficients, 40, 42
Time variable, 28
Transformation, 33
Transport equation, 83, 218 ff
Triangle inequality, 29
Triangular net, 208
Tri-diagonal system of equations, 183, 213, 347
Truncation error, 19 ff, 44, 175, 179, 187, 188, 211
Twenty-one-point formula, 210
Two-level difference equations, 72

Unconditional stability, 19, 187, 210, 211, 212, 213, 216, 217, 230, 233, 240, 246, 264, 275
Uniform boundedness, 34, 174
Uniform convergence, 29, 31
Uniformly diagonalizable, 84
Unstable scheme, 262

Van Norton, work of, 257
Variable coefficients, 91 ff, 114, 193, 214
Vibration of a bar under tension, 282 ff
Von Neumann condition, 70 ff, 81, 84, 85, 86, 90, 96, 153, 165, 232, 239, 263, 274, 303

Wave equation, 72, 260
Weak solution, 337
Weak stability, 95, 96, 99
Well conditioned, 158, 160
Well-posed problem, 147, 155
Wick-Chandrasekhar method, 233 ff, 237
Wiener-Hopf factorization, 165